# Wytham Woods

# Wytham Woods
## Oxford's Ecological Laboratory

EDITORS

P.S. SAVILL
Oxford Forestry Institute
Department of Plant Sciences
University of Oxford

C.M. PERRINS
Edward Grey Institute
Department of Zoology
University of Oxford

K.J. KIRBY
Natural England
Peterborough

N. FISHER
Conservator of Wytham Woods
University of Oxford

# OXFORD
UNIVERSITY PRESS

Great Clarendon Street, Oxford OX2 6DP

Oxford University Press is a department of the University of Oxford.
It furthers the University's objective of excellence in research, scholarship,
and education by publishing worldwide in

Oxford  New York

Auckland  Cape Town  Dar es Salaam  Hong Kong  Karachi
Kuala Lumpur  Madrid  Melbourne  Mexico City  Nairobi
New Delhi  Shanghai  Taipei  Toronto

With offices in

Argentina  Austria  Brazil  Chile  Czech Republic  France  Greece
Guatemala  Hungary  Italy  Japan  Poland  Portugal  Singapore
South Korea  Switzerland  Thailand  Turkey  Ukraine  Vietnam

Oxford is a registered trade mark of Oxford University Press
in the UK and in certain other countries

Published in the United States
by Oxford University Press Inc., New York

© Oxford University Press 2010

The moral rights of the authors have been asserted
Database right Oxford University Press (maker)

First published 2010
First published in paperback 2011

All rights reserved. No part of this publication may be reproduced,
stored in a retrieval system, or transmitted, in any form or by any means,
without the prior permission in writing of Oxford University Press,
or as expressly permitted by law, or under terms agreed with the appropriate
reprographics rights organization. Enquiries concerning reproduction
outside the scope of the above should be sent to the Rights Department,
Oxford University Press, at the address above

You must not circulate this book in any other binding or cover
and you must impose the same condition on any acquirer

British Library Cataloguing in Publication Data
Data available

Library of Congress Cataloging-in-Publication Data
Savill, Peter S.
Wytham Woods : Oxford's Ecological Laboratory / Peter Savill, Christopher Perrins, Keith Kirby.
 p.  cm.
ISBN 978-0-19-954320-5 (hardback)
1. Forest ecology—Research—England—Wytham Wood.
2. Naturla history—England—Wytham Wood.   3. Wytham Woods (England)
 I. Perrins, Christopher M.  II. Kirby, K. J.  III. Title.
QH138.W95S28 2010
577.309425'7—dc22      2010000246

Typeset by SPI Publisher Services, Pondicherry, India
Printed in Great Britain by
the MPG Books Group in the UK

ISBN  978–0–19–954320–5 (hbk); 978–0–19–960518–7 (pbk)

1 3 5 7 9 10 8 6 4 2

# Contents

| | |
|---|---|
| Preface | ix |
| Acknowledgements | xi |
| Contributors | xii |

1. **Introduction** — 1
   *C. M. Perrins*
   - 1.1 History of British woodland — 3
   - 1.2 The forest returns — 3
   - 1.3 A green mantle — 4
   - 1.4 From forest to farmed landscape — 5
   - 1.5 Brief history of Wytham — 7
   - 1.6 The ffennels and Wytham — 9
   - 1.7 The Woods today — 13
   - 1.8 Conclusion — 17

2. **The Physical Environment** — 19
   *M. E. Taylor, M. D. Morecroft, and H.R. Oliver*
   - 2.1 Location — 19
   - 2.2 Topography — 21
   - 2.3 Climate — 21
   - 2.4 Geology and Soils — 24
   - 2.5 Hydrology — 26
   - 2.6 Pollution — 27

3. **Woods ancient and modern—land use history** — 29
   *C. W. D. Gibson and K. J. Kirby*
   - 3.1 Introduction — 29
   - 3.2 The early times — 29
   - 3.3 The landscape up to the late eighteenth century — 32
   - 3.4 Exploiting the Woods — 35
   - 3.5 The legacy of the fifth Earl of Abingdon — 36
   - 3.6 Stasis and decline in the Woods — 38
   - 3.7 The development of a research estate — 38

4. **The Woods in the modern landscape** — 43
   *K. J. Kirby and C. W. D. Gibson*
   - 4.1 Introduction — 43
   - 4.2 Wytham in the context of Oxfordshire's woods — 43

| | | |
|---|---|---:|
| | 4.3 Grassland | 48 |
| | 4.4 Fens and other wetland | 50 |
| | 4.5 Movement within the Woods and between the Woods and the surroundings | 51 |
| | 4.6 Changes in the relationship between the Woods and its surroundings | 52 |
| | 4.7 Conclusion | 53 |
| 5 | The Trees in the Woods 1945–2007 | 57 |
| | *K. J. Kirby* | |
| | 5.1 Active forestry management 1945–1963 | 57 |
| | 5.2 Minimum intervention and 'near-natural' stand development | 60 |
| | 5.3 Long-term studies of stand structure and composition | 60 |
| | 5.4 Tree physiology | 69 |
| | 5.5 Composition and structure of the shrub layer | 70 |
| | 5.6 The changes in woodland structure and its possible causes | 70 |
| | 5.7 Old growth and open space | 72 |
| | 5.8 Twenty-first century changes | 74 |
| 6 | The flowers of the forest | 75 |
| | *K. J. Kirby and M. D. Morecroft* | |
| | 6.1 Introduction | 75 |
| | 6.2 The Woods in a wider botanical context | 75 |
| | 6.3 Vegetation patterns—soil and woodland history | 76 |
| | 6.4 The woodland flora and stand dynamics | 79 |
| | 6.5 Other changes in the woodland flora 1974–1999 | 81 |
| | 6.6 Changes in Wytham Woods compared to trends elsewhere in the country | 83 |
| | 6.7 Bramble as a key species | 87 |
| | 6.8 Conclusions | 88 |
| 7 | The ecology of Upper Seeds—an old-field succession experiment | 91 |
| | *C. W. D. Gibson* | |
| | 7.1 Introduction: patterns in space and dynamics | 91 |
| | 7.2 Patterns in time: the nature of change | 91 |
| | 7.3 The experimental system | 92 |
| | 7.4 Colonization and species pools | 95 |
| | 7.5 Early and mid successional communities during an English old-field succession | 97 |
| | 7.6 Invasion and decline of *Arrhenatherum elatius* (false oat-grass) | 99 |
| | 7.7 How does the grassland on Upper Seeds compare with other grassland? | 102 |

|  |  |  |
|---|---|---|
| 7.8 | Structure, plant taxonomy, and invertebrates | 104 |
| 7.9 | Invertebrate movement between patches | 106 |
| 7.10 | Conclusion | 108 |

## 8 Invertebrates 109
*C. Hambler, G. R. W. Wint, and D. J. Rogers*

|  |  |  |
|---|---|---|
| 8.1 | Introduction | 109 |
| 8.2 | Pioneers of ecology and ecological genetics | 110 |
| 8.3 | The Wytham Ecological Survey | 113 |
| 8.4 | Invertebrates, habitat specialisms, and landscape management | 114 |
| 8.5 | Studies of cover and succession | 114 |
| 8.6 | Understanding animal abundance and density | 116 |
| 8.7 | Stability of complex systems | 120 |
| 8.8 | Ecological Energetics | 121 |
| 8.9 | The origins of Behavioural Ecology | 121 |
| 8.10 | Farm wildlife | 122 |
| 8.11 | Wytham's invertebrates and British conservation management | 123 |
| 8.12 | Invertebrate ecology in teaching | 126 |
| 8.13 | Improved sampling methodology and indicator groups | 128 |
| 8.14 | Present and future work | 129 |
| 8.15 | Notable invertebrates in Wytham and their requirements | 132 |
| 8.16 | The winter moth—an unlikely superstar | 134 |
| 8.17 | The 'life table' of a species | 137 |
| 8.18 | From moths to generalities | 143 |
| 8.19 | Acknowledgements | 144 |

## 9 Birds 145
*C. M. Perrins and A. G. Gosler*

|  |  |  |
|---|---|---|
| 9.1 | Introduction | 145 |
| 9.2 | Bird species in Wytham | 145 |
| 9.3 | Changes in species | 146 |
| 9.4 | The Wytham Tit Study | 149 |
| 9.5 | The future—where next? | 163 |
| 9.6 | Other Bird studies in Wytham | 165 |
| 9.7 | Conclusion | 171 |

## 10 The Mammals of Wytham Woods 173
*C. D. Buesching, J. R. Clarke, S. A. Ellwood, C. King,
C. Newman, and D. W. Macdonald*

|  |  |  |
|---|---|---|
| 10.1 | Wytham's small mammals | 173 |
| 10.2 | Deer | 182 |
| 10.3 | Weasels | 184 |

viii Contents

|  |  |  |
|---|---|---|
| 10.4 | Foxes | 185 |
| 10.5 | Wytham's Badgers | 187 |
| 10.6 | Hedgehogs | 194 |
| 10.7 | Squirrels | 195 |
| 10.8 | Amateur Volunteers as Biodiversity Monitors | 195 |

11 **Conservation Management of Wytham Woods** 197
*N. Fisher, N. D. Brown, and P. S. Savill*

|  |  |  |
|---|---|---|
| 11.1 | Conservation management before 1900 | 197 |
| 11.2 | Twentieth century conservation | 198 |
| 11.3 | Management by Oxford University | 200 |
| 11.4 | Wytham's role in national conservation policy | 201 |
| 11.5 | Conservation management: the last 60 years | 202 |
| 11.6 | Woodland Management | 203 |
| 11.7 | Veteran trees | 205 |
| 11.8 | Deer | 206 |
| 11.9 | Grey squirrels | 209 |
| 11.10 | Grasslands | 211 |
| 11.11 | Management of rides and other woodland edges | 212 |
| 11.12 | Ponds and Fen | 212 |
| 11.13 | Conflicts between Research and Conservation | 213 |
| 11.14 | Promoting Public Access | 213 |
| 11.15 | Monitoring the effectiveness of conservation management | 213 |
| 11.16 | Future Challenges | 214 |
| 11.17 | Climate change | 215 |
| 11.18 | Conclusion | 215 |

12 **Wytham in a changing world** 217
*M. D. Morecroft and M. E. Taylor*

|  |  |  |
|---|---|---|
| 12.1 | The Environmental Change Network | 217 |
| 12.2 | Upper Seeds climate change experiment | 221 |
| 12.3 | Tree growth and interactions with the atmosphere | 223 |
| 12.4 | Conclusions | 228 |

References 231
Index 253

# Preface

John Krebs (Lord Krebs, of Wytham)

If there were a Nobel Prize for Ecology, and if you could award it to a place rather than a person, Wytham Woods would surely be a prime candidate. It is almost certainly unmatched anywhere in the world as a place of sustained, intensive ecological research extending over nearly three quarters of a century.

The ecological concepts and discoveries based on work carried out at Wytham over the past seven decades read like a text book of ecology. Here are just a few examples: density-dependence and population regulation, the evolution of reproductive rates, the heritability of ecologically important traits, the response of populations and communities to climate change, natural selection in action, the relationship between stability and complexity in ecosystems, and the chemical defences of plants against insects.

How did this come about? Not as a result of a long-term strategic plan, but rather as a result of a lucky concatenation of circumstances. These included the donation of the Woodlands to the University, giving long-term security, the vision of a small number of key individuals, including David Lack, Charles Elton and George Varley, who collectively and individually not only realised the opportunities and benefits of long-term ecological research, but also had the scientific vision to create, in large part, the new scientific discipline of ecology. Inspired by these intellectual giants, many generations of talented research students have worked at Wytham. But, although more than 150 D.Phil students, and a comparable number of other researchers, post-docs, research assistants, and visitors, have carried out ecological research at Wytham, there has never been a substantial core of long-term funding from the Research Councils.

When I began my research career, in 1966, as a DPhil student working on population regulation of great tits in Wytham, it already felt as though I was following a long line of dauntingly important and clever predecessors stretching back to 1947, when David Lack started the tit studies. Now, I realise that I was, if not in the pioneer group, at least one of the early settlers. In fact the question that Chris Perrins and Andy Gosler pose in their chapter, "surely there can't be anything more you can learn after all these years?" would seem to require an answer. The answer is simple. On the one hand, as we now appreciate more than ever before, long term ecological records have an intrinsic value that grows over time, yielding unique and crucial insights into the response of the natural world to climate change. On the other hand, as new techniques are developed: DNA fingerprinting, radio-tracking, satellite imagery, new questions can be asked and answered. When I built the first population model of the Great Tits, with the help of Mike Hassell, it was run on a Facit electromechanical calculator, nowadays on show in museums around the world, the computing equivalent of travelling by horse-drawn carriage.

But for those of us who have spent many hours, weeks and years in Wytham, it is not just a source of ecological data, but also a place of great beauty. For me, it is hard to match the sensation of inhaling the scent of a carpet of damp moss on a February morning and chewing 'bread and cheese'—the first, pale green, buds of hawthorn that foretell the arrival of spring, and another tit breeding season, in Wytham.

# Acknowledgements

The area known as Wytham Woods consists of ancient woodland, modern plantations, and some grassland. Ecological research has been conducted in the woods since the 1920s, providing very rare levels of continuity. This has contributed to Oxford University's worldwide reputation in the biological and environmental sciences. The purpose in producing this book was to tell the story of Wytham in a way that is accessible to the interested general public. We have tried to provide an overview of what the Woods are like, their history, composition, how their wildlife has been changing, and also their significance in terms of the work that has gone on there. Contributors include some of the most respected field biologists in the UK.

In this book we have normally used the common English names of birds and mammals, but have usually included the scientific names of insects and plants as well. Some of the latter two groups have no common names anyway.

We are grateful to and thank many people who have contributed information or illustrations for this book. They include Colin Newbury, David Field, and George Peterken for information about Raymond ffennell and H.C. Dawkins and E.W. Jones respectively. Roger Mills, Catherine Dockerty, and Stella Brecknell the Librarians of the Departments of Plant Sciences, Zoology and Natural History Museum for help in locating material, Philip Powell for commenting on Chapter 2, Rachel Thomas for unpublished historical material in Chapter 3, Darren Mann for help with illustrations for Chapter 8, Jesus Cordero, Mrs K. Southern, Mrs Sherry Kendall, and Charlie Krebs have provided other illustrations.

Photographs have been contributed by Fred Topliffe, St. John's College, Merton College, Andrew Lack, George Peterken, Ian Walsh, and Mrs Margaret McIvor among many others.

The daily maintenance and other work in the Woods has been has been organized over the last 58 years by only three managers (currently called Conservators), Herbert Probitts (32 years), Michael Day (26 years), and Nigel Fisher (from 2000). They have the gratitude of many researchers whom they have assisted in numerous practical ways in the Woods. The willing help that is frequently given to students, University staff and the public by the two Woodsmen, Kevin Crawford and Nick Ewart is also gratefully acknowledged.

Finally we pay tribute to Dr Charlie Gibson who contributed three chapters to this book and who died from cancer shortly before its publication, and to Dr Robin McCleery, who helped many to understand the tit populations in Wytham, and whose sudden death in 2008 is sorely felt.

# Contributors

**Nick Brown**
Nick Brown is a lecturer at Oxford University in the Department of Plant Sciences, and currently (2010) Principal elect of Linacre College. Nick is chairman of the management committee of Wytham Woods and a trustee of the Sylva Foundation. His research looks at the impact of natural and man-made disturbance on trees and forests. He is a co-author of an independent report commissioned by the Forestry Commission into the ways that trees and forests can help us tackle climate change. Nick is also working with the Woodland Trust investigating changes in tree cover outside woods over the last 50 years and on the success of methods for restoring plantation sites in ancient woodlands.

**Christina Buesching**
Christina Buesching is a Senior Research Associate with the Wildlife Conservation Research Unit at Oxford University. Christina has a M.Sc. from the German Primate Centre, Göttingen, on the reproductive physiology and behaviour of the female lesser mouse lemur, and a D.Phil. from Oxford University investigating mammalian sociality and communication in badgers. Currently, she is investigating the socio-political and biological implications of involving volunteers in ecological monitoring. She is a founder and committee member of the UK government's Tracking Mammals Partnership, and advises this committee on volunteer engagement and deployment. Christina has worked as a Science Officer with Earthwatch Europe, and has been a Principal Investigator with the Earthwatch Institute since 2001. Christina is a Canadian citizen and currently resides primarily in Canada, where she is a member of the committee of the Tobeatic Research Institute of Nova Scotia.

**John Clarke**
John Clarke did an honours degree in Zoology, University of Western Australia. His thesis was an introductory study of the anatomy of the Quokka, *Setonyx brachyurus*, a small wallaby found in Western Australia, and particularly on Rottnest Island, 10 miles off the coast at Freemantle. A Rhodes Scholarship brought him to Oxford where he did an undergraduate degree in Zoology, during which he made a preliminary study of the hippopotamus in the Gambia, West Africa. For his D.Phil he investigated experimentally the aggressive behaviour of the field vole, *Microtus agrestis*, and its physiological and demographic effects. This work was suggested and supervised by Dennis Chitty (see Chapter 10), and comprised a first attempt to investigate experimentally, under laboratory conditions and also in the Woods, some inferences that could be drawn from Chitty's very important hypothesis concerning the cyclic population changes of field voles, with its implications for population regulation in a variety of other animals. Since then John Clarke has established laboratory breeding colonies of the field vole, the bank vole,

*Clethrionomys glareolus*, and the wood mouse (*Apodemus sylvaticus*) which had been caught in Wytham Woods. Using animals produced by these breeding stocks, as well as wild type specimens captured in the Woods, he and his co-workers have investigated their reproductive physiology, partly as a contribution to comparative mammalian reproductive biology which might also bear upon explanations for rodent population fluctuations. He was a lecturer in mammalian reproductive physiology in Oxford University, at first in the Department of Agricultural Science and then in the Department of Zoology. In retirement, he continued to analyse the results of experimental work on reproduction in field voles, bank voles, and wood mice. He died in 2010 shortly before publication of this book.

**Stephen Ellwood**
Stephen Ellwood has worked for the Wildlife Conservation Research Unit (Oxford University) for ten years on aspects of Wytham badger and then on deer ecology. Concurrent to this research he has pursued an interest in the application of monitoring technologies to wildlife biology. He obtained his D.Phil on the efficacy of deer census techniques and the density dependence and independence of deer skeletal size and body condition.

**Nigel Fisher**
Nigel Fisher is currently employed as the Conservator of Wytham Woods. He has held this position since July 2000. He has 22 years of countryside management experience after working for English Nature, the British Trust for Conservation Volunteers and Countryside Management Projects. He has a Geography BSc and an MSc in Environmental Forestry.

**Charles Gibson**
Charlie Gibson started his studies in Wyham over 30 years ago, in the 1970s. After a degree in Zoology at Oxford, he learnt the importance of plant ecology the hard way, by researching insects that ate grass and were a great deal more discriminating than the casual observer. This led to an interest in how animal and plant communities changed over time, and after some time wrestling with the dynamics of the impossibly long lived giant tortoises and their interactions with vegetation on the Indian Ocean island of Aldabra, he returned to Oxford, where he taught ecology and statistics and set up the long term experiments on grassland dynamics. His projects have spread and widened to include work on restoration and management across the country with bodies such as Butterfly Conservation, Natural England and the National Trust. Starting about 1994 he ran an environmental consultancy company based in Oxford, but kept up a strong relationship with the Wytham estate. He died in 2008 soon after completing his contributions for this book.

**Andrew Gosler**
Andrew Gosler is a University Research Lecturer in the Oxford University's Department of Zoology, and Human Sciences Lecturer in Biological Conservation. He has studied

birds in Wytham for more than 25 years, and has a particular interest in the population of great tits that breed in nestboxes there. Consistent with his commitment to ornithology and conservation locally as well as internationally, he is both President of the Oxford Ornithological Society and a member of the International Ornithological Committee of the I.O. Congress. He is a former Editor of the journal *Ibis*, and a Trustee of *A Rocha UK*. He has published more than 50 scientific papers, mostly from his ornithological studies in Wytham.

**Clive Hambler**
Clive Hambler has worked on and off in Wytham for about 30 years, starting during his undergraduate degree in Zoology. Within Wytham he has worked particularly on invertebrates such as spiders in the Upper Seeds limestone grassland restoration project. His main interests are in global conservation, including measuring and reducing extinction rates. Some of this research has application in restoration ecology including reintroductions and 'rewilding' in Britain. He has worked on a range of organisms and habitats including giant tortoises and sea turtles, endemic birds and endangered plants—but all inspired by the teaching in Wytham!

**Carolyn King**
Carolyn (known to friends as Kim) was the last graduate student to slip in through the doors of the BAP before Charles Elton retired. She ignored warnings that choosing to study the field ecology of weasels would be too risky a topic for a D.Phil. Thanks to the close collaboration with colleagues made possible by working in Wytham Woods, she managed to complete the first ever full-length thesis on weasels (1967–71), before radio-tracking became the standard field methodology, by improving ways to observe these inconspicuous little animals using conventional live traps and by swapping data with fellow-student John Flowerdew. She is the only person in the history of Oxford University to persuade the Zoology Department to buy a pony as a research vehicle. In 1971 she moved to New Zealand to join DSIR Ecology Division as a scientist specializing in introduced carnivores, especially stoats. Her two best-known books, *The Natural History of Weasels and Stoats* (1989) and *The Handbook of New Zealand Mammals* (1990), both went into second editions with Oxford University Press (in 2007 and 2005 respectively). Since 1995 she has been with the Department of Biological Sciences at Waikato University, Hamilton (New Zealand) where she is now a Senior Lecturer. She lives in a country house outside Hamilton with her husband, Ken Ayers.

**Keith Kirby**
Keith Kirby has a BA in Agricultural and Forest Sciences 1970–73 and a D.Phil. obtained for his studies of growth, production and nutrition of bramble 1974–76. He has been Woodland Ecologist with Nature Conservancy Council from 1979 (which became English Nature and then Natural England). He has carried out survey work in Lake District 1977–78 and in many other areas since. His research at Wytham has been largely based on the permanent plot system established by Colyear Dawkins.

## David Macdonald
David Macdonald holds a D.Sc. from Oxford. He has been Director of the Wildlife Conservation Research Unit at Oxford University since founding it in 1986. He won the 2005 Dawkins Prize for Conservation, and in 2006 was awarded the American Society of Mammalogists' Merriam Award for outstanding contributions to research. He has published over 300 refereed papers on aspects of mammalian behaviour, ecology and conservation and has served on, or chaired, a variety of committees. He has twice been awarded the Natural History Author of the Year.

## Michael Morecroft
Mike Morecroft started his research career at Cambridge University with a PhD on the effects of climate on mountain plants (1987–90). After postdoctoral work at Manchester University on the effects of air pollution on upland grasslands, he joined the Institute of Terrestrial Ecology (ITE) in 1992. He was out-posted to the Field Station at Wytham to establish monitoring and research for the UK Environmental Change Network programme. Over the next sixteen and a half years, Mike worked on a wide range of issues around climate change—its ecological impacts, mitigation, and adaptation. ITE was integrated into the Centre for Ecology and Hydrology and Mike's office moved to Wallingford, but he continued to work actively at Wytham. He has been Principal Climate Change Specialist in Natural England's Evidence Team since April 2009 and is a senior research associate of the School of Geography and the Environment in Oxford University.

## Chris Newman
Chris Newman is a Senior Research Associate with the Wildlife Conservation Research Unit at the University of Oxford and a Principal Investigator for Earthwatch. He undertook his D.Phil. on Population Ecology, Demography and Parasitology at Oxford University and now co-manages (with Christina Buesching) the Mammal Monitoring and Badger Research Projects for the Wildlife Conservation Research Unit. He also serves on the executive committee of the UK government's Tracking Mammals Partnership. Chris's primary residence is in Canada, returning to Oxford for periodic badger research and the supervision of graduate students, where Chris is a member of the committee of the Tobeatic Research Institute of Nova Scotia.

## Howard Oliver
Howard read Physics at Oxford University and obtained an MSc in Meteorology at Reading in 1969. He then joined the hydrometeorology section of the NERC's Institute of Hydrology investigating forest energy and water balances. He obtained an external PhD from Reading in forest aerodynamics in 1974. He was later involved in a range of applied meteorology, hydrological and environmental research projects in the UK and overseas. During the latter part of his career Howard was a scientific coordinator for the major NERC programmes, including the 'Terrestrial Initiative for Global Environmental Research', and was a college lecturer in Geography at Oxford University. After 'retirement', in 2001, he became a CEH Fellow at Wallingford and a Supernumerary Fellow at

Harris Manchester College, Oxford. He is currently continuing with some research, writing and lecturing on applied meteorology and environmental topics.

## Christopher Perrins

Prof. Perrins LVO, FRS first started studying the tits in Wytham, under the supervision of David Lack, in the autumn of 1957, just over half a century ago! In addition to woodland birds, he has also made studies of seabirds—especially those on the Pembrokeshire islands of Skokholm and Skomer—and of mute swans on the Thames and at the colony at Abbotsbury, Dorset. He was deeply involved in the discovery that decline of swans in the 1980s was in large part due to the ingestion of lead angling weights, which led to bans on the use of most of these weights. He has been President of the British Ornithologists' Union, the International Ornithological Congress and the European Ornithological Union and is an Honorary Fellow of the American, Dutch, German and Spanish Ornithological Unions.

## David Rogers

Between 1964 and 1970 David Rogers did first (Zoology) and second (Entomology) degrees in Oxford, was taught during the former *inter alia* by Charles Elton, E.B. Ford, David Lack, Niko Tinbergen, and George Varley and was supervised for the latter by George Varley, and examined by George Gradwell. By the late 1960s much of the observational work on the winter moth in Wytham had finished (Mike Hassell had just completed his doctorate on *Cyzenis albicans*) but the analyses were still on-going and the book on much of this work, *Insect Population Ecology* (by Varley, Gradwell and Hassell) was still to be written. The attic of the University Museum was a wonderful place for a D.Phil. student to construct various bits of apparatus and constant temperature/humidity boxes for experimental work on insect parasitoids, well away from the judgemental gaze of the Hope Librarian. Towards the end of this period David began talking with John Ford (a New College student of Charles Elton in the 1930s but, in the late 1960s, in the Hope Department writing his seminal work on the African Trypanosomiases) about tsetse flies in Africa, he helped produce the maps for that book and eventually ended up on a Wellcome Trust grant to study tsetse in Uganda. The ultimate aim of this work was to try to apply Varley and Gradwell life table techniques to a species (tsetse) with continuous, overlapping generations rather than the winter moth with convenient, synchronized generations. This work eventually resulted in models for both tsetse populations and for the transmission of the African trypanosomiases and the first application of remotely sensed satellite data to predicting the distribution and abundance of tsetse and the incidence and prevalence of the African trypansomiases, later extended to many more vectors and diseases. David returned to a Lectureship in the Hope Department in 1972 and moved to the Zoology Department in 1979 where he still works as Professor of Ecology.

## Peter Savill

Peter Savill started his career by working for the Forest Services in Sierra Leone for four years, and in Northern Ireland for 12 years, and retired in 2006 a Reader in Forestry at the

Oxford Forestry Institute, Department of Plant Sciences, Oxford. His main interests are in silviculture and tree breeding. His Wytham-based research has been into the potential invasiveness of sycamore in the Woods; other research has mainly been concerned with broadleaved trees. He was Chairman of the Wytham Woods Research and Management Committee up to 2006, and is Chairman and a founder member of the British and Irish Hardwoods Improvement Programme.

**Michèle Taylor**
Michèle Taylor is a Higher Scientific Officer (Ecologist), at the Centre for Ecology and Hydrology (formerly the Institute of Terrestrial Ecology (ITE)) and Site Manager for the UK Environmental Change Network programme at Wytham. She originally studied agriculture and first worked for a plant breeding company, before joining ITE in 1994 to carry out long-term environmental monitoring at Wytham, completing an MSc in Habitat Creation and Management in 2002. Her special interests are ecological impacts of climate change and land use and long-term ecological monitoring.

**William Wint**
William Wint carried out field research on the determinants of arthropod community structure on native and introduced trees in Wytham, for eleven years (1975–1986), six of which were with Sir Richard Southwood. Since 1986 he has been Managing Director of the Environmental Research Group Oxford (ERGO) carrying out extensive field work and development of air and ground based survey methods for environmental and agricultural resources in Africa, and more recently developing remote sensing and Geographic Information System based methods providing high resolution global maps of livestock populations and animal diseases for the UN, EU, and the UK government.

# 1
# Introduction

C.M. Perrins

This is the story of a piece of woodland, now known as Wytham Woods, and its history and natural history. It is, in many ways, very ordinary—a piece of ash–maple–hazel woodland typical of central England. However, therein lies one of the story's strengths, for what happens here is also a reflection of what occurs through much of our remaining lowland English woodland.

There are local elements in what is otherwise a general story and two quirks of fate in particular have affected Wytham Woods. The first of these is largely geological. After the retreat of the last glaciation, during which the ice-cap reached as far south as the northern edge of Oxfordshire (but not quite so far south as Wytham), the rapid development of two local rivers, the Cherwell and the Thames, resulted in a large northerly loop in the modern River Thames—around the hills on which Wytham Woods are situated. Shallow areas of the Thames at both the western the eastern ends of this loop were much used by people as crossing places (see Fig. 1.1).

Oxford City grew up beyond the bridge at the eastern crossing and the crossing at the west was the main route to Cheltenham, to Gloucester, and to Wales. This western crossing, Swinford Toll Bridge, is still a toll-bridge to this day (5p for a car). Two Bronze Age shields, unearthed from the river at this point, provide evidence of the track's antiquity; whether the soldiers lost them in defence of the crossing, or while finding it deeper than they bargained for, we shall never know. The toll bridge was built by one of the Earls of Abingdon and opened in 1769. The original route—authorized as a turnpike road in 1751—ran, and still runs, along the higher, and so drier, ground through the Woods. Its name, at least according to local custom, reflects an earlier period when medieval monks on pilgrimage from Cirencester to Canterbury got their first view of Oxford (Plate 1)—with its promise of a meal and a roof over their heads—and so broke into song: the 'Singing Way'.

Even as late as the early 1900s, this area was poorly drained and Wytham was almost an island cut off by the river and the dry route along the Singing Way and connected to the east and west only by poor, low-lying routes across the floodplains. Now, with better, modern drainage the east–west route runs across this lower land along the southern side of the Wytham Estate. Later still, in the 1960s, a ring road around the city was completed, cutting the Woods off from the floodplains on its eastern flank.

## 2 Wytham Woods

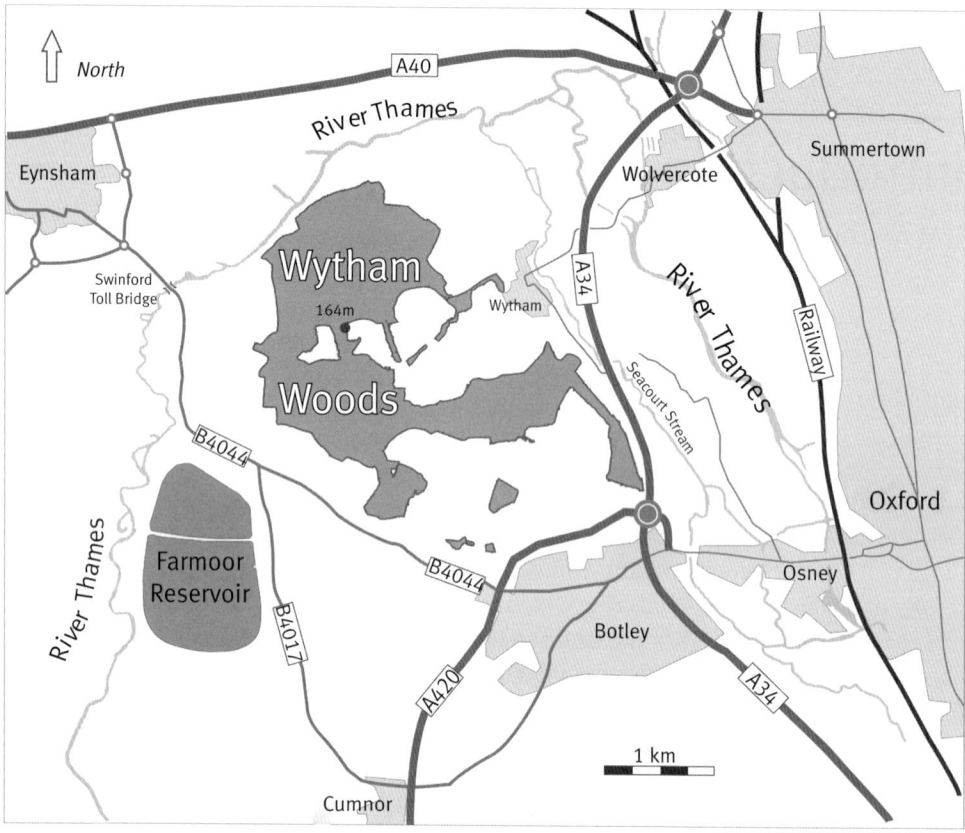

**Figure 1.1** Map showing the location of Wytham Woods.

The second quirk of fate that influenced this story stems from the series of decisions which lead to Oxford becoming a city where many places of learning were founded that later merged to form the University of Oxford. In 1942 the estate, of about 500 hectares of farmland and 350 hectares of woodland, became the property of the University and an obvious and handy choice for many of the scientists who were developing the new fields of ecology and behaviour and who needed 'outdoor laboratories' for this work. These pioneers included Dennis Chitty, Charles Elton, E.B. Ford, David Lack, H.N. (Mick) Southern and George Varley. Their studies were major contributions, milestones in the foundation of ecology as a respected science. The studies that they started, and many others that have followed since, made Wytham one of the world's most famous ecological sites. The Woods, despite their ordinariness, have become a site of unequalled ecological knowledge. This synthesis, drawn from the work of so many expert ecologists, will give the reader a greater understanding of what makes an English woodland 'tick'.

## 1.1 History of British woodland

The ancient history of the Wytham woodland is similar to that of other areas of lowland Britain (Godwin 1975; Rackham 2006). Changes in the vegetation since the last Ice Age have been reconstructed from careful examination of the pollen record; the outer cases ('exines') of pollen are extremely tough and can persist in the soil for millennia if the soil conditions are right. Being variously patterned, the species of plant from which they came can be recognized. By examination of dated soil samples—especially aquatic sediments—botanists have been able to obtain a good impression of the changes in the vegetation since the ice cap retreated. Increasingly a second strand of evidence, the remains of insects—particularly beetles—are being used to cross-check and expand on conclusions drawn from the pollen record (Robinson 2001).

Over the course of several hundred thousand years, the polar ice cap has expanded many times, covering large areas of Europe—including much of the British Isles—in huge sheets of ice several hundred metres thick; in inter-glacial periods these ice caps melted away as the ice retreated northwards. Sometimes the ice has covered virtually the whole of the British Isles, but in the last glaciation (which ended around 14,000 years ago), the ice extended only as far south as the northern edge of Oxfordshire. The county and probably almost all of southern Britain would still have been largely devoid of forest; south Oxfordshire, and our hill, would have looked much like arctic tundra.

## 1.2 The forest returns

After the ice cap started to retreat, about 12,000 years ago (12,000 BP, before the present, or 10,000 BC), early 'forests' started to reappear and, as the ice moved progressively further north and a warmer climate returned, so the vegetation became richer. First came birch and pine, forming forests perhaps similar to those still common in Scotland. By about 9000 years ago a more mixed, deciduous woodland of oak, elm, and lime was well established, with hazel in the understory (but also at times and in certain places hazel seems to have been a dominant 'tree'). The fauna was also changing; for example reindeer moved northwards and eventually became extinct in Britain, but were replaced by red and roe deer in the woods (Yalden 1999). The pattern of reforestation was initially similar to that in Continental Europe. However, around 7000 years ago, partly as a result of rising sea levels because of the melting ice cap, the English Channel opened up. This created a barrier to movement with the result that some of the trees that had been moving northwards more slowly—conifers such as spruce, firs, and larch conspicuously amongst them—reached northern France too late to spread into the British Isles. Although they spread much farther north on the continent, they failed to cross the Channel. So our English lowland woods were poorer in tree species than their continental counterparts. Another tree species that appears not to have crossed the Channel is the sycamore, until it was introduced to Britain at least 450 years ago; it is now one of the commonest woodland trees in Wytham Woods. Hence, by a chance of timing, our ancient forests became rather different from those of the continent at similar latitudes.

## 1.3 A green mantle

By about 8000 BP broadleaved forests were well spread and were probably dominant in areas such as south Oxfordshire. The pollen record for grasses and other species of open ground is very scarce (Rackham 2006), again suggesting a heavily wooded landscape. It is difficult to imagine what the forests looked like at this time since almost no natural forests exist today in Western Europe. Bialowieza Forest in eastern Poland may approach the natural condition, although even this has been influenced by management to some degree. There, huge trees dominate the forest; there is much dead timber, both fallen and still standing. The dense canopy prevents much light from getting to the ground so that over many areas it is sparsely vegetated.

Godwin (1975) describes the English countryside of 8000 years ago: 'closed deciduous forests covered the lowland landscapes with vast continuous expanses broken only by river valleys, mires, lakes and steep local scarps of rocks'. This is not to say, however, that the forest was wholly stable in either its structure or composition. Peterken (1996), building on the work of Oxford ecologist, Eustace Jones, has emphasized the variety of structures (including large and small clearings) that can occur in natural woodland. In particular the river valleys that we know today were then very different; their winding nature encouraged the collection of huge masses of floating logs and other debris that held the waters back and flooded wide areas of adjacent land. The site of Oxford City would have been under water almost all the time and the 'channel' of the Thames hard to find amongst the swamps. Getting from Oxford to Wytham would have involved struggling through several miles of alder- and willow-covered swamp, perhaps interspersed with beaver ponds.

However, the story may not be quite as simple as sometimes portrayed. The Dutch ecologist, Vera (2000), suggests that the natural forests were much more open, with many clearings where large grazing herbivores such as deer and aurochs lived. Was the future site of the city of Oxford once the haunt of large herds of wild oxen? Vera attributes the lack of grass pollen in the pollen record to heavy grazing by these animals—and the many smaller grazers too—which prevented the grasses from flowering. This idea is challenged by Mitchell (2005) who has compared the pollen records from Continental Europe with those of Ireland, where almost all the large mammalian grazers were absent. Mitchell found that the pollen profiles in the two regions were not significantly different and therefore concludes that the forests at that time were not open. Vera's hypothesis does not seem to be sustained as a general proposition (Hodder *et al.* 2005). Nonetheless, we do now have a large suite of plants and animals that only really flourish in open country, many of which were also present during the 'forest maximum'. There must have been some glades and edges where these could exist—when the forest was mostly closed. Possibly some of them depended on the temporary clearing made when the forest giants died and fell down; perhaps the forests were more open on very thin soils, such as those on the top of Wytham Hill, and perhaps our Mesolithic ancestors were already starting to create clearings in the forests—a process that expanded rapidly with the development of Neolithic farming.

## 1.4 From forest to farmed landscape

Pollen of grasses—so common today—was rare before about 4500 years ago (Godwin 1975), but after that it becomes slowly, but progressively, more common. This gives us a good guide to the fact that humans were already making small clearings by this time, presumably for pasture for animals. The signal from beetle remains is similar (Robinson 2001)—those associated with dung start to increase from the Neolithic period onward. The landscape was changing.

The archaeology of the Thames floodplain west of Oxford has been particularly well studied (Miles *et al.* 2007) and may reflect what was happening more generally in the land around Wytham. Mesolithic flints have been found scattered along the Corallian Ridge, and the early Neolithic settlements (from about 5000 years ago) are similarly concentrated here and on other chalk and limestone hills of the Berkshire Downs and the Cotswolds. For the middle and later Neolithic periods there is more evidence for activity on the gravel terraces lower down.

On the floodplain at Drayton, south of Oxford, there appear to have been a number of episodes of Neolithic clearance and woodland regeneration; elsewhere the clearance seems to have been more-or-less permanent. So Ingrem and Robinson (2007) suggest a landscape on the gravels of the middle and upper Thames that was a mosaic of clearings of various sizes and duration set against a background of old woodland. Gathered woodland food plants, such as hazel nuts, were being eaten alongside cultivated cereals during the Neolithic.

The species composition of the forests of this time was also changing (Rackham 1980). Lime and elm were common components of the lowland English forest, but both often show marked declines in the pollen record. This decline in elm is widespread across Europe. In Britain it seems to have occurred about 4900–5300 years ago and has variously been attributed to the cutting of elm shoots for cattle fodder or to outbreaks of Dutch Elm Disease. The two are not mutually exclusive because disease spread may have been facilitated by the proliferation of small branches and stems following lopping of elms. Lime tends to decline in the pollen record later than elm and the timing is more variable, but again seems to be linked to the spread of agriculture. By contrast other species increased, such as ash—which still colonizes abandoned farmland rapidly—and beech. Beech has been present in Britain for about 9000 years, but was initially scarce, only becoming widespread in Neolithic times. It may have benefitted from the early clearances and has since become a dominant member of many southern English forests, especially those on the chalk and limestone ridges.

Elsewhere in the country, forest clearance during the Bronze Age was extensive—some of our heathland seems to have its origin in this period. In the Upper Thames Valley the degree of clearance during the early to middle Bronze Age remains uncertain (Ingrem and Robinson 2007). There is some evidence that suggests areas of permanent open ground, for example around the Devil's Quits at Stanton Harcourt, but also for areas of wooded floodplain. By the later Bronze Age, however, much of the Upper Thames floodplain may have been largely cleared.

Iron tools would have made cutting and clearing forests easier. Ingrem and Robinson (2007) suggest that, by the middle Iron Age, the landscape around what is now the Cotswold Water Park was a predominantly open agricultural country, perhaps primarily used for raising domestic animals, but with arable fields on the drier ground. At Claydon Pike, near Lechlade, around AD 25–125 (late Iron Age, early Roman period) the evidence from pollen and beetle records suggests a landscape of heavily grazed grassland. Any woodland appears to have been distant from the site.

Roman Britain, at least in lowland England, was a densely settled country: to the north of Oxford, for example, the Ordnance Survey shows a dense cluster of villa sites around Wychwood. Woodland resources would have needed to be conserved, not only to supply fuel to heat the water in their famous baths, but also for their industries, such as brick and glass making, iron and lead smelting (Rackham 1976). Just to the south-east of Oxford there is evidence for local pottery and tile manufacturing sites. The Romans also added to the species mix of our woods through the introduction of sweet chestnut, still a very important tree in northern Italy for its wood and nuts. In parts of south-east England it is now treated as an 'honorary native' tree; in Wytham though it is only a minor component. Occasionally bones of fallow deer are found amongst Roman remains (Yalden 1999), suggesting that the Romans may have introduced it to Britain (rather than it coming in with the Normans). If so, then this is a legacy of their occupation that is still having a major impact on our woods today.

The 'Dark Ages' are now thought to have been more a period of reorganization of parts of the countryside, rather than large-scale abandonment of Roman farms to woodland, followed by fresh Saxon clearance. Certainly, by the Norman Conquest, the woodland cover may only have been about 15 per cent, at least in the counties covered by the Domesday survey (Rackham 1980). However, this probably underestimates the contribution of scattered trees in fields and commons, along streams etc. to the landscape and to the local economy. Rackham places Oxfordshire as about average for its woodland cover at this time. The Oxfordshire Domesday Book entries record that: 'the King's demesne forest of Shotover, Stowood, Woodstock, Cornbury and Wychwood were nine leagues (c 27 miles) in length and breadth'. This would have given a forest of some 750 square miles. Wychwood was still about 10 square miles in the mid-nineteenth century, parts of what is now Blenheim Park forming the southern edge. However a 'forest' meant a place for deer, not a place for trees. These forests were often more like open, heavily wooded parkland, or forest, with extensive glades rather than true continuous seas of trees. The Oxfordshire forests certainly included some agricultural land, and how much was actually wood is unknown (Rackham 1980). Nevertheless the Oxfordshire forests do seem to have retained more woodland than the surrounding countryside and their locations are often still reflected in clusters of ancient woods.

In the medieval and post-medieval world, trees and woodland were now very much a key element of local and national economies. However, this did not stop them being cleared: if about half the original forest cover of Oxfordshire had gone by Roman times, by Domesday (1086) only 15 per cent remained and this had been reduced to as little as 7 per cent by 1300 (Rackham 2006). The structure and composition was more and more coming to be determined by deliberate management. Lime, a major component of the natural woods, had become scarce. Trees and shrubs were regularly coppiced or pollarded, producing much small and easily handled timber and at the same time producing rich growths of herbaceous plants profiting from the extra light let in onto the forest. Wood was widely used as fuel for cooking, heating, and in 'local industries'. Bark was in demand for the tanning industry.

Meanwhile, indirect human influences on the woods included the increasing numbers of introduced animals, notably rabbits introduced by the Normans in the twelfth century. Rabbits were originally used for meat and fur, but doubtless soon escaped and established the wild populations that we see today. Rabbits in numbers are destroyers of young trees, barking saplings especially in cold winters when other food is scarce.

During the nineteenth century the next big shift occurred with the gradual replacement of wood fuel by coal, and the shift in some woods away from coppice to high forest management. This latter was facilitated by the Enclosure Acts that allowed owners to exclude the stock belonging to local villagers from their land. The fashion for landscaped estates and parks and interest in shooting provided other reasons for keeping trees and woods at a time when most of the wood and timber being used was imported. This reliance on imports nearly proved disastrous in the two World Wars in the twentieth century. This triggered a renewal of interest in forestry and woodland management and the creation of a state forest service—the Forestry Commission—that has overseen a doubling of England's woodland cover over the last hundred years, albeit mainly with introduced conifers. However, we can now move from a general account to the specifics of what has happened at Wytham.

## 1.5 Brief history of Wytham

We do not know a great deal about Wytham Woods after the ice retreated. However, there is a small bog in Marley Wood from which cores containing pollen have been extracted (Hone *et al.* 2001). Unfortunately it was not possible to determine the exact dates of the layers in these. Nevertheless the pattern is so similar to dated cores in other localities not far away that it seems safe to assume that the changes in Wytham would have occurred at the same times. In the lowest parts of the cores, pollen of Scots pine was abundant, as was pollen of hazel and grass; oak and elm pollen were present, but not in large amounts. After this, grass and pine pollen became scarcer and the samples were dominated by lime and hazel, with elm, oak, and alder also present. Lime pollen dominated for some time: using dates from elsewhere, the increase in lime started about 6800 years ago and the decrease dates from 3800 to 3100 years ago. Otherwise the pollens present seem similar to those on many other sites, although alder pollen remains quite common, presumably because this species flourished around the bog.

Rackham (2006) has pointed out that most English woodland exists not because it was on good ground for growing trees, but rather that it was on land that was bad for cultivation. In the case of Wytham the ancient woods tend to survive best on wet, north-facing slopes.

The history of ownership of the Woods is reasonably well known, though doubtless small changes in ownership of some small areas are long forgotten (Lamborn 1943; Grayson and Jones 1955). In the tenth century King Eadwig granted a large tract of land, possibly as much as 1300 hectares, at Hinksey, Seacourt, and Wytham to Abingdon Abbey. Seacourt at that time was a large village on the Botley–Wytham road, but by 1439 it was deserted. The land passed out of the control of Abingdon Abbey with the dissolution of the monasteries (c. 1535). Much of it was acquired by Lord Williams of Thame and was inherited from him by Lord Norreys, who in turn passed it on to the Berties. In 1682 the Berties became the Earls of Abingdon (there are both Norreys and Bertie roads in Cumnor). Although it was a large estate by that time, it is fairly certain that by 1761 only a little woodland remained. In addition there were many freeholders who owned small properties or rights which prevented the Earls from doing what they might have liked with the whole estate.

Many of these were bought up as the years went by, but 'it was only the passing of the Enclosure Acts (1814) which enabled Montagu Bertie, the fifth Earl, to complete the process and acquire absolute control of a large enough area to create a domain worthy of his wealth, rank and ambitions' (Grayson and Jones 1955).

The fifth Earl was the person who made the major alterations to the estate and who left it largely in the state that it is today. It was he who planted the now large beeches along the Singing Way in the early 1820s and many of the larger oaks seem to date from about the same period. According to Grayson and Jones, beech, lime (*Tilia* x *vulgaris*), sycamore, and wych elm were also widely planted then. The fifth Earl died in 1854, and the sixth and seventh Earls (1854–84 and 1884–1928 respectively) took little interest in Wytham. Indeed the seventh Earl became impoverished, eventually selling the estate to Raymond ffennel in 1920. The impact of these changes in land use are discussed in Chapter 3.

Though the name 'Wytham Wood' was used in 1623, in a purchase of timber by the University, it is a relatively recent name. For most of their history, the Woods were referred to as Cumnor Wood, a not surprising name in view of the fact that the southern parts of the woods fall within the parish of Cumnor. The part now known as Marley is not often referred to and this seems to be because it was for much of the time separately owned, falling as it did in the parish of Seacourt. The Abbey itself is no longer part of the estate, having been leased to a private individual in 1991 and the freehold subsequently sold as of right.

Introduction 9

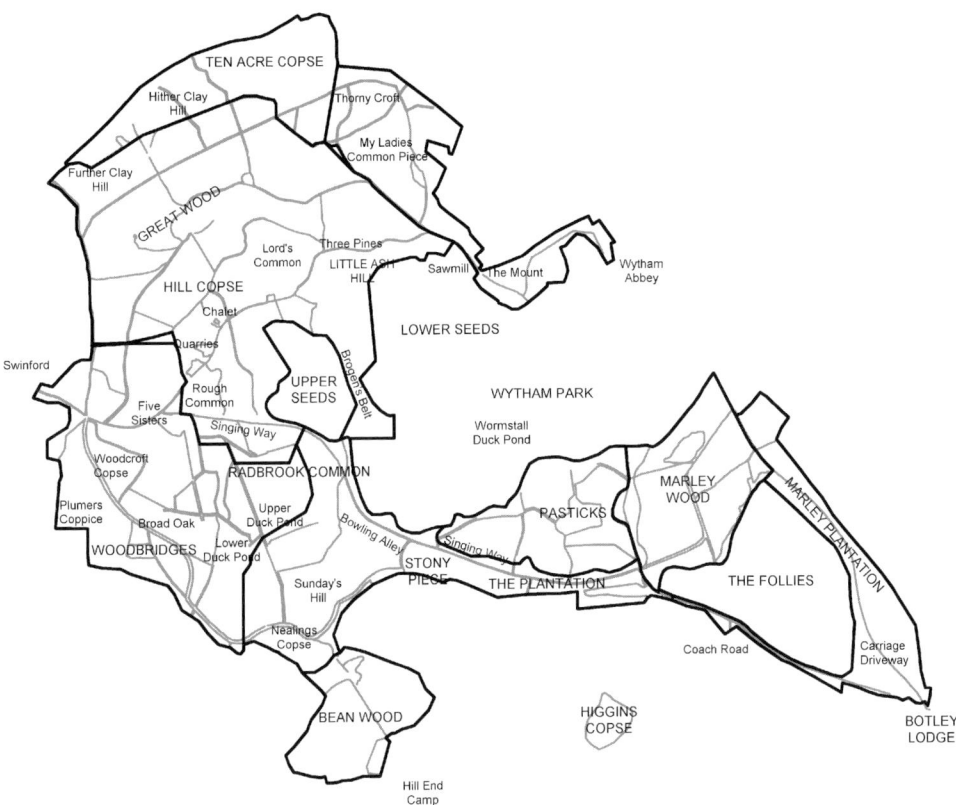

**Figure 1.2** Place names in Wytham Woods that are referred to in this book.

Today's visitor to Wytham can find many indications of the way the woodland has been shaped by the past: there are remnants of field boundaries and woodland rides that have been there for many centuries. The areas which were formerly wood pasture may be indicated by veteran ash and oaks. Almost all of them have the spreading build of trees planted with plenty of space around them, in contrast to those grown in plantations with their characteristic straight trunks without side branches.

## 1.6 The ffennels and Wytham

### 1.6.1 Raymond ffennell

Raymond William ffennell (1871–1944) was obviously quite a character. Born in London, he was the second son of Erwin A. Schumacher and Lucy Harvey ffennell. He was educated at Eastbourne and Harrow, studied banking and finance, and in 1894—through his father's business connections—joined the Johannesburg mining finance house of H. Eckstein & Co. at the boom period of deep-level gold-mining investment on the Rand. In the late 1890s his business career advanced rapidly with directorships in Ecksteins and in the industrial conglomerate, Central Mining and Investment Corporation, formed by Wernher, Beit & Co. By 1910, when Central Mining absorbed its parent house and all subsidiary mining companies and investments into the biggest industry in South Africa, Raymond Schumacher, at the age of 41, was a senior director and ran the organization in the absence of Sir Julius Wernher and Lionel Phillips as a major shareholder and Rand millionaire.

As a member of the Reform Committee, set up by mining interests for the overthrow of the Transvaal government by the Jameson Raid at the end of 1895, Raymond was arrested and questioned but released without charge. With a strong sense of British patriotism he served in the South African War as a captain in the Rand Rifles, rising to the peacetime rank of colonel in the Transvaal Light Infantry with an interest in the promotion of challenge cups for cadets throughout the Empire.

Although the family moved backwards and forwards between South Africa and England, by about 1915 he had decided to settle in England. Two years after his return, he adopted his mother's family name, together with the ffennell coat of arms (a number of British South Africans of Continental origin changed their names about this time to avoid the hostility directed at German names in wartime).

He had by then made a fortune. And it must have been quite a fortune, because he was able to buy a 3000 acre estate, which included most of the village of Wytham, where he lived and entertained in a lavish style. But although wealthy, he was also a philanthropist. Before his return to London in 1915 Schumacher gave his Johannesburg home to the city to house convalescent children, naming it after his wife, Hope Weigall, whom he had married in 1903. He also set up Hill End Camp (see p. 13). Raymond ffennell also wrote two books, one on the Chinese labour question in the Transvaal and the other: *Oxford As It Was, Now Is, And Never Should Be* (Blackwell, 1930). There is an oil portrait of him in the Hope Home, and much of his early correspondence is in the Barlow Rand Archives, Johannesburg.

The ffennell's acquisition of the Wytham Estate did not happen all at once. On a visit to England in about 1916, the ffennells found Wytham and rented the Abbey, falling in love with the area. However they returned to South Africa and when, in 1920, they had the opportunity to purchase the estate there was a sitting tenant in the Abbey who stayed on there for the remainder of the lease. During this time they commuted from a large house in London. When in Wytham, they lived in 'The Camp', a set of tents on top of the hill near where the chalet (Fig. 1.3) now is. 'Camp' is not a wholly adequate description of the establishment; a set of marquees might be nearer. There were servants, the tents were carpeted and heated, and the dining tent was 40 feet long. 'The dining-room tent is set up, and, when I entered it at the tiny door, I was literally amazed, a perfectly lovely room confronted me—yellow walls, red carpet, refectory tables and a side board and other furniture.' Entertainment was on a grand scale and at least one undergraduate who accepted an invitation to dinner

**Figure 1.3** The chalet in Wytham Woods that was originally built by Raymond ffennell as a hunting lodge.

and cycled to the top of the hill was embarrassed to discover everyone else in formal evening attire.

It was ffennell who built the Swiss-style chalet as a more permanent summer residence ('brise-block and waney elm-board'). Mysteriously, the chalet incorporates a rather fine wooden staircase which must have been imported from some other building since it does not quite fit!

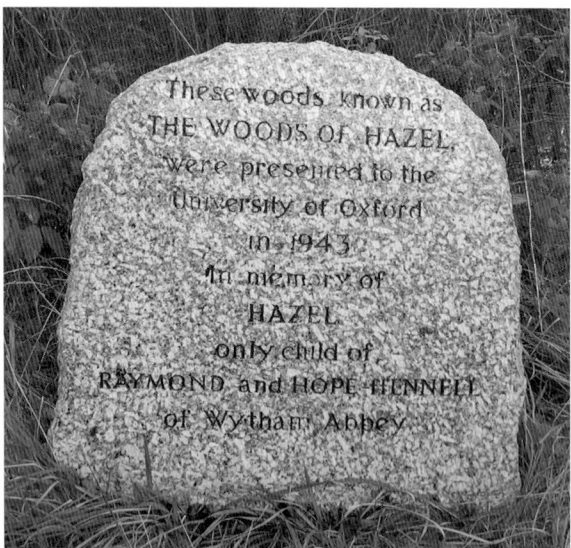

**Figure 1.4** Memorial to Hazel ffennell on the Singing Way.

### 1.6.2 Hazel ffennell

The light of the ffennell's life was their only child, Hazel, born in the Transvaal in 1905. Hazel died, after some years of illness, in 1939, aged 34. Much of what is known about her (and the family) comes from a book about her. Called, simply, *Hazel*, the book contains a miscellany of bits about her by friends, notes from her diaries etc., and also an interesting collection of photographs of the Abbey and the village. The book is not easy to come by; it was privately published (by Oxford University Press) and probably few were printed. Further, locating copies is perhaps not helped by the fact that the book lacks an author and a date of publication.

Like many children of the very wealthy, she was privately educated and travelled widely, including several lengthy tours of the Mediterranean and Middle East based on the family's 250 ton yacht 'Halcyon'. Despite the proximity of the University and the steady stream of undergraduates invited to the ffennell's, Hazel did not take a degree. She does, however, seem to have attended some of the degree course at the Department of Agriculture ('Hazel began an Agricultural course at Oxford for a year so that she should know something of Estate Management', but 'Not having sat for Responsions she could not, of course, take the examination'.).

Hazel had many talents; in particular almost everyone points to her great love of animals and affinity with them. She obtained much enjoyment from exploring the estate and its natural history. In notes from her diary she refers to finding a number of nests in Wytham including several nightingale nests, one of which was in Marley Wood. These seem to have been found in the company of Mr Tickner, presumably George Tickner, a founder-member of the Oxford Ornithological Society. Although, as with so many other records, the book does not give dates, this may have been in 1922 (when Hazel would have been 18). In that year the OOS Report states that George Tickner (and others) noted that the nightingale was unusually numerous all over the county particularly in north-west Berkshire (where Wytham, at that time, was).

### 1.6.3 The ffennell bequest

Hazel died. Having no one else to whom they could leave their estate, the ffennell's gave the Woods to the University in 1942.

The agreement between the University and ffennell, recording the gift of the Woods, sets out the ffennell's wishes concerning the management of the area:

> The Grantor being desirous that eventually the whole of the Wytham Estate should come under the protection of the University has agreed to make the gifts to the University of the Woodlands at Wytham described...... and to sell to the University the farmlands...... the total area dealt with in this Agreement being approximately three thousand acres.

He also requested that:

> .... every care should be taken to preserve their present state of natural beauty.
> .... the Grantor leave the guardianship of the Woodlands to the kindly sympathy of the University. The University will take all reasonable steps to preserve and maintain the woodlands and will use them for instruction of suitable students and will provide facilities for research.

Another of fennell's interests was the instruction of the young, especially those from deprived areas. He set up Hill End Camp, off the Eynsham Road. This was a camp designed to enable under-privileged children, many from the East End in London, to come and live in, and experience, the pleasures of the countryside. The site is now leased by the University to Oxfordshire County Council as a field study centre, and so in broad terms the intentions of Raymond ffennell continue.

ffennell requested that the Woods should be known as the 'Woods of Hazel' after his daughter and this is stated on the memorial stone on the Singing Way (Fig. 1.4). It is, however, not easy to ordain what name local people should use for an area, and this has never really caught on.

## 1.7 The Woods today

### 1.7.1 Management

When the University acquired the Wytham Estate, the whole was reckoned to be 1250 hectares of which some 400 comprised the woodlands (this measure included some open areas within the woodland boundary). The estate has remained largely unchanged in these dimensions since that date except that some farmland was lost on the eastern boundary when Oxford's western by-pass was completed and, a further parcel of land became separated from the estate with the completion of the Cumnor Hill by-pass in the mid-1970s.

The Woods were originally put under the management of the Forestry Department of the University, a natural enough decision. In 1951, in an agreement with the Forestry Commission, the University 'dedicated' the Woods to the production of timber. Most of the younger plantations still present date from the 1950s and early 1960s. The Department's management of the Woods as a commercial forestry operation involved the gradual felling of the old mature trees and planting areas with young trees. In the younger plantations, such as Radbrook Common and the Pasticks, there are the remnants of rows of conifers in between rows of hardwoods—a legacy of the forestry practices of the 1950s and 1960s. The conifers were there as 'nurse' trees to shelter the growing hardwoods and to provide a cash crop when the areas were thinned. They are being gradually removed as they become large enough, while the hardwoods with which they were inter-planted will be allowed to grow on. Immediately after the thinning more light can reach the forest floor and the ground flora can respond vigorously. Dense thickets of bramble or tall grasses and herbs may develop.

In other parts of the Woods evidence of the former coppice management can be seen as old stools with multiple stems. Attempts in recent years to re-establish a coppice regime, so as to maintain the flower-rich coppice habitats, have largely failed because the increased numbers of deer have prevented the coppice from regenerating (see Chapter 5 Section 5.6).

However, as the value of the Woods for research grew, pressure to maintain the old stands in the Woods increased. Eventually the whole of the woods came to be managed as a biological reserve rather than as a commercial forest. All felling of big, old trees ceased. The Woods were removed from dedication in the 1960s. Archived correspondence shows debates on the intensity of management, flower picking, etc., were occurring between foresters, ecologists, and others in the University during the 1950s, with ecologists such as Charles Elton pressing for reduced intervention. Wytham Woods were designated a Site of Special Scientific Interest (SSSI) by the Nature Conservancy (Council) (now Natural England) in 1950. This was confirmed in 1999 in a Site Management Statement in which the general presumption is towards high forest development and non-intervention; however there are clearly defined, localized areas that are or will be actively managed for conservation. The University's objectives have been in line with Raymond ffennell's wishes (see above) while at the same time attempting to maintain the nature conservation value of the site. Specifically, this latter involves the following aims:

- To restore and maintain broadleaved high forest woodland.
- To maintain scrub habitats.
- To maintain and promote veteran trees.

- To restore and maintain grassland habitats.
- To maintain other minor habitats, especially wetland features (fens, ponds, stream sides etc.).
- To maintain reasonable access to the various parts of the woods by keeping rides open and maintaining roads.

Hence the aim is to maintain a forest canopy over much of the woodland area and to allow the trees to continue to mature. At the same time, small areas of other habitats—such as the grassland on Rough Common and Upper Seeds—are maintained in their present form in order to have areas of other habitats where research can be carried out.

### 1.7.2  The Woods are not an island

It is one thing to decide to maintain a broadleaved high forest woodland, as requested by ffennell, but quite a different matter to achieve such an aim. This is because things that happen outside the Wood have major effects on the Woods themselves. The most striking are those that have affected the abundance of some of the Wytham mammals and the effects that they have on the vegetation.

One of these is the grey squirrel. Grey squirrels were introduced into several places in Britain during the period 1876–1929, spreading rapidly during the 1920s and causing the local extinction of our native red squirrel as it did so. They are an abundant member of the Wytham fauna where they seriously damage young beech trees, by girdling them either near the base or in the lower crown. It remains to be seen if any of those planted since the 1950s will survive to be replacements for the fine, old beeches that we admire along the Singing Way when they die. If not, the Woods will be the poorer.

The deer (Chapters 5, 10 and 11, Sections 5.6, 10.2 and 11.8) are also a major management problem, and their numbers are no longer influenced by native predators such as wolves. There are three species. The native roe, the fallow (introduced by the Normans), and the muntjac (introduced by the Duke of Bedford—who was also largely responsible for the grey squirrel!). In the 1950s, it was a rare event to see a deer, and there were only fallow in the Woods. During the very cold winter of 1962–63, the warden located the only herd in the Woods, then numbering 13, followed their tracks through the snow and saw them cross the frozen Thames towards Cassington. They later came back!

The deer started to increase towards the end of the 1960s and during the 1970s and by then all three species were present. The reason for this big change is still debated, but one view is that the increase was strongly linked to the change from sowing wheat in spring to sowing in the autumn. Formerly the fields around the Woods were brown plough all winter; afterwards they were lush green grass on which the deer came out to graze—and flourish. The University responded to complaints from the neighbouring farms that the Woods were harbouring all the deer that ate their crops, by surrounding the Woods with a 10.2 km long deer fence, completed in 1989. This restrained the deer population and possibly resulted in them making larger inroads into the vegetation than they might otherwise have done. Almost everywhere a browse line was visible; young growth on hazel was largely absent. Although deer numbers are now much reduced, their presence at even the current levels may mean that future tree regeneration will be reduced.

In contrast to the increases of deer and squirrels, rabbits have declined in recent years. Some idea of their past impact can be gained from a quote from the University's Working Plan (for 1949/50–1959/60) for the Woods:

> The Condition of the crop…was of…little value over much the greater part of the area. Rabbits which have existed in very large numbers have prevented satisfactory regeneration…..and there was….much loss of coppice.

However, the rabbit disease *Myxomatosis*, introduced to Britain, reached Wytham about 1955. The once-common rabbit was almost exterminated and many of its predators, such as foxes, were also severely affected and stoats virtually disappeared from Wytham. Mick Southern (1970) later commented that 'it is tempting to believe that *Myxomatosis* caused the big upheaval, lasting over several years, before the situation settled down again'. Even today, rabbits are nowhere near as abundant as they were before the advent of *Myxomatosis*. The legacy of the rabbit decline was that everywhere the little patches of open grassland, together with special floras, soon became overgrown with scrub. In the face of the drastic decline of one of its main prey, the weasel changed its hunting behaviour with serious outcomes for some nesting birds (see Chapter 10 Section 10.3).

Other arrivals from outside have also been unwelcome. The possible role of Dutch Elm Disease in the prehistoric elm decline has already been mentioned, but the more recent outbreak in the 1970s made a major impact on the British countryside and also left its mark in Wytham. Within the woodland areas, elms were not a major part of the forest cover, though there were exceptions (the top corner of Marley and the 'Five Sisters' on the corner of Rough Common). However, there were large specimens along the edges of the Woods and along the hedgerows in the farmland. Nowadays, all that we have left are little groups of small trees, which once they reach about 15–20 cm in diameter are large enough to become infected and die. Some of these trees survive to become just large enough to set seed, and so one could imagine the species surviving as a very minor component of the woodland for a very long time—perhaps until some of them develop immunity to the disease. Many of the larger elms were hollow and provided good nesting or roosting places for many birds and bats, but the effects—if any—of their disappearance are not known.

Even the oaks are at risk. The knopper gall, which destroys the acorns and so prevents the trees from setting seed, has now spread to Wytham. Even if the acorns germinate, the seedlings seem to struggle to survive: there are very few naturally regenerated seedlings within the woodland areas. Until the early 1900s, young oaks seemed to be able to regenerate in open woodland as well. Rackham (2006) suggests that with the arrival in Europe in 1908 of a silvery bloom, the fungus *Microsphaera alphitoides* common on almost all young oaks, the oak's competitiveness has been reduced such that in shaded areas the young plants can no longer survive. However, oak does still regenerate in the more open grassland areas, so it may be here that we will need to look for the future generations of old oak.

Other outside influences include the agricultural pesticides which destroyed the sparrowhawks (Chapter 9 Section 9.4). Doubtless there will be many others. Wytham is definitely not an island!

## 1.8 Conclusion

Like all English woods, Wytham is like a book which records the activities of people over many centuries. The visitor can easily see traces of the last 400 years of human activity interacting with natural processes; in places even older relationships can be found. The challenge for researchers is to document these, to try to understand what they mean, both for Wytham and woods elsewhere; but importantly also to try to predict how matters will change in the future. That challenge has been taken up by Oxford ecologists over the last 60 years and continues to drive the coming generation. The fruits of their work can be found in the rest of the book.

# 2
# The Physical Environment

M.E. Taylor, M.D. Morecroft, and H.R. Oliver

## 2.1 Location

Wytham Woods are located 5 km north-west of the city of Oxford, (National Grid Reference SP 462083). The woodland and associated grassland cover 415 ha in a loop of the River Thames (Fig. 2.1). The Anglo-Saxon for a bend or loop (wiht), gives Wytham its

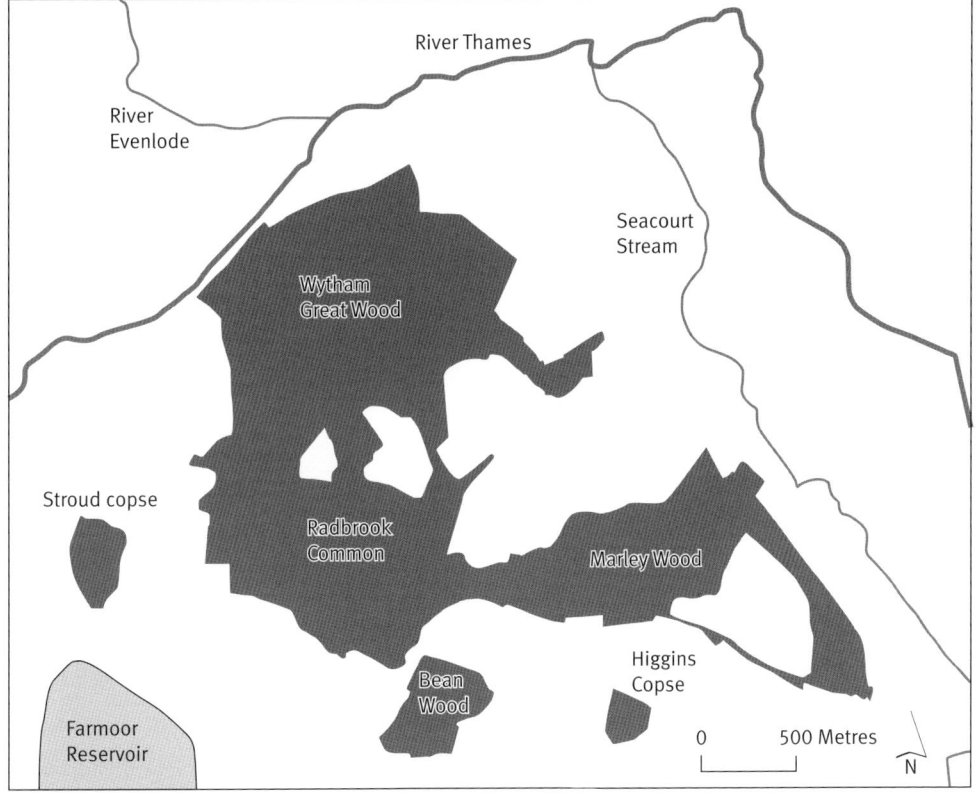

**Figure 2.1** Wytham Woods and surrounding features.
Source: Michèle Taylor.

20 Wytham Woods

name (Arkell 1945). Most of the site is in one main block of 386 ha (including areas of semi-natural grassland bounded by the Woods), with Bean Wood (18 ha) to the south attached now through an area of scrub. In addition, there are four smaller detached woods (Stroud Copse, Higgins Copse, Cammoor Copse, and Stimpson's Copse), totalling 12 ha, to the south and west. There are 26 ha of grassland within the Woods.

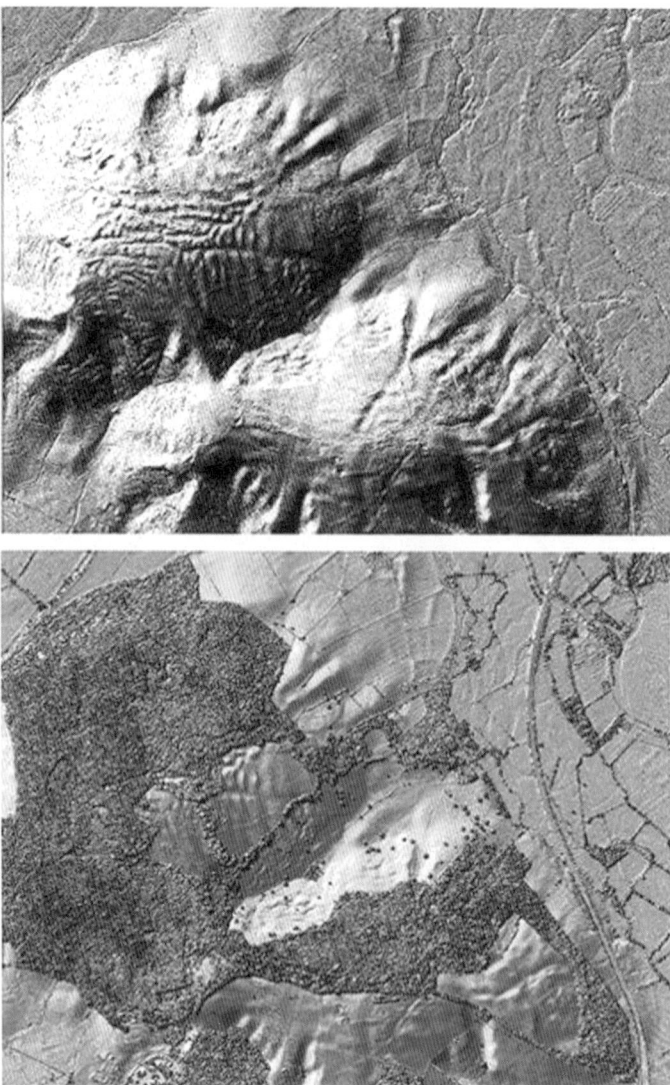

**Figure 2.2** Shaded relief images of Wytham showing the topography (top image) and overlying woodland (bottom image). These images are derived from airborne LiDAR data which were acquired in summer 2005 by the NERC Airborne Research and Survey Facility. These LiDAR data were processed by Ross Hill of Bournemouth University whilst based at CEH.

## 2.2 Topography

The land rises from 60 m above sea level on the river floodplain, to 164 m on Wytham Hill, with a ridge running east to Seacourt Hill (148 m) (Fig. 2.2). This high ground is an outlying area of the Oxford Heights limestone ridge to the south-east, part of the Mid-vale ridge stretching from Wiltshire to Buckinghamshire (Arkell 1945). Beyond Wheatley the limestones and sandstones that form the ridge pass into clays so the ridge dies away. This high ground forms a barrier to the River Thames flowing eastwards, causing it to flow north before turning east and then south, encompassing Wytham.

## 2.3 Climate

The Oxford area lies at the very heart of England, as far as it is possible to be from the climatologically ameliorating influence of the sea. The Oxford University Radcliffe Meteorological Observatory in Oxford boasts one of the longest data sets available and lies 5 km to the east of the Wytham site. The main Wytham meteorological observation site (Fig. 2.3) was established in 1992 as part of the Environmental Change Network (ECN) monitoring (Box 2.1). It is located near the top of Wytham Hill on grassland known as Upper Seeds. It is at an altitude of 160 m, in comparison with 63 m at the Radcliffe Meteorological Observatory.

An average temperature index for Central England—the 'Central England land surface Temperature data set' (CET)—allows the Oxford and Wytham records to be put in a regional context. Analysis shows consistent results, with the Radcliffe Observatory being 0.5 °C higher than CET and Wytham 0.3 °C lower. The Wytham site is significantly cooler than Oxford city due both to its greater altitude and its rural as opposed to urban location. Figure 2.4 shows the relatively recent Wytham air temperature data set placed into a long-term context with the Radcliffe Observatory data. Over the data period from 1993–2007 the Wytham monthly mean temperatures ranged from 1.6 °C in January 1997 to 20.3 °C in July 2006, with the long-term annual mean being 10.0 °C.

Rainfall comparisons between Oxford and Wytham show considerable variation as they depend on the spatial differences between localized storms (especially summer thunderstorms) as well as more consistent longer term differences due to altitude. However, over the entire data collection period, the Wytham site has a rainfall approximately 4 per cent higher than the Radcliffe Observatory. The annual rainfall at Wytham from 1993–2007 varied from 499 mm in 1996 to 923 mm in 2000, with an annual average value of 717 mm.

Woodland has long been known to experience a different microclimate to open sites (Geiger 1965). In addition to the standard meteorological recording on Upper Seeds, a variety of other measurements of climate and microclimate have been made at other locations at Wytham. Measurements of the woodland microclimate were carried out for 10 years at Ten Acre Copse and on a canopy walkway in Wytham Great Wood. Comparison of long-term mean air temperatures in the woodland and an open site showed that, in

**Figure 2.3** Photo of Automatic Weather Station, Upper Seeds.
Source: Michèle Taylor.

the winter, the near-surface soil and air temperatures were close to the grassland air temperature but, in summer, average air temperatures were up to 1 °C cooler and soil temperatures up to 3 °C cooler under the forest canopy. Maximum air temperatures followed a similar but more pronounced pattern. More marked differences were found in diurnal temperatures: for instance, in the middle of a sunny day in summer, woodland air temperatures could be up to three degrees lower than grassland. Soil temperatures are considerably less variable in their daily fluctuations than air temperatures, with temperatures at 100 mm depth in woodland varying by 1 °C, as opposed to 7 °C on grassland (Morecroft *et al.* 1998).

The greatest differences in air temperature between the grassland of Upper Seeds on Wytham Hill and the woodland at Ten Acre Copse (altitude 65 m) were found for minimum

> **Box 2.1** Environmental Change Network
>
> The Environmental Change Network (ECN) is a network of terrestrial and freshwater sites monitoring a range of physical, chemical, and biological measurements in order to detect and understand changes in ecosystems resulting from climate and other environmental changes. Wytham Woods and the farmland to the north and west (Northfield and Home Farms) are one of the 12 terrestrial sites with the Network.

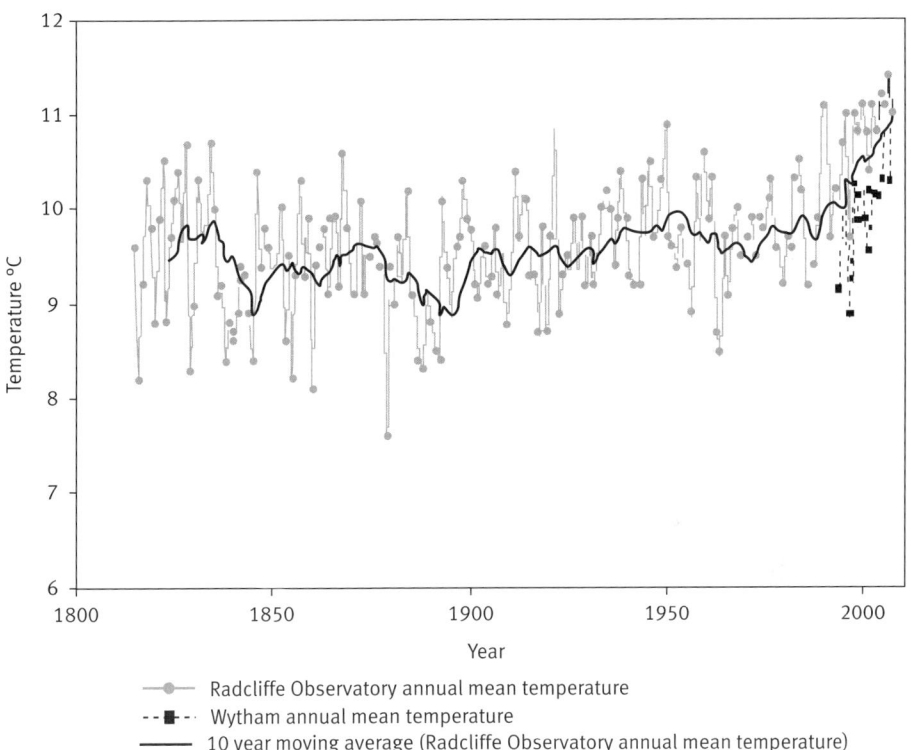

**Figure 2.4** Graph of annual mean air temperature at the Radcliffe Observatory (1815–2007) with a ten year running average and at Upper Seeds, Wytham. Data source: School of Geography and the Environment, University of Oxford and Centre for Ecology and Hydrology.

temperatures. During clear and calm winter nights the air temperature at Ten Acre Copse could be up to 4 °C lower, due to katabatic cooling (cold air flowing down the slope).

For short grass areas there are very marked differences in soil temperatures associated with different slopes and aspects, as all the solar radiation input is intercepted by a shallow layer. This effect is far less marked for forested areas because solar energy absorption is spread over a far deeper vegetated volume and the soil surface is shielded from direct radiation.

The direct solar radiation input to the ground surface is very small during the summer when the leaves are present but shows a proportionate increase before leaf development and after leaf fall. Similarly, the relatively light below-canopy wind speeds showed a small increase when the leaves are not present.

## 2.4 Geology and soils

The majority of the Woods are located on three Jurassic strata—Corallian Limestone, Calcareous Grit, and Oxford Clay—which tilt gently in a south-easterly direction. On the lower slopes, up to approximately 110 m, is the Oxford Clay, a Jurassic sediment laid down about 164 to 154 million years ago when the area was a muddy sea. The depth of the clay is recorded as being 91 m from a boring made in 1829, whilst searching for coal (Arkell 1945). Bordering the Thames, and just encroaching into the woodland on the northern edge, is a strip of clayey river alluvium overlying calcareous gravels.

Above the clay, on the higher ground, are sand and limestone deposits, part of the Corallian Formation laid down when sea levels fell leaving a warm shallow sea during the Upper Jurassic period. Approximately 159 million years ago the hills would have been coral reefs, with sand depositing beyond the reef. The sand is light yellow and very loose, and is classified as Calcareous grit. At Wytham this sand is predominantly exposed on the steep western slopes and the saddle between the two hills.

The Coral Rag limestone that caps the high ground is a rubbly limestone, of about 13 m depth. It has been extensively quarried in the past with the stone being used for building, road foundations, and lime. The stone is typified by large amounts of fossil corals with good examples of the 'honeycomb' *Isastrea* regularly found in the limestone outcrops. The limestone dome is irregular in shape, overlying the sand layer to the north and, in places, reaching as far as down as the Oxford clay.

On the higher land, in the south-east part of the woods (around the area known as the Plantation), Plateau Gravel occurs. This is a stony drift deposit made up mainly of quartzite cobbles and fragments of vein quartz. It was brought from country to the north-west by Pleistocene rivers, or by a precursor of the modern Evenlode some half a million years ago.

There has been considerable movement and displacement of the Coral Rag down the slopes of Wytham Hill due to springs washing out the underlying sand until a large block of unsupported limestone collapses parallel with the line of seepage. Spring sapping continues at a higher level and eventually a new block founders. The blocks then form a series of step-like ridges or terraces (see Arkell 1947 pp. 147–8). Consequently, owing to the removal of the sand, the limestone lies in contact with the Oxford Clay, notably on the eastern side of the hill.

The soils of Wytham were extensively surveyed in 1993 by the former Soil Survey and Land Research Centre (now National Soil Resources Institute at Cranfield University), as

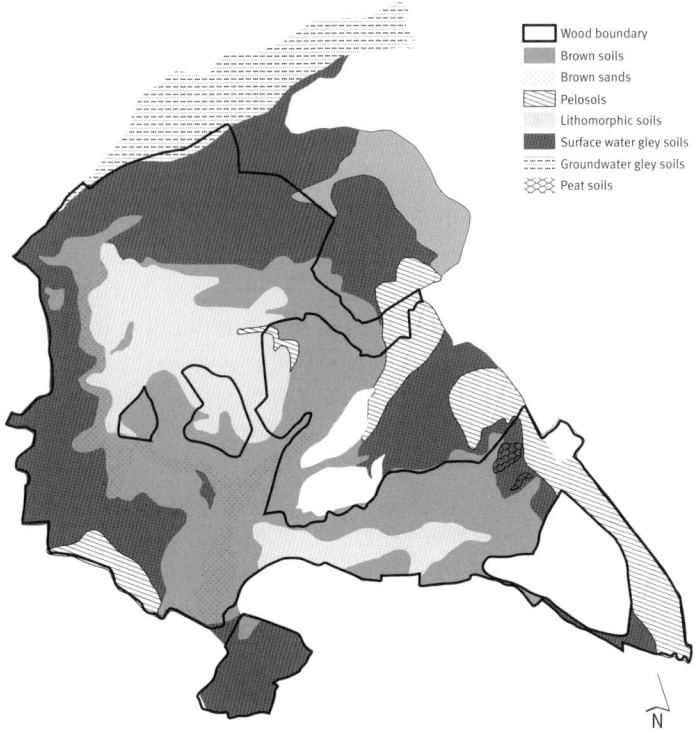

**Figure 2.5** Map of major soil groups at Wytham.
Source: National Soil Resources Institute.

part of the baseline information for the Environmental Change Network. Figure 2.5 shows the major soil groups according to the classification of the Soil Survey of England and Wales (Avery 1980), where soils are classified on the basis of horizon features such as stoniness and permeability.

Generally the soils reflect the underlying parent material. The Lithomorphic soils that occur above the Coral Rag limestone are a Rendzina soil, of the Sherbourne series. This is a shallow, stony, well drained calcareous clay soil, lying directly on the calcareous bedrock at 30 cm depth. Towards the edge of the limestone, soils are deeper at about 50 cm. Again they are stony, well drained calcareous clay soils containing medium and large limestone fragments, but are classed as a Brown earth soil rather than a Rendzina.

The Calcareous Grit deposits give rise to a Brown sand soil, one that is deep (greater than one metre) and generally well drained, although some seasonal water-logging may occur depending on fluctuating groundwater. Another Brown soil, of the Berkhamsted series, occurs on the small area of Plateau gravel. This is a moderately stony loam over a clay at 70 cm so, although it is permeable in the upper layers, downward movement of water is restricted. The greatest area of Brown soils are from the Oxpasture series, which are slightly stony and seasonally waterlogged and are found on the slopes below the Calcareous Grit. The other Brown soils (small areas of Burlsdon and Aberford series) are clay loams varying in stoniness and soil water regime, except for two of patches of Fyfield series, a stoneless loamy sand.

The most common soil at Wytham on the Oxford clay is the Denchworth series, a surface water gley soil (Stagnogley soil). These soils, together with small areas of Pinder and Wickham series, are deep and poorly drained clays, with rainwater unable to drain through quickly so that they become seasonally waterlogged, but also characteristically shrinking during dry periods, creating vertical cracks extending deep into the subsoil. The soils are mottled (patchy colouration), most obviously in the lower B (subsoil) horizon, and appear bluish-grey due to the lack of oxygen penetrating into the soil, causing the iron to be reduced. The Denchworth series, unlike the Pinder and Wickham series, are stoneless and non-calcareous to a depth of at least 60 cm.

The proximity of the River Thames to the lower lying land along northern edge of the Woods gives rise to groundwater gley soils that are seasonally waterlogged, but here it is caused when the water table rises with the river water level. The soil is of the Fladbury series, a stoneless clay with an appearance similar to the Stagnogley, but the soil material is moderately permeable.

The small areas of Pelosols on the western and eastern boundaries are, again, clay soils developed on the Oxford Clay, but while they can be seasonally waterlogged, they have slowly permeable subsoils with less obvious gleying than truly gleyed soils. However, in summer they will often develop cracks due to drying.

In Marley Wood (the eastern part of the Woods) there is a calcareous lowland fen which has formed where a spring occurs but drainage has been impeded and an area of peat built up to at least 2 m in depth (Hone *et al.* 2001). The Marley Wood Marsh is an unusual feature, which gives rise to a distinctive wetland vegetation type, quite unlike anything else in the woods (see Chapter 3 Section 3.1) and covers approximately 2 ha.

## 2.5 Hydrology

The limestone and sand on the top of the hills are permeable and the soils are consequently well drained. In contrast, the clay is impermeable and the soils become waterlogged during wet conditions, particularly in winter. Clay is also prone to shrinkage and cracking during dry weather in summer. Although the clay is deep, the effects of drought on vegetation, such as the early loss of leaves from trees, are often seen first on the clays.

Soil water content is monitored at the Upper Seeds weather station and has also been intermittently recorded on clay soils at the bottom of the northern slope. In general, soils are saturated for most of the winter and dry out from late spring onwards. A dry winter or a wet summer may modify this pattern, but the seasonal cycle remains apparent. The presence of the forest canopy affects soils' hydrology and rates of water loss from the soil increase once leaves emerge in the spring and transpiration begins. Water loss can be up to 50 per cent higher at the woodland edge (Herbst *et al.* 2007) than the centre, because the trees receive high light levels from the side, as well as above, leading to the development of a greater leaf area. This causes measurably faster drying of the soil at the edge.

A series of small spring fed streams rise within the woods at a seepage line, where water percolating through the limestone and sand reaches the clay. These watercourses are mainly dry in summer and only flow from late autumn, although the soil at the seepage line typically remains wet for most of the year and supports small patches (typically a few square metres) of wetland vegetation dominated by sedges and rushes. One semi-permanent stream flows south-west from Radbrook Common, with two manmade ponds occurring along its course, which were created for duck shooting in the nineteenth century.

The streams flowing north and west from the high ground drain into the River Thames, although, once on the farmland, much is through artificial channels, and the natural flow is altered greatly due to the presence of Farmoor Reservoir. Streams flowing eastwards from the Woods join the Seacourt Stream, which is a backwater of the Thames and ultimately rejoins it. The lowest lying ground at the northern end of the woods is on the floodplain of the Thames and flooding has occurred for periods of a few days, or even weeks, in some winters. It also, uniquely, flooded in July 2007 following extraordinarily heavy rainfall across much of central and southern England.

Weekly water samples, taken as part of the ECN monitoring from one of the ephemeral streams within the ancient woodland, had a mean pH of 7.8 between 1993 and 2007, reflecting the basic geology. The nitrate concentrations ($0.78 \pm 0.06$ mg l$^{-1}$) show a strong seasonal pattern, with concentrations peaking soon after the stream starts to flow in autumn and then quickly declining. The highest concentrations were found between 1993 and 2007 following hot, dry summers. The maximum weekly nitrate concentrations in autumn 1996 were 170 per cent higher than the maximum in autumn 1994 (Morecroft *et al.* 2000). This is consistent with experimental results (Jamieson *et al.* 1998), demonstrating an increase in the release of inorganic forms of nitrogen during the process of organic matter mineralization following drought, as well as a concentrating effect of lower water volumes.

## 2.6 Pollution

The main air pollution issues for Wytham, and other similar forests in much of Europe in the late twentieth century, were acidification and the fertilizing effects of nitrogen deposition. Although nitrogen is a nutrient, many plants typical of natural and semi-natural habitats are adapted to low levels of supply and when nitrogen availability increases, they are out-competed by other species more able to use the additional nitrogen to enhance growth.

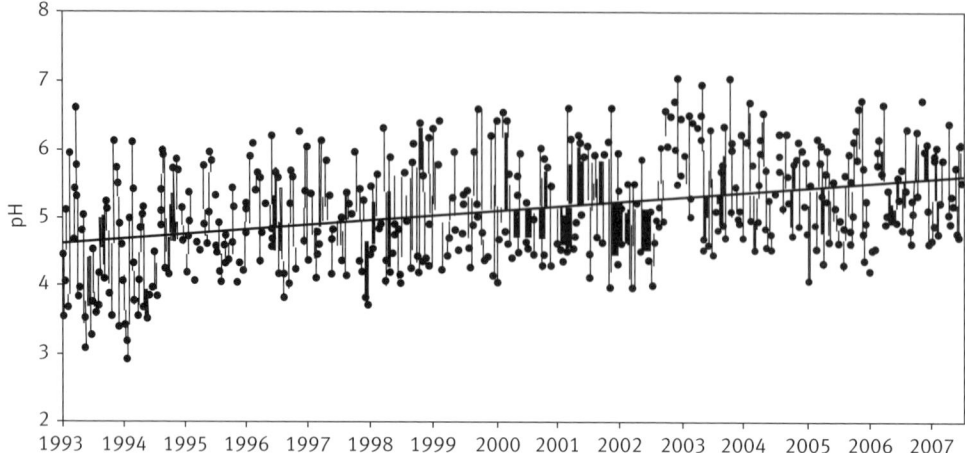

**Figure 2.6** Change in pH of rainwater 1993–2007.
Source: Centre for Ecology and Hydrology.

Oxides of nitrogen and sulphur are produced from fossil fuel burning in, for example, vehicles and power stations and dissolve in the rain to produce nitric and sulphuric acids. They were responsible for the concern over 'acid rain', which came to public consciousness across Europe in the 1980s. As a result emission controls were introduced. ECN monitoring at Wytham has shown a rise in the pH of rainfall following this (Fig. 2.6): mean pH in the years 1993–1996 was 4.7; from 2004 to 2007 it was 5.4 (the pH of unpolluted rainwater being approximately 5.6); this is a substantial change when it is remembered that pH is a logarithmic scale. In parallel with this, nitrogen dioxide levels, measured using diffusion tubes sited on Upper Seeds, have declined 50 per cent from 1996 to 2006, no doubt largely due to the compulsory introduction of catalytic converters on all new petrol cars from 1993. They also show a seasonal trend with higher concentrations in winter.

Although the problems of acidification are diminishing, nitrogen deposition continues to be a cause for concern. Most ammonia is produced from agricultural sources, particularly intensive livestock rearing, and these emissions remain high. Ammonia has been monitored on Upper Seeds as part of the UK National Ammonia Monitoring Network since 1997. Mean calculated ammonia level from 1997 to 2005 was 1 µg m$^{-3}$, a typical value for much of the UK, although relatively low for the lowlands of England (CEH 2007). Concentrations are, however, likely to vary over the Woods, depending on proximity to local sources of ammonia.

Dawkins and Field (1978) sampled the soils in forest monitoring plots in 1974 and Farmer (1995) re-sampled a sub-set of these plots 16 years later. These two surveys showed a large increase in soil nitrogen across the woods and some declines in pH, although, because of the calcareous geology, the soils at Wytham are buffered against the worst effects of acidification.

# 3
# Woods ancient and modern—land use history

C.W.D. Gibson and K.J. Kirby

## 3.1 Introduction

Wytham Woods are a typical enough few square miles of countryside to have lessons for ecology elsewhere. There is also enough preserved of their history (although not as much as in some places) to know how this landscape has developed over the centuries. As elsewhere the 'best' land has been farmed most intensively while the 'marginal' land has been used for common grazing, the provision of fuel and other woodland products. The nature and intensity of farming and woodland management has varied over the centuries: coppice management in the eighteenth and nineteenth centuries gave way to high forest and conifer plantations in the mid-twentieth. Today the conservation of biodiversity is a prime object of management.

This chapter draws heavily on the work of Grayson and Jones (1955), but sets it in the context of developments in our understanding of general woodland history since (Rackham 2003). However, there is undoubtedly scope for a more comprehensive review.

## 3.2 The early times

Sometime between 955 and 958 (see Fig. 3.1) King Eadwig of Wessex granted land to Abingdon Abbey. His charter survives, setting boundaries between mysterious places whose names generally have not. An exception is however Plum Leage, temptingly similar to the eighteenth century name 'Plumers' Coppice. This copse still stands on the western border of the Woods. If there is a real link, this charter is our earliest written reference to Wytham (Grayson and Jones 1955).

There are, however, even earlier glimpses of Wytham's ecological history, because Wytham is relatively unusual in lowland England in having a well-preserved pollen record covering the whole of the period since the last retreat of the ice (the Holocene) (Hone et al. 2001). Under the fossil corals of the limestone on the top of the hill lie sandy calcareous grits and below that clays. At some point during the retreat of the ice, a spring line at the junction of the grit with the Oxford clay fractured and split, making a wide

| Time period | Signficant events |
|---|---|
| **2000–present** | |
| 2005–2008 | 2009 Deer numbers estimated at 20 fallow, 20 roe, virtually no muntjac.<br>Majority of grassland now grazed, spring and autumn, by sheep.<br>2006 Flux tower established; increasing work on canopy interactions. |
| 2000–2005 | Fallow deer numbers reduced from about 440 to 40; muntjac reduced from about 200 to 15.<br>2002 Marley Wood separately fenced with the aim of keeping it deer-free.<br>2002 Restoration of Wormstall pond in Wytham Park. |
| **1900–2000** | |
| 1990 | 1998 Deer exclosures established within the wood.<br>1997 Cleveland's veteran tree survey.<br>1994–6 Severe drought effects monitored.<br>1994 Canopy walkway erected.<br>1992 accepted into Environmental Change Network. |
| 1980 | 1987–89 Main wood fenced.<br>1987 Initiation of long-term badger monitoring.<br>1987 Field margin study on Estate started.<br>1984 Upper Seeds experiment started.<br>1981 Last crop of wheat sown in Upper Seeds. |
| 1970 | 1974 Lloyd notes lack of management of the grassland and initial steps taken to clear scrub.<br>1973 Colyear Dawkins plots set up.<br>1972 Fox and Badger studies commenced. |
| 1960 | 1966 Elton's *A pattern of animal communities* published<br>1963 Woods removed from Dedication Agreement with Forestry Commission. Plantation creation stopped.<br>Last commercial coppicing. |
| 1950 | 1953 Myxomatosis largely eliminates rabbit population.<br>Varley and Gradwell Winter Moth surveys initiated.<br>First 10-yr forestry management plan.<br>Elton initiates the Wytham Survey. |
| 1940 | 1947 Study of tit populations begins.<br>1947 Southern's study of tawny owls begins.<br>1942 The Estate passes to the University.<br>Temporary cropping of parts of Radbrook Common. |
| 1930 | 1939 Death of Hazel ffennell. |
| 1920 | 1920 The Estate is bought by Raymond ffennell; some small replantings; chalet built. |
| 1910 | Fellings during the First World War.<br>Practice trenches dug on Sundays Hill. |
| 1900 | Estate rather neglected during lifetime of Seventh Earl. |
| **0–1900 AD** | |
| 1800 | Major restructuring of Estate<br>1814 Enclosure Acts allow fifth Earl to gain full control. |
| 1700 | 1799 Fifth earl, Montagu Bertie, succeeds to title. |
| 1600 | 1632 First reference to botanical record for Wytham. |

| | |
|---|---|
| | 1623 First reference to name Wytham Wood. |
| 1500 | 1538–1546 Disposal of monastic property at Wytham. |
| 1400 | 1439 Seacourt Village abandoned. |
| 1300 | |
| 1200 | |
| 1100 | Abingdon Abbey feudal overlords for the Wytham Estate: Henry 1st granted right to take roe deer and red deer from Cumnor (Wytham) Woods. |
| 1000 | 1086 Domesday Book, no direct reference. |
| 0 | 955–58 the Charter of King Eadwig. Roman occupation of Wytham Hill. |
| **10,000–0 BC** | |
| 1000 BC | Lime declines; increasing clearance of landscape. |
| 2000 | Neolithic activity in Upper Thames Valley; some woodland clearance apparent in pollen record. |
| 3000 | Elm declines. Mesolithic and early Neolithic activity on Corallian ridge. |
| 4000 | Lime starts to increase. |
| 6000 | Oak, elm, hazel woodland dominant. |
| 8000 | Open pine–birch woodland, with some juniper. |
| 10,000 | Northern part of Oxfordshire glaciated. |

**Figure 3.1** A time line for Wytham.

seepage on the slopes of what is now Marley Wood. On the seepage, wet conditions prevailed, first depositing fine clay and marl and then forming a peat fen. Most such places in the lowlands have been destroyed by drainage, but the fen in Marley survives to this day and preserves a rare pollen record of the vegetation right back through the Holocene.

The deposit still awaits full investigation, but the first examination (Hone *et al.* 2001) found an initial spread of pine woodland. This was followed by establishment of mixed forest in which lime and elm were conspicuous, as they seem to have been over much of lowland England (Grieg 1982, Rackham 2003). Elm declined at Wytham in parallel with other English sites, but lime remained dominant until Neolithic times. Later Bronze Age farming led to a decline in woodland overall, as indicated by the increase in the pollen of grasses and plants of disturbed ground.

By Roman times there were people living on the hilltop, including in areas that have long been wooded. When the gales of the late 1980s caused havoc in a 1920s pine plantation around Holly Hill, within the 'ancient' part of Wytham Great Wood, up with the pine roots came Romano-British hearths, pottery, and other artefacts. Elsewhere in Britain and on the continent evidence of former occupation on ancient woodland sites is becoming common and in some cases the effects on the woodland soils can be very long-lasting indeed (Dupouey *et al.* 2002).

More 'accidental archaeology' took place at Wytham when experimenters on small mammals and their ticks cut into Upper Seeds with burrow-proof metal sheets. The narrow trenches revealed pits, ditches, and hearths, associated with mainly Romano-British pottery. A treasured find was a whetstone still bearing the traces of knife sharpening. There was probably cultivation around this settlement, shown by a persistent scatter of pottery fragments across a wide area, and the possible effects of this on the soil phosphorus levels may still be detectable (see discussion of *Arrhenatherum elatius* abundance in Chapter 7 Section 7.6).

The people chose a good place to live with a wide vantage from the hilltop but close above the spring line in the Dell where water still emerges. Other evidence scattered around the estate suggests that there may have been earlier occupation.

## 3.3 The landscape up to the late eighteenth century

In the eleventh century the Domesday Book refers to plough land and meadows in the manors of Cumnor, Seacourt, and Wytham, but is silent about the Woods. However, there is later evidence from Abingdon Abbey records of the large Cumnor Woods, which remained the more common name for the Woods until two centuries ago (Grayson and Jones 1955). It may be that Radbrook Common remained well wooded (the South Wood) until later inroads of common grazing, and the name 'Newlands' suggests further small clearances.

During most of the medieval period, wood was a valuable commodity that might often be in short supply at least locally (Rackham 2003). Before canals brought cheap coal, firewood and charcoal were the main fuels. Before cheap coal brought cheap iron and steel, parts of tools now made of metal were wooden or only edged in metal. Houses might be framed in timber but with walls of wattle (inter-woven thin rods such as hazel) and daub (mud mixed with straw); carts, boats, and ships were predominantly wooden. Thus wood in a bewildering variety of shapes and sizes was essential to life: no part of a tree was wasted. Two examples from Wytham illustrate this (Grayson and Jones 1955). Abingdon Abbey granted to Godstow Nunnery the right to 'fowre burdyns of thornys' every day from Cumnor Wood: later, in 1632 eight oaks were purchased for £11.6s.5d from Wytham Wood (the earliest known use of this name) to make the great gates of the University's Schools Quadrangle.

Before power tools became common, trees were normally cut when they were the right size for the job in hand, to save the effort of cutting up large timber. Most British native trees and shrubs then grow again, provided livestock or wild animals do not browse them. This forms the basis for coppicing, if the stems are cut close to the ground; pollarding when the trees are cut out of reach of browsing; or shredding, the high pruning of the branches from (usually) hedgerow or field trees. The coppice stool, pollard bole, or shredded trunk can live on through many cycles of cutting.

The oldest living things in Wytham are therefore probably coppice stools. Each time the stool is cut, new shoots arise from the rim of the stump. Given enough time and repeated cuttings, the shoots spread out like a fairy ring from the site of the original trunk, which has long since rotted away. The resulting 'stool' forms a circle, or part of a circle, from which growth springs after every cutting. Many field maple and ash stools in Ten Acre Copse, Marley Wood, and Plumers Copse, are as much as 4 m across, but in Higgins Copse there stands a connected ring of ash trunks forming a quarter circle arc, with over 6 m between the ends of the arc. The age of such a tree is impossible to guess accurately, but is likely to be in excess of a thousand years.

Sometime around or after the Black Death (1348) many villages in England were deserted. The settlement of Seacourt next to Marley Wood, which had been larger than Wytham village itself, was abandoned; there was still a vicar but he was virtually without parishioners by 1439. However, for much of the medieval period most land was exploited: the best was used for arable crops, pasture, and meadow; the common 'waste' provided more extensive pasture, bedding such as bracken, and was the source of nutrients that were transferred, via the livestock and their manure, to the arable land.

### 3.3.1 Coppice and pollards

The future Wytham Estate remained relatively rich in woods. Abingdon Abbey, and later those who acquired the land after the Dissolution, kept control over much of the High Wood (now Great Wood) and perhaps some of the other coppices. Wood-banks between wood and pasture or wood and common, still visible today, were part of the boundaries used to restrict grazing in the woods long enough for cut coppice to regenerate. Much of Cumnor parish was divided into tithings for administrative and land rights, and of these Stroud tithing was perhaps the worst off, with only small woods: Stroud Copse itself, Plumers Copse, Machins (now Nealings), and the tiny Fairbirds (or Fosberts) Copse.

In the woods, timber and coppice were harvested young, in comparison with the abundance of old trees and dead wood found in near-natural temperate forests, such as Bialowieza in Poland (Falinski 1986). Consequently medieval woods were poor places for animals and plants associated with old trees. Rotten wood was no use except for the small need for tinder for striking fire.

Outside the woods, for example in parks, royal forests and commons, and on wood edges, pollarded trees provided a resource for an astonishing variety of invertebrates, fungi, birds, and bats that need the structures or the sheltered decaying wood. The resulting bole is thought to be a good mimic of a primary forest veteran and may survive for even longer producing regular wood crops from regrowth above the bole.

Trees on the common and on field boundaries at Wytham helped ensure the survival of at least some of the special species associated with veteran trees and dead wood. When, much later, pollarding and common grazing declined as an economic practice, the fashionable creation of Wytham Park early in the nineteenth century (Grayson and

Jones 1955) provided another reason for retaining and fostering old trees. The importance of this aspect of the Woods' ecology is discussed in Chapter 8 Section 8.11.

### 3.3.2 Field and furrow

As important as the Woods to the medieval economy were the grazing rights on the unenclosed swathe of common land straddling Wytham and Cumnor parishes and lying to the east and south of the present Great Wood. This is likely to have maintained the hilltop as open limestone grassland, with trees surviving on the lower slopes, especially on the Wytham side. On these lower areas, the open ground would have been a mixture of neutral grasslands, marshy areas, and probably some bracken fields.

Away from the Woods and common, the detailed history of land use is not always clear, but arable land was widespread (Grayson and Jones 1955). There are hints of scrappy management and poor yields for the time, especially in the Cumnor tithings, but the remains of a 'classic' three field system appear to exist for Wytham itself. There was a North Field preserved in the name of Northfield Farm; Down Field between this and the village; demesne (retained by the abbey and later lords) immediately west of the village; and the Upper Field or Upfield occupying land above the valley of Wytham Park on the site of the present Lower Seeds and extending into part of Upper Seeds. Much of the land above the floodplain and outside the Woods and common thus appears to have been arable at least part of the time.

On the floodplain the soils tended to be too fertile for arable use until modern varieties of grain were developed. The crops grew tall and tended to lodge and fall over, making harvest by traditional methods impossible. Instead such land was used as seasonal pasture and, as the practice of hay conservation spread, was shut up for a hay crop during the spring and early summer. The estate had vast areas of these meadows, contained as it is in a bend of the Thames with extensive floodplain areas. Land divisions, however, produced some odd results here with, for instance, Summerford Meadow belonging to Cumnor rather than Wytham, despite its remoteness from the Cumnor side (Grayson and Jones 1955).

### 3.3.3 Stone

The final resource that shaped the landscape of the estate was stone. Although much of the Corallian limestone is of little use for building, it is associated with some better material, namely Wheatley stone. Also the Abingdon to Eynsham road, which survives as the Singing Way, needed maintenance and the quality of stone was probably less important than it being close at hand. Road stone was won from fields beside the Singing Way (Stony Pieces) and probably from Rough Common. Eighteenth century advertisements for timber sales (below) locate some woodland (the current Hill Copse) as near these 'stone pits'. More extensive quarrying took place on the common to the north, with a major excavation to the west of the present chalet. The Earls of Abingdon presumably stopped winning stone from beside the Singing Way when ornamental plantations were made there. Aside from this, it appears that quarrying around Rough Common continued into the nineteenth century but not beyond. This past quarrying remains, however, a major influence on the local ecology—the grassland that re-established on the quarries after they were abandoned was less likely to be colonized by scrub after rabbits declined in the 1950s.

By the middle of the eighteenth century there had been a change from the old open field system, but otherwise the broad distribution of woods, common, farmland, and meadow probably retained much of the pattern from the early medieval period.

## 3.4 Exploiting the Woods

By the 1770s, the Earls of Abingdon appear to have taken most of the coppice and timber rights in hand and most or all of the products were sold standing at auction in the White Hart at Wytham. For the period 1772 to 1840, Rachel Thomas painstakingly extracted the advertisement details from *Jacksons Oxford Journal*; the results show how the Woods were managed at this period in some detail (R.C. Thomas unpublished).

The advertisements give most detail at the beginning of this period, often with an acreage, the type of product being sold and the coppice name, sometimes even further subdivisions. For instance in 1774, 22 acres (1 acre = 0.4 ha) of coppice in Lords Copse and 1.5 acres in Nealings went up for sale. In the same year 145 maiden (single stem) oaks were sold from Wherren's ground quarter of Marley Wood and from Bean Wood. Like all advertisements, coppice was described for its quality and oaks were often promoted as being good for shipping and the Navy. More fragmentary records from the estate accounts show that these claims often have to be taken with a pinch of salt. A very large oak in the 1780s did sell for £18 but the average for woodland oak was only about £2 10s, with oak from the poorest clays (Lower Common) only fetching 14s per tree. Coppice generally sold for £10 to £12 per acre and elm and ash trees for less than a pound each.

Timber from a wood often seems to have been sold with the coppice: the precise numbers given suggest that the trees selected for felling would have been marked beforehand. Throughout this period the estate seems to have been aiming for a coppice cycle of ten years but this was not rigid: even the advertisements state between ten and fourteen years' growth and some gaps suggest a longer rotation in places, unless an intervening cut was reserved for estate use and not sold. The acreages given may not have been very accurate, but the data suggest an average of about 50 acres a year of coppice was sold, with some coppice sold in most years: only 11 of the 69 years lack an advertisement.

Oak initially appears to have been the only timber tree sold, with all other species being managed as coppice. There were massive sales of oak between 1772 and 1790, with about 2500 stems going. There was then a long gap, presumably because any good-sized trees had already gone, with no more oak advertised between 1795 and 1827. However, from 1790 the estate was selling ash and elm timber trees from hedges and fields as well.

From 1827 to 1840 there were five years that had timber (as opposed to coppice) sales which involved ash and/or elm as well as oak, with one walnut in 1831. Late in the period seven ash from the Woods were sold as timber trees.

Both timber and coppice were sold on occasion from well-wooded parts of the common land on the Wytham village side. This may reflect earlier enclosures made after the Dissolution, when more than one landowner was jostling for the Abbey lands and on several occasions appeared to have taken more than their due. Very few trees sold were pollards, suggesting that these parts were protected from grazing before the main enclosure of 1814. The only pollards sold from the Wytham estate were 70 ash in 1833 and an unspecified number, again ash, in 1826. The scarcity of pollard sales from field and hedgerow trees is puzzling given the survival of such trees today. Either the modern survivors are mostly from pollarding after this, or the pollard products were mainly kept within the estate. In contrast, there were frequent references to oak pollards from other parts of the Earl's estates in the south of Cumnor and near Bagley Wood.

The broad picture is that most of most of the Woods were run as coppice with mainly oak standards along fairly traditional lines, albeit the harvesting of timber might have been more exploitative (did they run out of good oak by 1790?).

## 3.5 The legacy of the fifth Earl of Abingdon

One of the greatest changes in the estate's history took place when the fifth Earl of Abingdon claimed his rights under the Enclosure Acts in 1814 and 1816 and a massive programme of restructuring and new planting took place (Grayson and Jones 1955).

The fifth Earl inherited in 1799 and lived until 1854. He appears to have been impatient to improve the estate according to his lights, as some of the plantings date from before the Enclosure Acts—presumably on land he already occupied. The changes he initiated took place throughout his lifetime, although the major parts were enabled by the 1814 Act. Land consolidation on the Cumnor side of the estate, however, continued. Maps made in 1808 set out various proposals, not all of which were implemented and some were later changed, but these maps, together with what survives on the ground now, give a good overall picture of the changes (Grayson and Jones 1955).

The greatest effects on the ecology of the estate were the end of common grazing rights, the widespread planting of new areas, the creation of an 'instant park' and probably changes in farming patterns of which unfortunately little evidence remains apart from major rationalization of field boundaries.

On the lower slopes, the common lands either quickly reverted to woodland or were planted. Some may already have contained scattered trees, such that the change was only one of degree of tree cover. In contrast much of the limestone hilltop seems to have stayed open, grazed probably by increasing populations of rabbits. Deer were absent from much of lowland England at the end of the eighteenth and beginning of the nineteenth century (Yalden 1999) so would not have contributed to keeping the grassland from becoming covered with scrub.

Woods ancient and modern—land use history 37

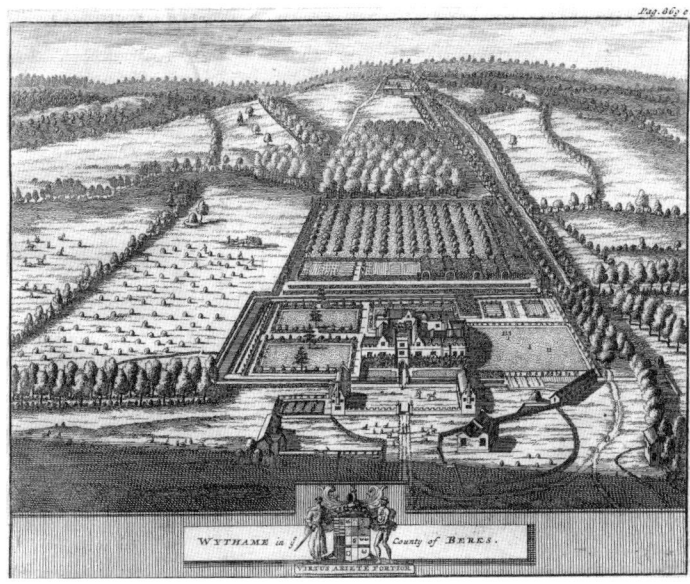

**Figure 3.2** Wytham Park from *Britannia Illustrata* (1st ed. 1720). Wytham Abbey originally had a moat around it. This was filled in by the fifth Earl of Abingdon.

Whatever the mechanism, the grasslands on limestone at least ensured a continuity of open habitats through this period. Other grasslands and bracken fields survived on a smaller scale on parts of Radbrook Common and the adjacent fields (the Woodbridges) that were reputed to be marginal for agriculture. These latter were taken into the Woods with the common enclosures but only planted in part (Grayson and Jones 1955).

The most obvious effects of the fifth Earl's activities to the present day observer come from his massive programme of planting. This was mostly of broadleaved trees, including species such as beech, wych elm, sycamore, hybrid limes, and sweet chestnuts that were previously rare or absent from Wytham. Conifers, European larch, and Scots pine, were used for the first time on a relatively large scale.

His aim was to create large scale sinuous belts and groups of planting to shield his new house (now Wytham 'Abbey') and to create landscaped drives and views all round the estate. He linked Marley Wood to Botley with a swathe of planting and carried it on round up the Singing Way through to the edge of Rough Common. With construction of the new valley road from Botley to Swinford, the Singing Way became a private drive. Further planted groups were laid around the fields of the Pasticks and in horizontal belts across the former Upper Field, of which Brogden's Belt survives. In the west end the scrawny remains of larches planted at the same time as the giant beech can still be seen. Still other belts and avenues linked across to the present chalet and along new sinuous broad rides that replaced the old straight ones in the Great Wood (Grayson and Jones 1955).

These plantations took time to achieve. The earliest appear to have been those closest to the house at Wytham. The size of the great beeches around the chalet might imply these were also amongst the early plantings. However, the great gale of October 1987 cracked one of them and left it teetering over the main chalet bedroom. This had to be felled, if only to soothe the inhabitants' nerves, and to everyone's surprise a ring count dated it to 1847, making it one of the last of the fifth Earl's plantings.

The Earl also created an instant park sometime in the first decades of the nineteenth century. The valley of the Earl's future Wytham Park encompassed the edges of Marley Wood and the small Mount Wood entirely on the valley sides. While some new trees were planted, the main parkland appearance was made by cutting out all of the underwood at the edge of Marley and in the whole of Mount Wood, leaving scattered large trees at a suitable density. The Earl planted a new Mount Wood as a backdrop to the park on the north side, which remains as the area beside the road leading from the village to the main entrance and car park (Grayson and Jones 1955).

## 3.6 Stasis and decline in the Woods

After the fifth Earl's death in 1854, the estate appears to have become more and more impoverished and his great impetus for plantation and estate improvement was not to be seen again. There are a few new features from this later period, for example the digging of flight ponds on Radbrook Common and Wormstall in the park (the latter possibly a refurbishment). Before this there had been no substantial water bodies on the hill. This also indicates that the sporting use continued; there were still four keepers on the estate between the First and Second World Wars.

During the First World War there was also a big area of ground disturbance from Sundays Hill down the slope to Nealings Copse, where raw recruits were trained in trench warfare before being sent to the front. The pattern of practice trenches, dugouts, and redoubts remains clearly visible (see Fig. 3.3).

In tune with the estate's impoverishment, there seems to have been exploitative felling of timber from the second half of the nineteenth century. There may also have been wartime fellings but by then there was little good timber left. Coppicing continued but with less intensity and in fewer places. In the 1920s two areas were planted with pine: around Thorneycroft and near Holly Hill in the Great Wood.

## 3.7 The development of a research estate

In 1920 the seventh Earl sold the estate to Raymond ffennell. In general the ffennells preserved the *status quo* in land management rather than initiating major changes, although in common with much of the country all open land possible was brought into cultivation in the Second World War. This included temporary cropping of part of Radbrook Common.

**Figure 3.3** Aerial photographs of Radbrook Common, (a) in April 1946 and (b) in June 1953 as it was beginning to be planted with trees. The circles in the bottom right hand of both indicate trenches dug for infantry training in the First World War. By 1953 strips of new plantations can be seen.

ffennell's philanthropy, besides new building, however, included setting aside Hill End for education and outdoor activities. This incidentally removed the surrounding land from intensive farming. Although it continued to be grazed, it escaped the later agricultural intensification that affected almost all the rest of the estate outside the Woods.

Ecological studies, or what would at the time be called natural history, had started in a small way during the nineteenth century and in the beginning of the twentieth century when the botanist G.C. Druce was active. His flora is deposited in the Herbarium in the Plant Sciences Department. The studies really got under way, however, after the University of Oxford was bequeathed the Woods and the end of the war allowed normal activities to resume. The University's tenure of the estate initiated a further period of major change.

The purpose of the ffennell bequest (see Chapter 1 Section 1.6) was for amenity, landscape, and education to ensure that the Woods were preserved. The remainder of the estate (acquired in 1942), aside from Hill End with its educational function, could be used for research but would also provide an income from farming. The estate thus became divided into the experimental farm at Northfield, the remaining tenant farms, and the Woods.

The experimental farm inherited the land between the former common, the Thames, and Wytham Village. Oxford University had a renowned agricultural school and the farm, while being productive itself, served as a focus for the research into husbandry, agrochemicals, and mechanization that drove the post-war boom in agricultural yields and

**Figure 3.4** Young broadleaved tree plantations with conifer 'nurses' on Radbrook Common, from Singing Way. Picture taken in May 1958.

productivity. The effect was the same as in most of lowland England, but perhaps more rapid and intense due to the research use. Pasture and meadow were converted to improved grassland or arable, weed species were greatly reduced and nutrient levels soared, influencing not only the land but the watercourses that drained it. By 1980, the only semi-natural habitats remaining were two small fields, the hedges and fragments on the edge of the old meadows, the riverbanks, and the nearby ditch system.

Events in the tenant farms, now merged into a few large units, paralleled this process although they varied according to the particular tenant. Again by 1980, semi-natural habitats had become restricted to a few hedgerows and verges, and places like Beacon Hill where the tumbled ground made agricultural improvement difficult.

In the Woods the research and education that is the subject of this book also gathered pace; forestry management was according to the good practice of the time. Conifers were used as nurses in mixed plantations mainly of broadleaves such as oak (see Fig. 3.4), ash, and chestnut. Large areas were planted, importantly removing open habitats on Radbrook Common and filling in the fields on the Pasticks with new plantation around the fifth Earl's boundary belts.

These changes were contemporary with the demise of rabbit populations after myxomatosis in the early 1950s, removing the last effective grazing from the remains of the open common land (Southern 1955). While ecologists, led by Charles Elton, secured a number of small areas as 'biological reserves' left out of forestry and intensive agriculture, these areas would not survive without management. Consequently, after the removal of Wytham from dedication to a Forestry Commission scheme in 1963 a period of neglect set in where management of plantations, woods, and the open habitats was less than that needed to maintain the wildlife of either coppice woodlands or the open areas.

However, we do not need to speculate about the effects of these more recent changes. They have taken place against the background of intensive research; and the findings from these studies form the major part of the rest of the book.

# 4

# The Woods in the modern landscape

K.J. Kirby and C.W.D. Gibson

## 4.1 Introduction

Wytham Woods, as described in Chapter 2, lie mostly on two limestone-capped hills that rise to 164 m above the Thames. The site is almost an island, within a great loop of the Thames, approached only from by-roads and its greater part kept in seclusion by Wytham Park (Plate 2). Nevertheless the Woods are linked physically, functionally, and historically with the landscape around them.

There is increasing interest in trying to understand how ecology works at the medium- to large-landscape scale: the diversity of individual woods or meadows is likely to decline if they remain isolated, and changes in climate over the next century will mean that species distributions will need to change to accommodate this (Hopkins *et al.* 2007). Wytham Woods and the surrounding landscape provide a test bed for how species might use this changing environment.

## 4.2 Wytham in the context of Oxfordshire' woods

Modern Oxfordshire has about 7 per cent woodland cover (18,235 ha), of which about 59 per cent is broadleaved. About 41 per cent of the total (c. 7600 ha) is ancient (Forestry Commission 2002; Spencer and Kirby 1992; unpublished data Natural England), of which about 63 per cent (c. 4700 ha) is classed as semi-natural (Box 4.1). The remainder of the ancient woodland is occupied by plantations, often established during the period 1945–85, when across the country landowners and the Forestry Commission were very active in clearing broadleaved woodland and replacing the stands with conifers (Tsouvalis 2000).

The distribution of woodland generally, and ancient woodland in particular, is neither random, nor evenly spread across the county (Wilson and Reid 1995; Forestry Commission 2002). As in much of southern England, woodland (hence surviving ancient woodland) tended to remain on land that was poor or difficult to farm, albeit what is considered poor or difficult changes over time as farming patterns change. At Wytham the ancient woods tend to be on heavy, intractable clays.

Woods are retained and managed where they have a purpose; through much of history even small woods were valued because they could provide local supplies of posts,

**Box 4.1.** Wytham Woods and the ancient woodland inventory

Peterken (1977) and Rackham (1976) developed the concept of ancient woodland—sites believed to have been wooded since at least 1600 AD—as a way of separating woods likely to be of high nature conservation and historical importance from more recent, less valuable sites. In 1981 the Nature Conservancy Council started to produce inventories of such ancient woodland across the country. Ancient woods were then divided into 'semi-natural' stands—those composed predominantly of native trees that had not obviously been planted—and plantations, which were mainly of conifers (Spencer and Kirby 1992). These fairly crude divisions of woodland according to origin and composition have proved very useful and have subsequently been incorporated in planning and forestry policies (Defra/FC 2005). However, while adequate on most sites, there are some places where the divisions are not clear-cut; Wytham is one such site.

Many of the old coppice areas are clearly ancient, the stands on Radbrook Common are recent, but areas that might have consisted of scattered trees on common land, i.e. wood-pasture, are less easy to classify. Wood-pasture was identified by Peterken (1977) as ancient woodland, but was only partially captured during the inventory process. In the north-east of Wytham, My Lady's Common Piece is an area of former common with scattered trees that was taken into the woods no later than the early nineteenth century; it was included within the 'ancient' category, but the other common lands that remained relatively open into the late nineteenth and early twentieth century were not.

Areas with a clear planting date and an even-aged high forest structure, were classed as 'plantations' even though some of these are of native species (for example in Further Clay Hill). Areas in the Great Wood with much sycamore were, however, classed as semi-natural because of their structure, albeit more disturbed than the copses on the lower slopes.

The boundaries shown on the inventory map for Wytham should therefore be considered as indicative of origins and composition, which are useful in broad-scale comparisons. However for any detailed work in the Woods reference should be made back to the work of Grayson and Jones (1955) and the Wytham Atlas.

poles, and timber for repairs to fences, gates, styles etc. (Edlin 1949) and in earlier centuries, fuel. The sawmill (Plate 3) at Wytham was still operating even into the 1980s.

Finally there are the accidents of history and land ownership. Some areas were within royal hunting forests and so tended to survive; elsewhere an estate, or a particular owner, was interested in trees; another was not. The influence of the fifth Earl of Abingdon on the structure and composition of Wytham Woods has already been discussed (Chapter 3 Section 3.4).

These and other factors produce the broad pattern of ancient woods in Oxfordshire today (Fig. 4.1). To the north of Wytham are the woods and parkland of estates such as Blenheim and Wychwood, and the pattern continues over the border into Gloucestershire as the hills and woods of the Cotswolds. As elsewhere in the country, ancient woods are not necessarily primary. For example, some of the woods in Wychwood overlie an area

The Woods in the modern landscape 45

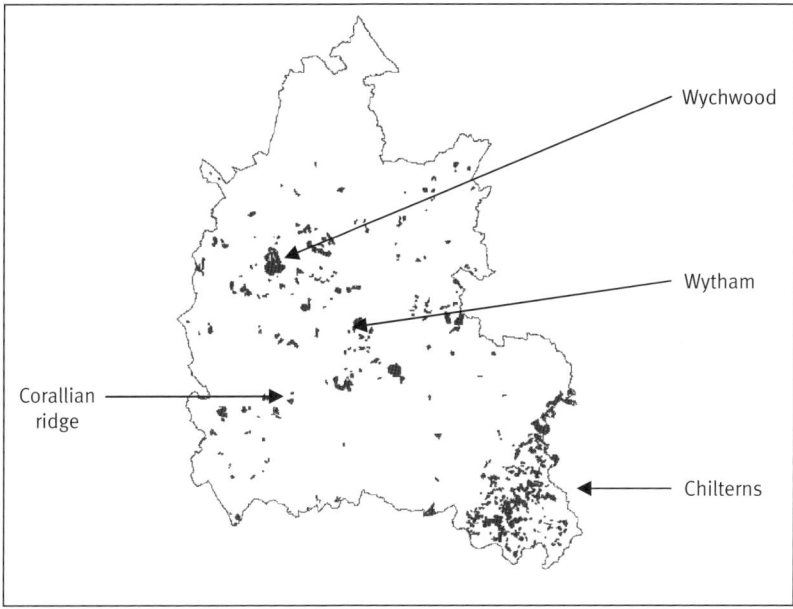

**Figure 4.1** Distribution of ancient woodland in Oxfordshire (Wilson and Reid 1995, unpublished data Natural England).

in which there was extensive Roman occupation (Copeland 2002), as do parts of Wytham (Chapter 3 Section 3.2). To the south are the Chilterns, again one of the most wooded parts of England. By contrast the fertile soils of the floodplains of the River Thames around Oxford and its tributaries, such as the Cherwell in the north around Banbury, have very little ancient woodland.

However there is a band of scattered ancient woods, with Wytham more or less at its centre, which runs from Faringdon, past Oxford, and continues through the clay vales, north-east into Buckinghamshire. In part this picks up the ridge of corallian limestone including the hills at Wytham (Fig. 4.1). It includes parts of former Royal Forests at Shotover and Bernwood (Steel 1984; Thomas 1987), and some substantial areas that survived on private estates such as at Middle Claydon, north-east of Bicester (Kirby and Wright 1988). More recent woodland tends to follow a similar pattern, reinforcing the parts of the county already rich in ancient woods. The overall area of woodland in the county has increased since 1990 (Forestry Commission 2002).

Across Oxfordshire there are also about 1.4 million trees outside woodland (Forestry Commission 2002). The comparative lack of woodland immediately to the north of Wytham is to some extent made up for by the frequency of non-woodland trees, in parks, on commons, in hedges, and along rivers and canals (Fig. 4.2). While there have been losses through hedgerow clearance in the recent past there are also new trees being put in, for example along the verges of the A34 and more recently on the University

**Figure 4.2** View across Wytham Park to the north-east of Wytham showing well-treed landscape.

Farm as part of a trial involving chickens foraging amongst trees (Jones *et al.* 2007). Many woodland species are thought of as relatively immobile, but they must be able to spread rapidly in some circumstances, or they would not have recolonized Britain as conditions warmed up after the last ice age. Small patches of trees, scrub and hedges may prove important for enabling species to shift their ranges in future in response to climate change (Hopkins *et al.* 2007).

Most ancient woods are small; nationally about 80 per cent are between 2 and 20 ha in area. Ancient woods less than 2 ha were not included in ancient woodland inventories (Spencer and Kirby 1992). Although there are a few very large blocks at Wychwood and in parts of the Chilterns this pattern is broadly true for Oxfordshire as well. There are 178 ha of ancient woodland on the Wytham Estate (including the outlying copses), and this includes some relatively large individual blocks such as the Great Wood and contiguous coppices, and Marley Wood. Woodland size is important because it is often linked to species-richness (e.g. Peterken and Gane 1984, but see Rackham 2003) and this pattern is seen in the individual blocks of ancient woodland in Wytham (Gibson 1988). However, whereas in much of the country ancient woods survive, isolated from each other, in Wytham some have become linked through the increase in woodland cover over the last two centuries.

Such woodland expansion and linkage is encouraged under recent forestry policies (Defra/FC 2005, 2007). The Wytham Estate can be used to test whether the expected benefits do arise. Certainly many woodland species, including some 'ancient woodland indicators' (Rose 1999) have spread into the more recent woodland (Chapter 6 Section 6.3). However, perhaps counter-intuitively some woodland plants have been lost from the Woods over roughly the same period as the area and connectivity of woodland habitats have increased (Gibson 1986): patch size is important but not everything. This decline seems to be linked to changes in 'woodland habitat quality' as a result of changes in management, particularly the conversion of some ancient coppice stands to plantations and reductions in open woodland as the coppice regime was abandoned. These are trends mirrored in national woodland surveys (Kirby *et al.* 2005).

The hills of the Cotswolds and the Chilterns are noted for their beech woodland, but beech is a relatively recent and limited component at Wytham (Chapter 3 Section 3.5; Kirby *et al.* 1996). Rather, the ancient coppices with ash and oak in the canopy and field maple, hazel, and thorns in the understorey (Chapter 6 Section 6.4) are more similar to those found commonly throughout the Midlands and into East Anglia, for example at Monks Wood in Cambridgeshire (Steele and Welch 1973). Overall, therefore, the woodland vegetation generally falls in the ash-field maple-dog's mercury type (W8) of the National Vegetation Classification (Rodwell 1991; Annex 4.1)—the most widespread type of woodland on base-rich soils in the lowlands. This equates broadly to Peterken's (1993) ash-maple-hazel and ash-hazel woodland types. There are also small areas of the beech-dog's mercury and the beech-bramble woodland. More details on the woodland composition and structure are given in Chapters 6 and 7.

Scrub is common at Wytham Woods. Of 66 woody plant species at Wytham over 40 participate in scrub formation at some stage of their life cycles. Chalk and limestone scrub (National Vegetation Classification (NVC) W21d; Annex 4.1) tends to be the most species-rich, although its extent at Wytham is limited because it has been cleared to open grassland. More common are the species-poor blackthorn and hawthorn scrubs (NVC W21a–c; Annex 4.1). There are also extensive patches of elder, nettles, and bracken

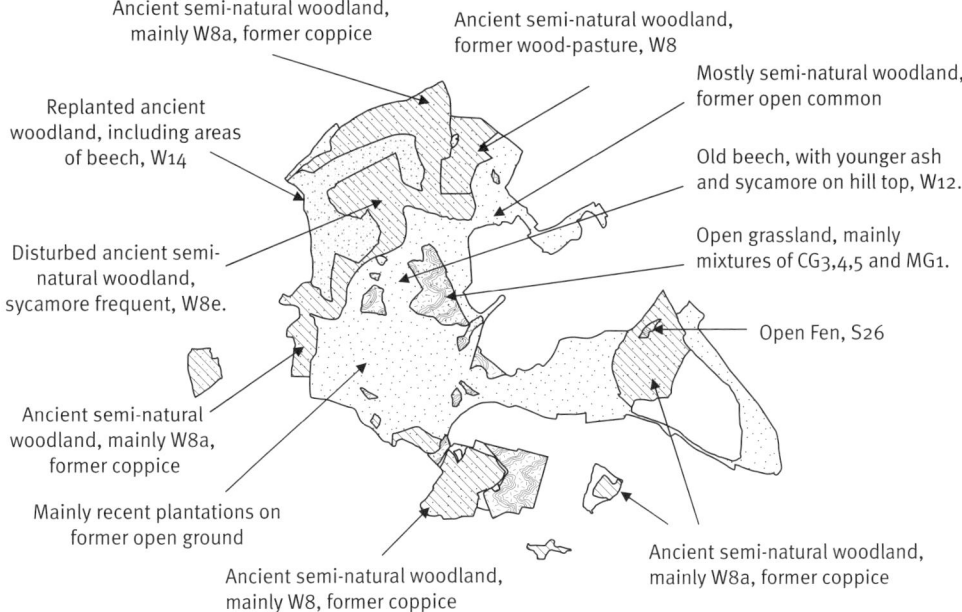

**Figure 4.3** The variety of woodland types and open habitats in Wytham Woods.

beds, often associated with badger setts, former rabbit warrens, and/or past elm invasion.

The role of different vegetation structures in providing opportunities for increased diversity in animal life was pioneered by Charles Elton and is particularly apparent at the edges between open habitats and woodland or where there is scrub. On thin soils many of the components of species-rich grassland may exist amongst low scrub with little loss of botanical diversity. The protection from grazing animals provided by the scrub may allow additional flowering and seed-production with concomitant increases in the associated invertebrates. Even bracken can sometimes function as a scrub species, sheltering species-rich vegetation, although at other times it becomes so dense as to exclude virtually all other plants.

On a slightly larger scale, scrub provides dense thorny structures and hollow dead stems used by birds and specialist invertebrates that shelter and/or nest in the scrub and forage on adjacent grasslands. Conversely, some grassland butterflies breeding in short grassland may obtain nectar from flowers that are protected by low scrub. Species that depend on the juxtaposition of one or more habitat structures for survival are legion. As Elton (1966) emphasized therefore, structure—both of the vegetation in a patch, of vegetation patches in a site, and across a landscape—is critical to maintaining the diversity of the countryside.

## 4.3 Grassland

Trees grow so well in the British climate that, without human intervention, there would be little open ground on the Wytham Estate. However, by medieval times much of the land was occupied by open habitats, because they were important to the local economy; the arable, pastures and meadows, land used for common grazing, and where fuel might be gathered. Even within the Woods open areas were not uncommon, either temporarily as in recently cut coppice, or more permanent spaces such as grassy rides. As a consequence many plants and animals in Britain now depend on open semi-natural habitats either for all or part of their lives, although today such habitat patches are often surrounded by intensively farmed countryside.

Among the most important open habitats are grasslands. Just beyond the boundaries of the Woods are the nationally important floodplain pastures and meadows on Port Meadow and in Pixey and Yarnton Meads, while further east are the expanses of wet grassland and associated habitats on Otmoor. The Chilterns and the Cotswolds are nationally important for their chalk and limestone grassland (Ratcliffe 1977).

The same principle applies to Wytham: the Woods are obviously dominated by woodland in area terms, but the diversity of the estate depends on the grassland, fen, ponds, etc. These smaller habitat types contribute out of all proportion to their sizes. The open areas of limestone grassland on the top of Wytham Hill are composed of mixtures of erect brome (*Bromus erectus*) and tor grass (*Brachypodium pinnatum*) (CG3, CG4, CG5 in the National Vegetation Classification terms, Rodwell 1992); Annex 4.1). The patches supporting the highest botanical diversity consist of mixtures of erect brome and tor grass on the skeletal soils of old quarries. These contain a great many other low-growing herbs and grasses characteristic of chalk and limestone grassland. However, even the best Wytham grassland lacks some of the typical species for these types nationally. Their absence is probably a consequence of this being a relatively small isolated patch of suitable geology. The effects of this lack ramify through the ecological community. One of the absences is the horseshoe vetch (*Hippocrepis comosa*), the only food plant of a number of insect species including the spectacular adonis blue and chalkhill blue butterflies (*Lysandra bellargus* and *L. condon*); consequently these butterflies are also not recorded from Wytham. Although small in extent compared to the grassland communities of the Cotswolds and Chilterns, the limestone grassland at Wytham may still be significant as a potential stepping stone in the landscape for species movement between these areas.

Across the country unimproved grassland has shown major declines over the last century through natural succession to scrub and woodland in areas where grazing (including by rabbits) has declined, and through ploughing and agricultural improvement by fertilization (NCC 1984; Natural England 2008). Similar changes occurred on the Wytham Estate, although the conversion of grassland to plantations was probably more significant than natural succession as a cause of species loss, because to some extent scrub clearance was successful in maintaining open areas when the rabbit populations declined (Gibson 1986).

Because of these changes, there are targets for restoring and recreating such grassland habitats under the UK Biodiversity Action Plan (HMSO 1994; English Nature 1998). Evidence to support how far this might be successful is coming from an experiment in recreating neutral grassland on the floodplain in the narrow strip of land between the Woods and the Thames at Somerford Mead (McDonald 2001 and personal communication). Cutting for hay at the end of June/early July and then not grazing reduces species-richness as tussocky grasses establish at the expense of herbs. Cutting for hay and grazing with cattle (the traditional management regime) results in diverse grassland within which both alluvial meadow plants and invertebrates flourish. Cutting for hay and grazing with sheep comes in between the other two regimes. These results have become more evident with time.

A parallel study to that on Somerford Mead is occurring on Upper Seeds within the Woods (Chapter 7). This experimental restoration to limestone grassland from intensive arable illustrates in particular how rich such successional areas can be in plants: the field had accumulated 247 species in less than ten years from the last crop being sown (1981) and 264 species over the 25 years to 2007.

The plants came from a number of different habitats. Some are specialists of chalk and limestone grassland, which is not surprising, but even some ancient woodland indicators (including *Acer campestre, Bromopsis ramosa, Carex pendula, Carex sylvatica,* and *Veronica montana*) moved into the field. There are others more often associated with neutral grassland (Annex 4.1). Although not obvious from maps or aerial photographs, mesotrophic/neutral grassland does occur in Wytham along some of the wider rides and ungrazed areas and in the woodland as well—often the colonizers of recently felled areas.

Another large group of species that have made use of the changing conditions in Upper Seeds are those associated with the middle years of succession to calcareous grassland, typically robust herbs such as field scabious (*Knautia arvensis*), perforate St John' wort (*Hypericum perforatum*) and wild parsnip (*Pastinaca sativa*). Many but not all are naturally short lived, but all generally take a few years to build up abundance, are common for a few decades, and then decline again (Gibson and Brown 1992).

Until the establishment of the experiment on Upper Seeds in 1984 (described in Chapter 7), these mid-successional plants had been sparse or even undetectable in Wytham Woods for many years. Likewise there were only a few intermittent records for specialist herbivores associated with these plants. However, some of these insect herbivores are very specialized, often in response to highly toxic secondary plant chemistry: they appear to be almost glued to their food plants, with very high powers of detection and dispersal. By 1988, only a few years after the start of the experiment, virtually all the specialist herbivores on these species within geographic and climatic range (some are coastal only) had appeared on Upper Seeds. These included species from groups such as Lepidoptera, Coleoptera, and Diptera. The life-styles of Diptera (true flies) are similarly diverse, including leaf-miners and other internal feeders, leaf-chewers, and seed-eaters.

## 4.4  Fens and other wetland

Scattered around Oxford are various small fens, including to the south Cothill Fen National Nature Reserve and a small area in Sydlings Copse to the east of Oxford. Other patches occur on the Wytham Estate; one patch even survives within the farmland (Watt and Kirby 1983). Relatively little work has been done on their composition, but they appear to correspond broadly to S26 tall herb fen in NVC terms (Rodwell 1995; Annex 4.1). Analysis of pollen remains from cores taken from the fens have contributed much to our understanding of the long-term history of the landscape (Day 1991; Hone *et al.* 2001, Chapter 3 Section 3.2).

The largest blocks of fen habitats are on the slopes in Marley Wood, and together cover several hectares. In contrast to the grasslands, the vegetation is botanically simple, with most areas dominated by the reed (*Phragmites australis*). There are few associated species and species richness at the scale of 2 × 2 m quadrats is much lower than in grasslands. Other open fen types cover a smaller area but are likewise botanically simple, being dominated by the great horsetail (*Equisetum telmateia*) or the sedges *Carex acutiformis* and *C. riparia*. Smaller areas of reed swamp survive on the remainder of the Wytham Estate, in riverside swamp and on ditch banks. Here pure reed is often juxtaposed with shorter fen vegetation and with open, still and flowing water.

Reed has a rich associated fauna of specialist herbivores, with the Diptera (true flies) being well represented, including species that also form galls on other tall wetland grasses. The other main group, the Lepidoptera, tends to have more species confined to reed as a food plant. Of these, only two moth species occur in Marley Fen: the southern wainscot (*Mythimna straminea*) and twin-spotted wainscot (*Archanara geminipuncta*). The remaining fen and swamp habitats add four more, giving approximately 46 per cent of the possible British reed specialist fauna. This is little different to the overall proportion of the British Lepidopteran fauna supported at Wytham, and may reflect the small total area and small patch size of these habitats. By contrast an equivalent estimate of the oak specialist fauna (excluding a few leaf-miner groups that await checking) gives 38 species or 67 per cent of the possible total, from Wytham Woods.

The River Thames and its tributaries run as muddy brown veins around and through the landscape—potential pathways for movement of species, nutrients, energy, and people. Ponds and small streams around the Woods tend to have been drained or culverted. The stream that leaves the wood to the north and joins the Thames at Hagley Pool, for example, runs largely in straight ditches and its course has been altered by new cuts. Nevertheless, it still supports locally uncommon species such as greater water parsnip (*Sium latifolium*), greater spearwort (*Ranunculus lingua*), and water violet (*Hottonia palustris*) (Watt and Kirby 1983). New open water habitats have been created, such as the reservoir at Farmoor to the east and the various gravel pits to the north.

## 4.5 Movement within the Woods and between the Woods and the surroundings

There can clearly be considerable movement between patches within the Woods and to some degree an interchange of species between the Woods and other habitats in the neighbouring areas. A red underwing moth (*Catocala nupta*) caught and marked by an entomologist in Kidlington was caught by a second entomologist in Wytham three nights later, a flight of about 6.5 km (Waring 1984).

Looking at the patches of grassland within a woodland setting, and the Woods within their wider landscape setting, can the theory of Island Biogeography (MacArthur and Wilson 1967) help to explain variations in species richness of different patches? Some of the patterns expected from Island Biogeography theory do appear if only the specialists of different habitats are considered. The large ancient woodland patches do, on the whole, contain more ancient woodland species than small patches, and large limestone grassland patches have more chalk and limestone grassland indicators. However, there is no evidence of an equilibrium between colonization and extinction on this scale as is predicted in Island Biogeography theory. It may simply reflect that if a larger area is sampled (as will be the case for a large patch) more species are found. In contradiction to Island Biogeography ideas the largest areas of reedbed at Wytham (Marley Fen) support only two specialist species among the reed feeding Lepidoptera, and is not the richest in such species. Here the most likely explanation is that Marley Fen lacks the variety of other structures and microhabitats, including additional plant species and open water patches, compared to smaller fen patches elsewhere on the estate.

So, while some patches may behave as islands for some specialist species the analogies are poor and heavily scale-dependent. At one end of the spectrum a leaf miner in rock-rose will perceive at most a small patch of plants within a patch of ancient grassland in a generation. At the other end of the spectrum, a buzzard will see the same patch as one of a number of potential foraging areas spread across the whole Woods and their surroundings.

There is also a risk in assuming how species 'perceive' habitats is the same as how we perceive them. The abundance of neutral grassland indicator plant species in Wytham was noted above despite the lack of real neutral grasslands: they are spread among the woods, limestone grassland, fens, etc. Thus, rather than species 'belonging' to particular habitats as perceived by ecologists, habitat patches are where notable concentrations of species with particular requirements come together. This does not prevent the same species being scattered through other habitats and therefore, in dynamic situations such as Wytham, perhaps much more likely to turn up in what appears to be the wrong place.

## 4.6 Changes in the relationship between the Woods and its surroundings

The landscape around Wytham has not been immune to the changes that have taken place elsewhere in the country since the 1940s (NCC 1984; Natural England 2008). The high tide of arable farming during the 1970s swept up to the fields above the sawmill, including Upper Seeds, and through much of the floodplain. Wetland plants and those of cultivated land (arable weeds and other casual members of the Wytham flora) were lost as a consequence (Gibson 1986). Some of these fields have been put back to grass since, but round the southern edges of the Woods they remain largely arable. There is still the potential for drift of pesticides and fertilizers into the Woods that could alter its composition, at least at the edges (Gove *et al.* 2007).

Modern farming practice has also tended to sharpen the division between woodland and non-wooded habitats across the landscape. There are fewer scrubby transitions between grassland and forest such as existed on the former common grazings. Many hedges—an extension of woodland out into farmland—have been removed to create larger fields, simplifying the landscape (Macdonald and Johnson 2000). Sufficient survive on the Wytham Estate, however, to illustrate the contribution they can make to the wildlife of the countryside through, for example, the presence of veteran trees and, at least until recently, woodland plants—such as bluebells—in some.

As farming has become more intensive, so more and more of the plants and animals that were previously widespread through the countryside have been pushed to the edges of fields. Whether field boundaries such as hedges really do function as 'corridors' for movement of species between larger patches of habitat is debatable. There is, however, no doubt that they increase the richness of wildlife in modern farmland compared to where they are absent. This importance was recognized in the late 1980s and 1990s in the various Countryside Stewardship and Environmentally Sensitive Areas schemes. Field margin creation and management options have continued under the Environmental Stewardship agri-environment schemes that currently operate. Part of the evidence to justify such options comes from studies on the Wytham Estate, where margins were allowed to develop naturally or deliberately sown with a variety of flowers and grasses (Smith *et al.* 1993; 1997).

## 4.7 Conclusion

Compared to the past, the habitats and species populations on Wytham Hill must be viewed as more isolated. To the east and south the A34 and A420 dual carriageways present potential barriers to species movement. There is less movement of stock, carts, and muddy boots between Wytham and its immediate surroundings than when the Hills were mostly open common and the coppices regularly cut. However, the analogy with species populations' changes on islands should not be pushed too far: management and habitat 'quality' changes seem to have been more important over the last 60 years than the balance between immigration and extinction processes in explaining plant species changes on the Wytham Estate (Gibson 1986).

The relationship between the Woods and the human communities around them has also changed. Formerly the Woods were valued for their physical products—wood, meat, stone, etc—now a different type of value is attached to them. For the researchers there is the intellectual challenge of understanding the ecology of the Woods. Volunteers from across the world come to Wytham to help under the Earthwatch programme; other groups (perhaps less willing) from drug and alcohol rehabilitation programmes contribute to work in the Woods. Importantly there are also the visitors who just enjoy the peace and quiet and the natural beauty that the Woods provide, whose maintenance is a condition of the ffennell bequest to the University.

To Charles Elton, Wytham Woods may initially have been a rather ordinary piece of countryside. However, they must increasingly be seen as a surprisingly large and diverse survival of semi-natural habitat in lowland England. The components of this landscape are analysed in more detail in subsequent chapters.

## ANNEX 4.1.

Main vegetation types in Wytham Woods in relation to the National Vegetation Classification. (Based on largely unpublished data from M.D. Morecroft, C.W.D. Gibson, K.J. Kirby, and others.)

## Woodland

W8 *Fraxinus excelsior, Acer campestre, Mercurialis perennis* type.
This community is widespread on nutrient rich soils in southern and eastern Britain. While ash, field maple, and hazel are characteristic the tree and shrub layer can be very variable, depending on the past history and management of the site. The ground flora can also vary considerably, but dog's mercury is usually abundant.

In Wytham this is the main woodland type overall. Dog's mercury is widespread, along with a range of other herbs such enchanter's nightshade, wood-avens, and violets. Bramble was formerly abundant.

The least disturbed ancient semi-natural coppice areas tend towards the *Primula vulgaris, Glechoma hederacea* sub-community (W8a) but the tendency towards increasing abundance of tufted hair-grass is more typical of W8c (*Deschampsia cespitosa* sub-community).

The more recently colonized ash woodland and some of the ancient woodland with plantations or sycamore dominance show similarities with W8e (*Geranium robertianum* sub-community) with frequent *Urtica dioica* and *Poa trivialis*.

The recent increases in abundance of wood false-brome which in Rodwell (1991) tends to be constant only in W8g may be an indication that the sub-community classifications are not necessarily fixed over time.

W12 *Fagus sylvatica, Mercurialis perennis* woodland.
This type includes beech stands on free-draining, nutrient-rich soils and is mainly found in the south-east lowlands of Britain. Beech tends to be dominant in the canopy, but ash and sycamore are often present, particularly in regeneration gaps. The ground flora may be sparse, because of heavy beech shade, but where present tends to be mixtures of dog's mercury, ivy, wood melick, sanicle, and sweet wood-ruff.

In Wytham this type is best developed amongst the beech on the top of the hill (the fifth Earl's plantings).

W14 *Fagus sylvatica, Rubus fruticosus* woodland.
This type is found on more acidic soils than the previous type and consequently the ground flora is more limited in species; bramble is the most common component, with bracken, honeysuckle, wood-sorrel, bluebell, and ivy.

In Wytham the type is restricted to small areas of 40–50 year old beech plantations scattered round the lower slopes of the hill. Some of these provide the most striking displays of bluebells in spring.

## Scrub

W21 *Crataegus monogyna, Hedera helix* scrub.
Hawthorn scrub is widely distributed across Britain, on abandoned fields, roadside verges, waste ground, and in linear form as hedges. It can be very variable but is most commonly

associated with bramble, blackthorn, ash, ivy, nettles, and cleavers. On chalk and limestone soils the *Viburnum lantana* sub-community (W21d) can be much richer in shrubs, including privet, wayfaring tree, dogwood, clematis, yew, and even locally juniper.

There are patches of the *Viburnum* sub-community on Wytham hill (though no juniper) but it has often been cut back to maintain/restore the open rich grassland. On the lower slopes, where scrub occurs it is less species rich; local stands of dense hawthorn with little on the ground below (*Hedera helix-Urtica dioica* sub-community) or slightly more open stands with much elder and nettles (*Mercurialis perennis* sub-community).

W24 *Rubus fruticosus, Holcus lanatus* underscrub.
This community is an untidy mixture of brambles, rank grasses (typically cock's-foot and false oat-grass), and tall herbs. It is typical of derelict land, run-down arable, and pasture but is also frequent around wood-margins and hedgerows.

In Wytham it occurs around the edges of grassland and rides and temporarily during the regeneration of recently cleared areas.

## Grassland

CG3 *Bromus erectus* grassland
CG4 *Brachypodium pinnatum* grassland
CG5 *Bromus erectus, Brachypodium pinnatum* grassland
These three types are common on the chalk and limestones of south and east England forming rich communities of grasses, sedges, and herbs. They tend to depend on close-grazing for their diversity.

These communities were probably widespread on the Commons at Wytham in the past, but then declined as grazing was abandoned, although a few areas survived on the thinnest soils at least while rabbit grazing continued. A priority for conservation management is now to retain the best surviving areas and to try to restore other areas through re-instating grazing (see Chapter 7).

MG1d *Arrhenatherum elatius* grassland, *Pastinaca sativa* sub-community.
*Arrhenatherum* grassland is virtually ubiquitous throughout the lowlands of Britain, but this sub-community is more restricted, by soil requirements, to the south and east. Tussock-forming grasses such as false-oat grass, Yorkshire fog, and cock's-foot are abundant, along with tall herbs such as wild parsnip, yarrow, ragwort, and ribwort plantain.

In Wytham this occurs in the ungrazed/lightly grazed areas, particularly on the limestone.

MG5 *Cynosurus cristatus—Centaurea nigra* grassland.
This grassland type is typical of grazed hay-meadows throughout the lowlands on circumneutral brown earths, although agricultural improvement has drastically reduced its extent. In Wytham it has been recorded from the 'Ant Reserve' at the top of the park, where the grassland is rather different to anywhere else on the site.

MG9  *Holcus lanatus, Deschampsia cespitosa* grassland
This type is characteristic of moist, gleyed soils through the British lowland and is noted as occurring on woodland rides. In grassland terms they are considered floristically dull. Tufted hair-grass, rough-stalked meadow-grass, and Yorkshire fog tend to be abundant,
   In Wytham it is common on the rides on the lower slopes, on the clay, but often in rather disturbed stands because of traffic along the rides.

**Fens**

S26  *Phragmites australis, Urtica dioica* tall-herb fen, *Filipendula ulmaria* sub-community.
This is a community of eutrophic, circumneutral-basic water margins and mires and springs. Common reed dominates with a range of other tall grasses and herbs, nettle, meadow-sweet, and hemp agrimony, and sprawlers such as woody nightshade and greater bindweed.
   In Wytham Woods this is mainly found in the open fens in Marley Wood, but similar vegetation survives in one of the fields outside the Woods, on part of the University Farm.

# 5

# The Trees in the Woods 1945–2007

K.J. Kirby

## 5.1 Active forestry management 1945–1963

The recent history of Wytham Woods encapsulates many of the trends and debates over the policy and practice of forestry in Britain (Richards 2003; Tsouvalis 2000). A national forestry school was established at Coopers Hill (Surrey) in 1885 by Schlich, but moved to Oxford in 1905, where initially Diploma courses were run with the first degree courses established in 1919 (Edlin 1966). There was a need to train foresters to serve across the Empire, particularly in India, but the case for a home forestry service was also increasingly recognized.

The First World War had highlighted Britain's dependence on imported timber and wood products and in 1919 the Forestry Commission was established. Its objective was to create a strategic timber reserve, largely through new afforestation, but these stands were still too young to be of much use in the Second World War. So, as in the previous conflict, the existing woods were searched for timber. Some reports refer to half the woods or half the trees being cut-over (Richards 2003).

The state of England's forests after the war was recorded through a woodland census in 1947 (HMSO 1952); large areas were described as scrub, felled, or 'devastated' where the best timber had been removed, leaving only a scattered remnant of the original crop. Wytham Woods had been heavily felled after the death of the fifth Earl of Abingdon in 1854, and possibly during and after the First World War, and so there was not much scope for further exploitation during the 1939–45 conflict (Smeathers 1939). Even so, the state of the existing woods in 1947 (Table 5.1) was probably typical of the state of broadleaved woodland more generally (Anon 1950). There were also large areas of open ground with various degrees of shrub cover that were deemed suitable for planting (see Fig. 3.3).

The quality of many of the stands was poor: the 1950 and 1959 management plans contain frequent references to short-boled, wide-crowned branchy oaks (Anon 1950; Osmaston 1959). This reflects the Woods' various origins as ancient coppice with standards, in-filled former wood-pasture, secondary growth from scrub on former commons, overlain by varying amounts of felling. However, there were indications that the site had a higher timber potential: beech, elm, lime, and sycamore 20–25 m tall in places; ash and elm butts of 150 years, 90 cm diameter and virtually sound; mature ash of excellent form; and some oaks felled that were 'near veneer' quality (Osmaston 1959). There were also good local markets, including for coppice products.

National forestry policy post-1945 was geared towards restoring the productive potential of existing woods as well as creating new ones. This tended to be by establishing plantations,

**Table 5.1** Wytham Woods post-1945 (Anon 1950).

| Forest type | ha | % |
|---|---|---|
| Broadleaved | 62 | 18 |
| Conifers | 5 | 1 |
| Mixed | 11 | 3 |
| Coppice/coppice with standards | 57 | 16 |
| Unproductive (open ground) | 177 | 50 |
| Biological and recreational reserves | 40 | 12 |
| Total | 352 | 100 |

often broadleaf–conifer mixtures or high-yielding pure conifer crops. The Forestry Department of Oxford University consequently took on the management of the Woods with enthusiasm and produced a management plan with the following objective:

> The woods are to be managed to improve, maintain and utilize the existing woodlands under an approved plan and in accordance with the practice of good forestry insofar as the special local conditions and conditions of ownership allow. The woods have a high amenity value which must be maintained and also their value for instruction and research.
>
> (Anon 1950)

The woods were 'not to be regarded primarily as a financial asset'. The bulk (277 ha) was to be worked so as to create an irregular mixed broadleaved high forest, not necessarily of uniform composition throughout, nor precluding the use of coniferous species on short rotations where their use would help to achieve the main objective. The remaining 34 ha were to be worked as coppice.

An active work programme was put in place with thinning of existing stands, planting of gaps, and bare ground. The new planting took place mainly in the Radbrook Common and Pasticks areas and there was some replanting of existing woodland within the Great Wood. Beech, oak, and ash were commonly used, sometimes with conifers such as larch or western red cedar. Small areas were, however, also planted with other species, such as poplar or Norway spruce. Ten years later, Osmaston (1959) noted that the area of 'unproductive scrub or bare land' had been reduced from about 177 ha to 102 ha. The 1959–69 management plan envisaged that over this ten-year period there would be further planting of the open ground and other unproductive areas. Continued thinning of the young, naturally-regenerated ash and sycamore stands would be needed.

All this activity would require an increase in the workforce from five and a half to at least seven men with one, possibly even two (!) power-driven chain saws, even if the trees were sold standing. Eight or nine men might be needed in the longer term. Had the second plan been fully implemented, the Woods would have been very different today.

Shortly after the 1959 plan was produced, a rift developed between the ecologists (primarily from the Department of Zoology) and the foresters as to the priorities for the

# Fitzwalter Camplyon 'F.C.' Osmaston (1901–1979)

F.C. Osmaston in about 1940, taken in Penang at the outbreak of war with Japan.
Photo by courtesy of Mrs M. McIvor.

F.C. Osmaston came from a widely known forestry family. His father had worked first in India, before becoming Deputy Surveyor first in the Forest of Dean and then in the New Forest. Osmaston was educated at St. John's College, Oxford, immediately after the First World War and studied under Professor Schlich. He took a Diploma in Forestry in 1921 and graduated in Botany in 1922 (Fairbairn 1980).

In the same year he was appointed to the Indian Forest Service in which he was to serve for 25 years. He was posted to Bihar and Orissa, where he later wrote the 1931–51 Working Plan for Orissa. In 1947 he was appointed a lecturer in the Department of Forestry at the University of Edinburgh and returned to Oxford in 1954 as lecturer in Forest Management in the School of Forestry. He was a founder member and Fellow of St. Cross College and retired in 1968.

It is in connection with planning and woodland management that Osmaston is known in Wytham. His revision of the management plan for the Woods was completed in 1959 and it provided the first complete inventory of trees. He drew attention to the fact that there was a serious imbalance in the age classes of oak at Wytham, with insufficient young trees to replace the old mature ones as they were felled or died. Every subsequent inventory has reiterated this point.

Osmaston's management plan proposed that the bulk of the woods should be managed along the lines, then being promoted quite vigorously nationally, of contributing to the creation of a 'strategic reserve' of trees to provide timber in case of another war. Only a few areas were excluded from this proposed management regime. As is evident, 50 years later, his plan to produce timber was abandoned (in 1963) in favor of ecological research. Had it been fully implemented, the Woods would have been very different today.

Osmaston is remembered more widely mainly for his classic book *The Management of Forests* (1968). Together with Professor Champion he also co-edited a fourth volume, covering the years 1925 to 1947, to Stebbing's three-volume *The Forests of India*. This was an outstanding collation of historical records for a difficult period in India.

Woods (see Chapter 11 Section 11.3). The 1950 and 1959 plans envisaged that the bulk of the wood should be managed along forestry lines, but with 9 areas excluded from the management plan; 40 ha as biological reserves and 1 ha reserved for a girls' youth organization. The ecologists felt that the balance between reserves and managed areas should be reversed, and in 1963 the Woods were taken out of the Forestry Commission's Woodland Dedication Scheme and active management declined. Responsibility for managing the Woods was removed from the Forestry Department and given to a Committee representing all the users of the Woods in the University.

The trees continued to grow, but now for the most part the dynamics of the tree and shrub layer were determined by the inherited age structure of the tree layer and various natural disturbance factors such as grazing, windthrow, and disease.

## 5.2 Minimum intervention and 'near-natural' stand development

It is.... clear that Wytham Woods have not for many centuries been 'virgin', though if given the chance to do so they might well return to something resembling a natural woodland, even if this would be different in composition from the original Saxon forest. What could be more fascinating than to watch this happen and record its progress over a hundred years or more, armed with the methods of modern ecology?

(Elton 1966)

Eustace Jones, the Oxford forester and botanist, co-authored the main historical account of Wytham Woods (Grayson and Jones 1955), and contributed much of the detail in the 1959 management plan. He was also among the first to suggest (Jones 1945a) that temperate, virgin forest might take various forms; with some areas having small-scale mosaics of different ages, whereas elsewhere more even-aged stands would predominate. The climax forest might be a concept only:

...never existing in practice either because of the catastrophic initiation of fresh seres, or because of the time lag — necessarily great where long-lived trees are concerned—in the adjustment of the vegetation cover to an ever-changing environment. Even if the climax condition is attained theoretical considerations give us no clue to what might be called its minimal area—the extent of the area within which the kaleidoscopic vegetation patterns remains unchanged in the aggregate, nor do they indicate the degree of precision with which the climax could be defined. There might for example be several variants which were equally stable and capable of replacing each other in time or space.

Jones makes virtually no mention in this paper of the role of large herbivores in temperate forests, although he was familiar with open grazed woods in the Balkans and in the tropics. Recently, Vera (2000) has proposed an alternative vision of the pre-Neolithic landscape which resembles much more an irregular mosaic of scrub, grassland, and dense woodland. The basis for this hypothesis is strongly disputed (Hodder *et al.* 2005; Kirby 2004a) and generally Jones's dynamic view of 'virgin forest', updated and expanded by Peterken (1996), still holds.

## 5.3 Long-term studies of stand structure and composition

In the 1950s, Jones set up, in Lady Park Wood in the Wye Valley, what has become probably the most recorded long-term woodland monitoring study in the UK (Peterken and Jones 1987, 1989). The results from this tend to confirm Jones's ideas on the dynamic nature of mature/old-growth woodland. He was involved in a number of other long-term studies elsewhere in the country, but he appears never to have set up anything similar in

# Eustace Wilkinson JONES (1909–1992)

E.W. Jones (L) in 1984 in Lady Park Wood with G.F. Peterken.
Photo by courtesy of G.F. Peterken.

After graduating in Botany from Cambridge he gained a PhD., also in Cambridge, for his work on plant physiology and in 1934 he took up a position as a lecturer in the School of Forestry at Oxford where he remained until he retired.

Eustace Jones will probably be remembered best by foresters who studied at Oxford. His silviculture lectures were well-attended, for he showed a good blend of theory and practice, frequently based on his own experience and careful observation. He conducted frequent excursions to Forestry Commission forests and to private estates in England and Wales.

He was a very competent botanist and knowledgeable ecologist who had an international reputation as a bryologist. His published work is considerable and wide ranging. He wrote the definitive accounts of *Acer* (1945b) and *Quercus* (1959) for the Biological Flora of the British Isles in the *Journal of Ecology*, and gave the introductory paper to the celebrated conference on the British Oak organized in September 1973 by the Botanical Society of the British Isles. He was one of the leading authorities on temperate forests, and in 1945 published an important paper 'The structure and reproduction of the virgin forest of the north temperate zone' (Jones 1945a).

A major focus for his work in Britain was the Wye valley and the Forest of Dean, which he used for teaching and furthering his considerable interests in bryophytes. Lady Park Wood in the Wye valley was also a focus for his research up to 1960, and his work there resulted in two papers in the late 1980s, in collaboration with Dr George Peterken. The original records for this were written on the back of the notes he made for the 1945 paper (mentioned above), probably due to wartime paper rationing, reflecting the conditions under which he operated. He was responsible for editing the second (1952) edition of R.S. Troup's *Silvicultural Systems*, modestly asserting that he did not need to alter much of that great work apart from bringing it up to date in a few places, and was an editor of *Forestry* from 1970 to 1978 (Stern 1993).

One of Jones's personal interests was local history, which he saw as intensely relevant to understanding the genesis of estate woodlands. It was this that led him to collaborate with Arnold Grayson in writing the 1955 paper *Notes of the History of the Wytham Estate with Special Reference to the Woodlands* which is much quoted in this book.

Wytham, even though he clearly knew the Woods well. Was this because of some residual rivalry between the Forestry and Zoology Departments?

Jones's young successor as forest ecologist in 1973 was Phillip Lloyd. Renewed interest in Wytham from the Botany and Forestry Departments led to the initiation of a series of D.Phil. studies on different aspects of the plant communities, and establishment of a series of permanent plots through the Woods. Unfortunately he never had a chance to fulfil his undoubted potential, dying in an accident in 1975. Though the data from the plots survive and the plot locations are known approximately, the plot markers have gone and the studies have never been followed up.

Survival of plot data, but not the plot locations and its converse, where plots survive but not the records, have proved to be all too common in nature reserves and other research forests across Britain. This challenge was addressed by Colyear Dawkins of the Forestry Department (Dawkins 1971).

> The inspiration for this project came from the toil and sweat of a futile attempt, in 1954, to find the exact position of one of the Davis and Richards.... plots in Morabilli Creek, Guyana, only 25 years after it was established....Thus one of the earliest and potentially most valuable of rain-forest studies was unavailable for precise re-observation...This experience...convinced us that vegetation ecology must equip itself with permanent plots with unambiguous addresses.
>
> (Dawkins and Field 1978)

Records from these plots from 1973–76, 1991–92, and 1999–2001 illustrate the changes in the composition and structure of the Woods (Kirby *et al.* 1996; Kirby 2004b), which can be extended back through comparison with the 1959 and 1950 enumerations.

The 1950 plan included a visual estimate of the composition of virtually the whole of the Woods and the stems were split into numbers of Small (<35 cm diameter at breast height) and Large stems (> 35 cm diameter) (Fig. 5.1a). The predominance of 'large oaks' is notable, although 35 cm is not really that large for an oak! The size classes for ash and sycamore are more even. Elm is almost as common as sycamore. Amongst the other species beech is represented almost entirely by large stems, whereas the converse is true for birch.

The 1959 plan gave a more detailed enumeration of 13 compartments, covering 111 ha, by main tree species and stem sizes. These data (Fig. 5.1b) have been summarized into the Large and Small categories used in 1949 to show the preponderance of small stems overall; the developing cohort of young ash and sycamore is very apparent and is repeatedly referred to in the compartment descriptions. This cohort was boosted by the death of rabbits from myxomatosis in 1954: whereas previously 1000–2000 had been being killed annually on the estate, Southern (1955) noted only two or three tracks in the snow.

The records from the Dawkins plots (from 1974, 1991, 1999) cannot be directly compared with the stem enumerations of 1949 and 1958 because tree abundance was assessed in a different way. The results (Fig. 5.2) do however indicate a continuation of the trends shown in these earlier surveys. The canopy cover overall has been fairly stable between about 70 and 80 per cent; basal area has doubled from about 14 m$^3$ ha$^{-1}$ estimated from

# Henry Colyear Dawkins (1921–1992)

H.C. Dawkins. Photo by courtesy of St. John's College, Oxford

Dr Colyear Dawkins MBE came from a family of foresters. He took degrees first in Botany and then Forestry at Oxford. During the Second World War he was sent to Uganda as an Assistant Conservator of Forests, then to the Northern district of Acholi, as a District Forest Officer, where he learnt not only the fundamentals of tropical forestry, but also the Acholi language with great skill. He was also able to identify more Ugandan tree species than almost anyone else there.

His work as Uganda Forest Ecologist, 1949–62, led Bob Plumptre to write of the Budongo Forest Project: 'The Ecologist…was Colyear Dawkins who had established numerous research plots; he usually inspected them at the double which left me well behind, mentally and physically'. Colyear's 1958 paper on 'The management of tropical high forest with special reference to Uganda' remains a classic on the subject. After retirement he started writing a comprehensive history of the management of tropical forests that was completed in 1998, after his death, by Michael Philip (*Tropical Moist Forest Silviculture and Management*).

He became Lecturer in Statistical Method at the Oxford Department of Forestry in 1971, and a Fellow of St. John's College. Offers of promotion did not interest him—he preferred to 'do something useful'. He found time to be Vice-President of the Lundy Field Society, visiting frequently; and also to be President of the forestry section of the British Association.

He taught statistics in a way that engaged biologists—as many will remember. His active interest in statistical quality meant he both tirelessly helped many with research design and analysis of results, and also co-developed with his wife, Barbara, computer programs for diverse analyses.

He was a man of robust integrity with a strong sense of justice and rightness; this, with his complete absence of pretension, meant he could relate to anyone, especially young children. He could only very rarely be persuaded to render his baboon bark—almost as startling as a first hearing of Barbara addressing him as 'petal', but as unforgettable as the man himself.

His main legacy to Wytham is the accurate map he made of the Woods and the grid system and its markers (orange posts at 100 m intervals) that he meticulously surveyed. There are 164 permanent 10 × 10 m sample plots in Wytham located on alternate intersections on a 100 × 100 m grid. In laying out the plots, Dawkins was well ahead of his time for detecting the influences of climatic change.

It was the trouble he had relocating forest plots in Guyana that led him to trial a system of permanent (and not subjectively sited) plots for long-term monitoring of woodland (Continuous Forest Inventory). These plots were set out and recorded in 1973–5 in three forests near Oxford. The reasoning for them is explained in his Chairman's address to the Forestry Section of the British Association in Leicester in 1972. Those in Wytham have been re-assessed several times in terms both of vegetation and of soil factors (as he hoped), and changes have been detected and commented upon elsewhere in this book. He also applied the same systematic plots approach in a different context—reclamation of slag heaps for the Coal Board.

> **Box 5.1** The 'Dawkins' Plots (Dawkins and Field 1978)
>
> The aim of the project was 'to devise a system of permanent low-fraction sampling of polyspecific uneven-aged woodland capable of diagnosing and predicting the long-term changes in specific composition and growth over large areas'.
> Plot distribution and identification:
>
> - 163 plots spread across the Woods;
> - 10 m x 10 m plots, located 14.1 m north-east of alternate points on 100 m marked grid (see Plate 4);
> - south-west and north-east corners marked by underground metal markers.
>
> Tree and shrub records:
>
> - diameters of the four largest trees in the plots; height estimate for the largest individual;
> - canopy (split by species) and shrub cover estimated across the plot diagonal;
> - basal area of the trees in the plot surroundings estimated by relascope sweeps.
>
> The ground flora and soil records are discussed in other chapters.
> No long-term study can be guaranteed to survive, but Dawkins and Field tried to increase the chances by ensuring:
>
> - that the information collected should be sufficiently useful and interesting to be worth looking at and to be positively sought by future researchers and managers of woodland;
> - that the plots should be easily found for many years to come by those concerned with recurrent observation, though unnoticeable to all other forms of life;
> - that the records also should be easily found, be intellectually as well as physically accessible, and be informative to any researcher or manager without any other source of knowledge of the project.

relascope sweeps in 1974, to 31 m$^3$ ha$^{-1}$ in 1999. Ash and sycamore have overtaken oak as the main tree species both in terms of canopy cover and basal area.

A third indication of the changing structure is the diameters of the largest ('leading') trees recorded in the plots. In 1974 the most common size class was 11–20 cm whereas by 1999 it was 21–30 cm. This reflects primarily the continued growth of the 1959 ash and sycamore cohort (also reflected in the basal area data), these two species being the commonest leading trees. However, while the survivors from this flush of regeneration have continued to grow, their numbers have been dramatically reduced and there has not been a concomitant initiation of a new cohort.

This is illustrated by a more detailed study of the area shown in the canopy photograph in Elton (1966, Figure 9). The photograph covers the boundary of the Great Wood and what had been common land with only scattered trees in the eighteenth century.

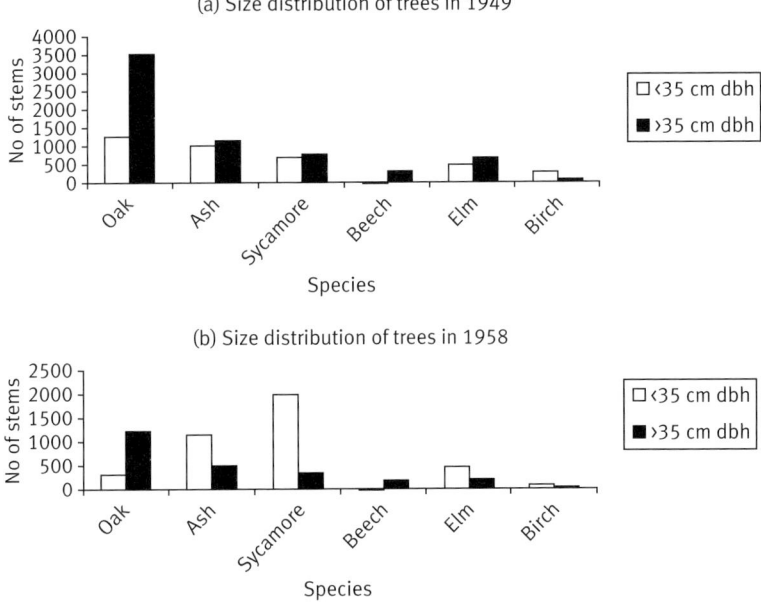

**Figure 5.1** Composition of the Woods, (a) c. 1949 (Anon 1950)—the main species were assessed visually as being above or below 35 cm dbh; (b) c. 1958 (Osmaston 1959)—stems were measured in 13 compartments; the data have been grouped to make them comparable with the earlier survey.

K. Paviour-Smith and C. Elbourn recorded the tree and regeneration density in 1968 as a grid of 10 x 10 m plots which were revisited in 2007 (Mihok 2007). By 2007 a few of the mature trees from 1968 had been lost and some of the pole-sized stems of ash and sycamore had joined the canopy. The most striking change, however, was the virtually complete disappearance of stems less than 10 cm diameter (Fig. 5.3).

The abundance of ash in relative and absolute terms is likely to continue to increase. Across the Woods some of the old oak are dying; there are young oaks coming on in the plantations established in the 1950s, but the contribution of oak to the canopy is likely to decline. Similarly, the beech surviving from the fifth Earl of Abingdon's plantings in the early 1800s are starting to collapse, although there were not the mass die-offs of beech that occurred in the New Forest and in Lady Park Wood after the 1976 drought. Beech is, however, also vulnerable to beech bark disease, air pollution, and attacks by the grey squirrels (*Sciurus carolinensis*) that seem to have reached Wytham in the 1920s (Elton 1966). Most of the stems in the young plantations show signs of damage.

Dutch elm disease became established in Wytham in the 1970s (Sheehan 1979). Consequently the large elms, which Elton described as the tall, flaming pillars of golden yellow in autumn, have disappeared, although locally suckers are still abundant. Rackham (2003) comments on the increasing amounts of birch in woodland during the

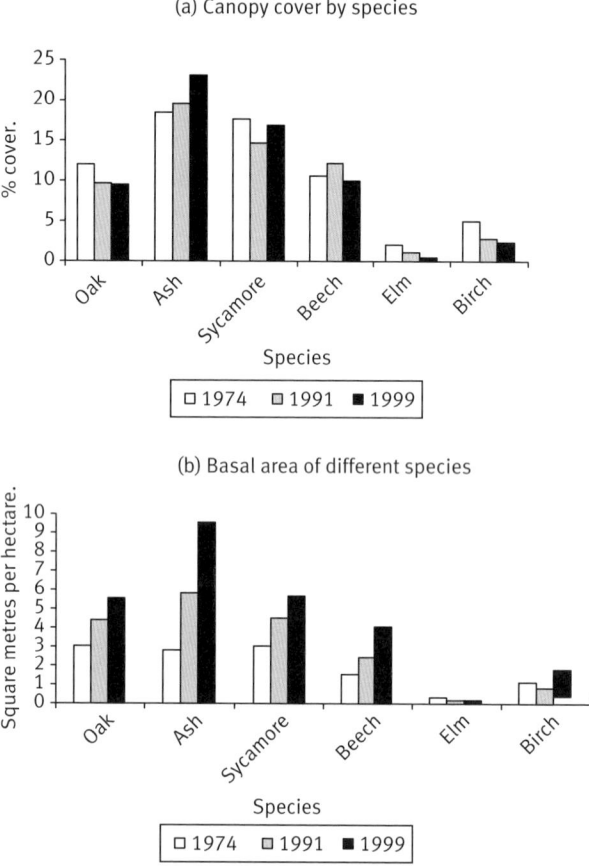

**Figure 5.2** Contribution of main species to (a) canopy cover (estimated across the plot diagonal) and (b) basal area (from relascope sweeps) for the Dawkins plots in 1974, 1991, and 1999.

twentieth century; this may be connected to the disturbance associated with fellings during or shortly after the two world wars. Birch is, however, a relatively short-lived tree. It has only been locally abundant in Wytham and has proved particularly susceptible to windthrow; its abundance in the canopy is now decreasing. A similar trend was seen at Monks Wood between 1966 and 1996 (Crampton et al. 1998).

The exotic conifers planted, mainly post-war, have matured and are gradually being removed leaving sycamore as the main 'non-native' tree in the wood. Green (2005) argues that it might be native to Britain, citing amongst other ideas the carving of sycamore leaf on St Frideswide's tomb in Christchurch Cathedral. However there is little positive evidence to go against Jones's (1945b) conclusion that it was introduced, from just across the Channel, probably in the early medieval period. Sycamore has been in Wytham at least since the early nineteenth century. In a large swathe across the north of

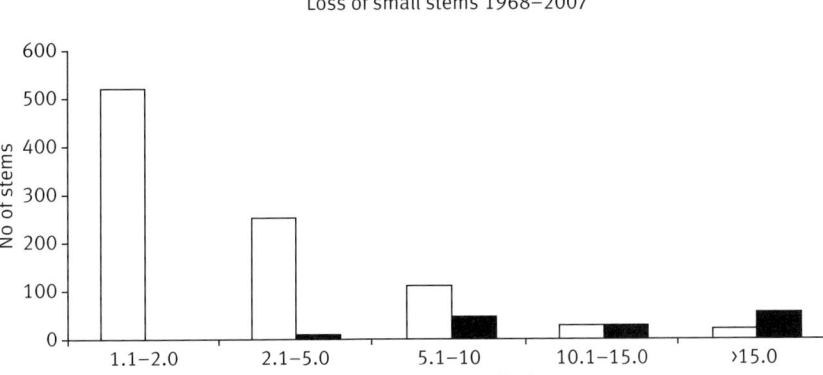

**Figure 5.3** Changes in the number of saplings and young trees (including recently dead stems) for 20 of the 10 x 10 m plots recorded in 1968 by Paviour-Smith and Elbourn (1993) and Mihok (2007).

the hill sycamore forms a major part of the canopy (Fig. 5.2) but its cover and the number of plots in which it is present (95 in 1974, 88 in 1999) show little change, (Fig. 5.4). Osmaston (1959) noted its regeneration on the limestone heavy loam soils, but that it rarely spread onto the clays. The clay areas also tend to be where abandoned coppice has persisted, largely unmanaged over the twentieth century. Morecroft *et al.* (2008) suggest that the failure of sycamore to colonize these areas reflects its inability to establish under shade.

Waters and Savill (1991) looked at the balance between ash and sycamore in the canopy and in regeneration. They found evidence that seedlings of one species might do better under the other's canopy, allowing for the potential alternation of species. Neil Riddle (summarized in Stiven 2007) working in the Peak District and Wye Valley concluded that soil disturbance appeared to be a more critical element for sycamore regeneration. The competitive balance between ash and sycamore can also be affected by grazing (Linhart and Whelan 1980), and by different growth responses to climate change. Sycamore is more drought sensitive than ash or oak and at present, in Wytham, is growing more slowly than ash (Morecroft *et al.* 2008). Elton (1966) may well have been right therefore to suggest that, left alone, sycamore would probably settle down eventually as a part of our future deciduous woodland composition.

Over the long term the relationship between different species can also be followed through detailed stand maps, such as that for the 1 ha square of the Environmental Change Network plot. Blackthorn and aspen, which spread clonally, are strongly clumped; hazel stools are spread fairly evenly across the whole area, as are the oak standards, though at a much lower density. This stand mapping has now been extended to a much larger area of the Woods in a new project, providing another type of baseline to study future change.

68 Wytham Woods

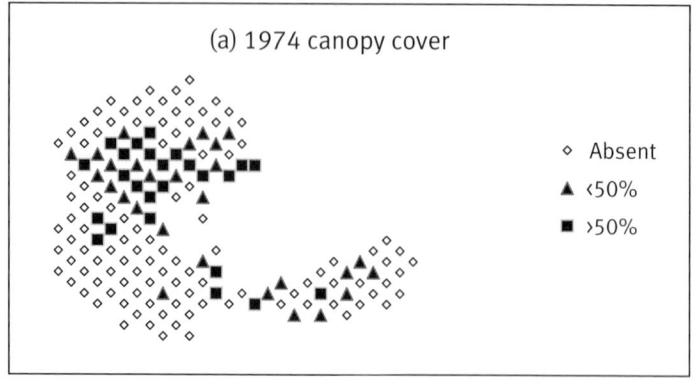

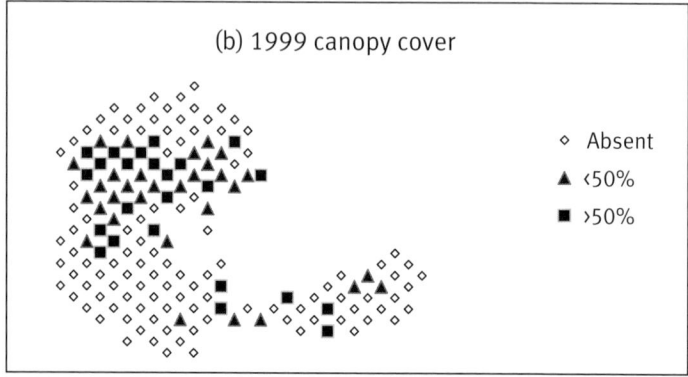

**Figure 5.4** Sycamore canopy cover estimates for 1974 and 1999 from the Dawkins plots.

Recording changes in structure and composition is time-consuming and tends to be carried out infrequently. Hence there is increasing interest in remote sensing techniques to try to identify changes in woodland structure and composition. These are starting to show positive results, for example in a trial looking at the distribution of ash and sycamore in the Great Wood (Atkinson *et al.* 2007).

Similarly LiDAR results, used in Chapter 2 to illustrate the topography of the Woods, also show up the variation in tree heights in different areas (Plate 5). The low heights in the ancient semi-natural areas in Ten Acre Copse, Bean Wood, and Marley Wood reflect that much of the canopy is composed of the former coppice layer. The former woodpasture of My Ladies Common with its tall maiden ash stands out in contrast. Other tall-tree areas include the well grown plantations from the 1950s on Radbrook Common, as well as those on the eastern edge of Jews Harp. Work in other sites such as Monks Wood (Cambridgeshire) has started to link such data to, for example, woodland bird habitats (Hill *et al.* 2003; Hill and Thompson 2005).

**Figure 5.5** The scaffolding walkway, erected in 1994. It allows researchers access to the canopy of the trees.

## 5.4 Tree physiology

From simple descriptions of the patterns and growth of trees and shrubs, research has increasingly focused on the underlying physiological processes. In the 1970s, detailed work was done on the processes of litter production and decomposition in the beech stand at Brogden's Belt (e.g. Phillipson *et al.* 1975). More recently attention has focused on canopy processes.

A scaffolding walkway (Fig. 5.5), erected in 1994, allowed researchers access to the canopy of the trees. This has presented a rare opportunity to measure the processes of photosynthesis and transpiration and to investigate the characteristics of leaves themselves. Results from this, dealing with the carbon balance of the trees, are considered in more detail in Chapter 12. Other studies have focused on the structure of the canopy itself and its relation to leaf properties. In particular Roberts *et al.* (1999) were able to quantify how the Specific Leaf Area of leaves (their weight per unit area; which is typically higher in sun leaves) decreased through the canopy. By comparison with leaves collected during the autumn leaf fall, Roberts *et al.* (1999) were able to show that leaf collection and measurement of Specific Leaf Area could allow the distribution of canopy leaves to be inferred from the leaf collection.

Herbst *et al.* (2007) and (2008) looked at rates of evapotranspiration from the woodland canopy, including comparison of rates in the woodland interior compared to close to the edge. The results, showing greater water loss close to forest edges, could have implications for estimates of the impact of creating new woodland in areas where water resources may come under stress in future.

## 5.5 Composition and structure of the shrub layer

Apart from hazel, the 1959 management plan gives little detail on the shrub layer, other than to deplore the areas invaded by elder and thorns—'useless scrub'. Gibb (1954) records elder, hazel, and hawthorn as most abundant in Marley Wood. Other species referred to by Elton (1966) include field maple, dogwood, privet, spindle, guelder rose, and wayfaring tree. Blackthorn was noted as being abundant chiefly on the clay soils low down on the hill. The hazel and maple tend to pick out the ancient coppices; the abundance of elder may be linked to ground heavily occupied by rabbits before the Second World War (elder is not eaten by rabbits). This diversity of species is important in providing a range of flower and fruit supplies across the year for birds and insects (Sorensen 1981).

The shrub layer appears to have been fairly dense prior to the mid-1980s. Osmaston (1959) commented that the few deer around were of little amenity value as they were not seen because of the density of low cover. Paviour-Smith and Elbourn (1993) describe a dense sapling layer in Great Wood in 1968. As late as 1987 the students producing a management plan commented that their inventory work was carried out in April and May before the vegetation became impenetrable.

However, the dense blackthorn thickets noted in the 1974 Dawkins plots (Dawkins and Field 1978) had started to break up by the late 1980s (to the relief of modern surveyors, who have not had to crawl and climb their way through them). The 'shrub layer' in the Dawkins plots (cover along the diagonal between 0.5 and 2.5 m high) in general has declined from 44 per cent in 1974, to 25 per cent in 1991, and 15 per cent in 1999. While this includes some tall field layer species such as bramble and bracken, it is primarily the shrubs and the lower leaves of trees.

## 5.6 The changes in woodland structure and its possible causes

A survey comparing the structure of 103 woods across Britain, based on plots recorded in 1971 and 2001, concluded that the loss of smaller stem sizes, such as seen at Wytham, could be linked to a shift towards more shaded conditions, largely due to stand growth (Kirby *et al.* 2005). Given the dominance of young growth in much of the Woods in the post-war period, some of the decline in the shrub layer at Wytham might be an inevitable consequence of young trees growing up and forming a dense canopy—the stem exclusion phase (Oliver and Larson 1996). However, this is unlikely to be the only reason, since banks of saplings would still be expected in more open areas or where the canopy has not changed greatly, for example at ride edges. Hence attention has increasingly focused on the impact of deer.

In the early medieval period, the Abbots of Abingdon Abbey granted Henry I the right to take roe deer and red deer in Bagley Wood and Cumnor Woods (the old name for Wytham Woods) (Grayson and Jones 1955). However, by the mid-nineteenth century both of these species were largely absent from lowland England, while fallow deer were often confined to parks (Yalden 1999). During the twentieth century these and an additional introduction, the muntjac, have increased in abundance and spread widely. Prior to 1950 numbers were very low. In the exceptionally cold winter of 1962–63 the one herd, of 12 or 13 animals could be tracked from their footprints in the snow. At one time they all actually crossed the frozen river towards Cassington, leaving the Woods temporarily deer-less. Until the 1970s it was still a rare event to see a fallow deer in Wytham. However in Wytham, as elsewhere, their impact gradually became apparent through the development of a clear browse line and the observed differences between areas inside and outside small exclosures in the Woods (see Fig. 5.6).

**Figure 5.6** Deer exclosure in Wytham Woods in March 2009. The fenced area on the right, which deer cannot reach, has prolific brambles growing in it. Elsewhere the ground vegetation is very sparse.

**Figure 5.7** Browse line caused by deer on a lime tree in Wytham Woods.

The following comments illustrate the changing appreciation of deer and their impacts in the Woods.

- Osmaston (1959) noted the need for fallow deer control, but estimated only one buck and five to seven does and young beasts. He was far more concerned about the effects of rabbits.
- Elton (1966) identified the potential threat of deer irruption, perhaps because the deer get much of their food from the surrounding farmland. He also noted that muntjac were not far away. There are, however, far more comments about rabbit impacts.
- Sheehan (1979) commented that there were about 60 fallow deer and a few muntjac. The deleterious effects of fallow were tolerable, provided that the herd remained at the current size; effects on adjacent farmland were minimal; muntjac might be encouraged on account of their amenity and scientific interest!
- Wardell (1987) considered that deer were now at an unacceptably high level; widespread damage occurred to regeneration and on neighbouring farms; there were about 200+/- 50 fallow, perhaps 300 muntjac, and roe also present. The numbers were based on estimates provided by Den Woods, the gamekeeper.
- In 1989 the Woods were fenced with the deer inside and impacts increased.

In 1989 the Woods were fenced to reduce the impact of the deer on the adjacent farmland and shortly thereafter management of the deer populations within the fence began in earnest, such that the numbers over winter may now be down to below 50 fallow and roe combined, with virtually no muntjac (see Chapter 11 Section 11.8). At the time of writing (2008) there are starting to be signs that the vegetation in the Woods is recovering.

## 5.7 Old growth and open space

Elton (1966) was amongst the first to stress and record systematically differences in woodland structure, and subsequent workers, particularly on birds, have noted the range of species using different levels for nesting or feeding. The loss of the low cover provided by the shrub layer has had a major impact on other groups of organisms (see Chapter 10 Section 10.1). Warren and Key (1991) stressed that, from a nature conservation point of view, the most interesting invertebrates in a woodland system tend to be those associated with the either the beginning (open space, young growth) or end (old growth, dead wood) of the cycle.

In the past, open space in the Woods would have been created by coppicing, but the extent of worked coppice declined through much of the twentieth century. Whereas England had some 134,000 ha of coppice in 1947, this had been reduced to 22,000 ha in 1998 (HMSO 1952; Forestry Commission 2001). In Wytham the 1950s plan did envisage retaining 34 ha as worked coppice, but this would probably not have survived the change in markets for woodland produce during the 1960s and 1970s and, as a relatively labour-intensive form of management, could not be sustained as the resources for management declined after 1963. Temporary open forest glade and shrub habitats in the Woods were consequently reduced.

The excellent ride system established in the nineteenth century—it was reputed that one of the Earls could ride round the wood in a coach/dog-cart—was neglected after about 1914. By 1950 several former rides had reverted to woodland and were lost, while those on clays and other poorly drained sites frequently became impassable to wheeled traffic. Since then the rides have gone through various phases of neglect and subsequent opening up. They tend, however, not to have the scrubby edges and overhanging bramble bushes previously reported, for example by Wardell (1987), and seen in photographs of expeditions to Wytham in the 1960s and 1970s. A study of the pollen sources used by hoverflies nonetheless noted over 50 plant species from a 750 m stretch of Broad Oak ride (Haslett 1989).

Elton (1966) was also concerned about dead wood and old trees. A process of cleaning-up the derelict woodland had begun after the estate came into the University's ownership, so Elton arranged that various decayed trees and fallen logs would be marked and left alone. He encouraged studies on the dead wood and associated microhabitats of old trees such as rot-holes and fungal fruiting bodies (Elbourn 1970; Fager 1968; Kitching 1971; Larkin and Elbourn 1964; Paviour-Smith 1960; Paviour-Smith and Elbourn 1993).

Since 1990, interest in dead wood and also in 'veteran' trees has increased considerably (Kirby and Drake 1993; Kirby *et al.* 1998; Read 2000). The biggest veteran in Wytham is an ancient hollow pollard oak by a field boundary on the Woodbridges (see chapter 11, Fig. 11.2), with a circumference of 520 cm in 1987 and great boughs arising from the last pollarding some time before enclosure in 1814. Not surprisingly Wytham Woods does not feature in the very top tier of sites important for dead wood invertebrates in Britain (places such as Windsor Forest, the New Forest, Epping Forest, and Moccas Park), but, in part because of the intensity of survey that has taken place in Wytham, it did feature amongst the next ten most important sites, based on an 'index of ecological continuity' (Harding and Alexander 1993).

Cleveland (1997) identified at least 85 veteran trees scattered through the Woods, largely in the areas of former wood-pasture, mostly oak, but also including old beech from the early nineteenth century plantings. On the thin limestone soils, the beeches with their shallow and wide root plates are unstable and the trees respond by sending out massive flying buttresses. A cross-section reveals not concentric tree ring circles but highly contorted shapes. With these come the early establishment of holes, cracks, and splits, which with beech especially often include small permanent water bodies. These are miniature ecosystems and are the breeding sites for mosquitoes amongst other invertebrates. The trees in Wytham Park were not included in Cleveland's survey, but Sheehan (1979) noted that about 50 of the original trees (mostly elm, but including some of the oaks) had died.

Warren and Key (1991) considered that the middle stages of closed forest were of less interest for their invertebrate populations, but Hambler and Speight (1995) caution that knowledge of canopy dwellers is still very limited. The development of new techniques such as 'fogging' of the canopy with insecticide and use of canopy cranes to lift researchers up amongst the branches is starting to address this gap (Ozanne *et al.* 1997; Ozanne 2005). Given that the closed canopy element of the woodland cycle has tended to increase during the last 50 years, this may be expected to produce exciting results over the next few years.

## 5.8 Twenty-first century changes

The early twenty-first century is half-way through Elton's hundred years of change. Ideas of what the natural woodland might have looked like have evolved. The time when the Oxford area was last wildwood has been pushed back long before Saxon times to the pre-Neolithic age.

The open, unproductive stands described by Osmaston have thickened up. Oak and beech—both deliberately encouraged by past management—have gradually been surrounded by a cohort of ash and sycamore from the 1950s. Deer have replaced rabbits as the main threat to future regeneration. The shrubs and smaller trees have tended to decline, partly due to the increased density of the canopy, partly the demise of coppicing, and particularly from increased deer browsing.

The physiological basis of growth of the tree canopy is better appreciated and competitive relationships between species are recognized as dynamic and unpredictable—the differential effects of disease, windthrow, and attacks by grey squirrels on trees will alter the balance locally. More substantial shifts are likely as the climate changes. What, for example, will be the long-term outcome of the ash–sycamore relationship? Should small-leaved lime be reintroduced? It does seem to have been the dominant shade-tolerant tree in the Woods in the distant past (Hone *et al.* 2001) and it is a species that may benefit from increased summer temperatures. How much intervention is needed or justified to create young open stands or to reduce the competition from young growth around the veteran trees? How low should the deer population be for there to be another burst of regeneration?

The interest in monitoring the Woods remains because fascinating times still lie ahead.

# 6
# The flowers of the forest
K.J. Kirby and M.D. Morecroft

## 6.1 Introduction

The archetypal image of English woodland—a swathe of bluebells (*Hyacinthoides non-scripta*) (see Plate 6), interspersed perhaps with a few primroses (*Primula vulgaris*)—can be seen in parts of Wytham Woods. More commonly there is a carpet of dog's mercury (*Mercurialis perennis*) or wood false-brome (*Brachypodium sylvaticum*). This chapter focuses on these plants of the forest floor; both species such as bluebell and primrose that tend to be more restricted to, or at least most abundant in, woodland and particularly ancient woodland (Peterken 1977), and 'woodland generalist' species such as bramble (*Rubus fruticosus*) and bracken (*Pteridium aquilinum*) that grow within woods but are common in other habitats as well (Kirby 1988).

The grassland vegetation is described more in Chapters 4 and 7. Relatively little research has been done on the lower plants and fungi at Wytham and so these are not considered further (Paviour-Smith 1971).

## 6.2 The Woods in a wider botanical context

'The wandering botanist is naturally mistaken for a tramp off the beaten track or for an officious inspector of something or other' (Church 1922). However, as a consequence of such 'tramps' Britain has one of the best-recorded, and best understood, floras in the world. The first botanical record for Wytham may be that in Johnson's edition of *Gerard's Herball* (1632); this refers to the plant now known as wood horsetail (*Equisetum sylvaticum*) being found in what is either the Great Wood or Woodcroft Copse (in Grayson and Jones 1955).

Gibson (1986) collated plant records for the Wytham Estate from the two then published floras of Berkshire (the Woods were in Berkshire until the 1974 county reorganization), various unpublished materials from Elton's Wytham survey, from other researchers, and his own searches of the whole estate during 1985. While there had been some losses over the previous century, over 500 species (excluding casuals) had been recorded since 1980. By comparison, about 1700 species are known for the whole of the West Berkshire vice-county for the period 1987–99 and a similar number for the vice-county of Oxfordshire, just across the Thames (Preston *et al.* 2002a). So Wytham has within its bounds between a quarter and a third of the flora of the whole county.

Much of this floral richness is associated with the grassland and wetland habitats within the Woods (see Chapters 4 and 7), but there are at least 190 of the 300–400 plants that in Britain are associated with woodland or wood-edge plants in at least part of their ranges (Kirby 1988). A further 19 wood or wood-edge species have been found in the past and some of these might yet turn up again. Wytham Woods have sometimes been dismissed as botanically not very distinguished, but the numbers of species are comparable to large blocks elsewhere: the flora includes very attractive species such as the meadow saffron (*Colchicum autumnale*), nettle-leaved bell-flower (*Campanula trachelium*), and various orchids.

## 6.3 Vegetation patterns—soil and woodland history

The woodland vegetation mostly falls in the ash-field maple-dog's mercury type (W8) of the National Vegetation Classification (Rodwell 1991), with small areas also of the beech-dog's mercury type (W12) and the beech-bramble woodland (W14) (see Annex to Chapter 4). The woodland species are therefore mainly those of free-draining to moist soils and of neutral to slightly base-rich pH, although there are a scatter of species associated with wetter or acid soils (Hill *et al.* 1999) (Fig. 6.1).

Osmaston (1959) noted differences in the vegetation according to soil type and these were generally confirmed in the frequency of species in the 1974 records from the Dawkins plots (Box 6.1, Table 6.1). For example, *Mercurialis perennis* is abundant through

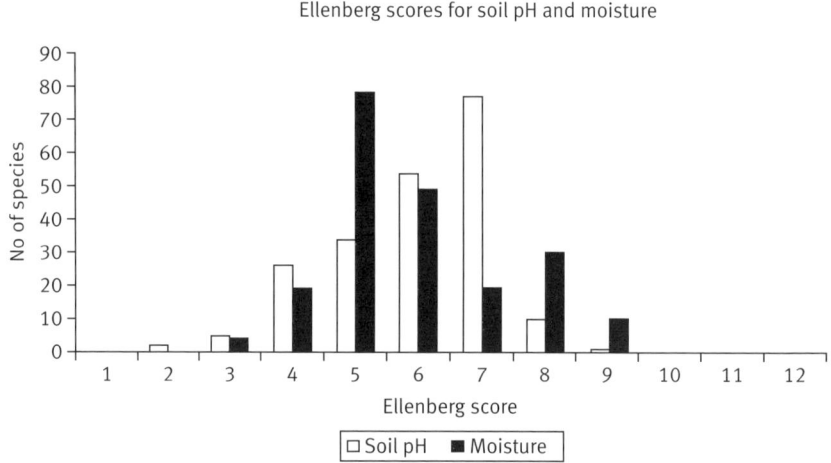

**Figure 6.1** Ellenberg scores for woodland species in Wytham. Moisture scores are from: 1 Plants of extreme dryness; 5 Moist site indicators; 7 Damp site indicators; 9 Wet site indicators; 10–12 Flooded/aquatic conditions. The soil reaction (pH) scores are from: 1 Plants of extremely acid conditions; 5 Indicators of moderately acid soils; 7 Indicators of weakly acid to weakly basic conditions; 9 Indicator of highly calcareous or otherwise high pH soils.

**Box 6.1** The 'Dawkins' Plots (Dawkins and Field 1978)

Plot distribution and identification:

- 163 plots spread across the Woods;
- 10 m x 10 m plots, located 14.1 m north-east of alternate points on 100 m grid;
- south-west and north-east corners marked by underground metal markers.

Ground flora records:

- a list of all vascular plants rooted in the plot;
- species occurrence in thirteen 0.1 m$^2$ 'circlets' evenly spaced across the diagonals;
- cover for ground flora (0–0.5 m high) estimated across the plot diagonal;
- cover over whole plot noted for seven species of known local significance bramble *Rubus fruticosus*, bracken *Pteridium aquilinum*, rosebay willowherb *Chamerion angustifolium*, bluebell *Hyacinthoides non-scripta*, nettle *Urtica dioica*, dog's mercury *Mercurialis perennis* (also for the shrub *Sambucus nigra*);
- in the 1991 and 1999 records the cover of each species was assessed using the Domin (1–10) range.

much of the Woods, though less vigorous on the clay. Species of more marshy sites such as *Carex pendula, Carex acutiformis*, and *Filipendula ulmaria* are locally common on the clay and rare on the limestone. One anomaly is that Osmaston associated *Sanicula europaea* with the clays, whereas all its seven occurrences in the Dawkins plots were on the limestone, which fits more generally with its national occurrence. Corney *et al.* (2008) in a more formal analysis of all three recordings of the Dawkins plots concluded that soil pH and fertility were key drivers of the floristic variation in the Woods.

Woodland vegetation is strongly influenced by historical factors. Eustace Jones (in Osmaston 1959) anticipated the work on ancient woodland indicators by Peterken (1974) and Rackham (1976) when he noted that *Paris quadrifolia* (Plate 7), a plant of damp shady places, was widely distributed in Marley Wood and in the Great Wood below the Calcareous Grit scarps, but was almost confined to 'ancient woodland'. Gibson (1988) mapped its distribution and confirmed that this was indeed the case.

Rose (1999) produced a list of 100 species believed to be strongly associated with ancient woodland in Oxfordshire and counties to the south, based on earlier work with Dick Hornby of the Nature Conservancy Council. Sixty-three species from this list have recently been recorded from Wytham (another 10 occur in older lists), which puts the Woods about tenth out of the 53 richest woods that Rose lists. The old coppices, the ancient woodland sections of Wytham Woods, tend to be the richest in 'ancient woodland indicators' but they are also found on the former commons and some even in the restored grassland on Upper Seeds (Chapter 4, Section 4.3).

Within Wytham Woods the distribution of these 'ancient woodland species' can be compared using lists based on occurrences in 12 ancient woods and 6 recent woodland

**Table 6.1** Occurrence of species in Dawkins's plots on different soil types in 1974. C = clay, S = sand, L = limestone. The ten commonest ground flora species in the plots as a whole, ten species more common on clay and ten species more common on limestone.

| Commonest species overall | C | S | L | Species commoner on clay | C | S | L | Species commoner on limestone | C | S | L |
|---|---|---|---|---|---|---|---|---|---|---|---|
| No. of plots | 62 | 46 | 56 | No. of plots | 62 | 46 | 56 | No. of plots | 62 | 46 | 56 |
| *Rubus fruticosus* | 57 | 40 | 45 | *Pteridium aquilinum* | 37 | 37 | 25 | *Geum urbanum* | 23 | 16 | 32 |
| *Mercurialis perennis* | 54 | 36 | 47 | *Dryopteris filix-mas* | 32 | 13 | 16 | *Brachypodium sylvaticum* | 19 | 13 | 31 |
| *Circaea lutetiana* | 54 | 35 | 45 | *Deschampsia cespitosa* | 31 | 7 | 11 | *Dactylis glomerata* | 8 | 9 | 19 |
| *Urtica dioica* | 46 | 40 | 36 | *Carex pendula* | 26 | 10 | 7 | *Heracleum sphondylium* | 5 | 12 | 15 |
| *Poa trivialis* | 41 | 32 | 37 | *Ajuga reptans* | 22 | 9 | 10 | *Silene dioica* | 2 | 7 | 10 |
| *Glechoma hederacea* | 37 | 31 | 31 | *Chamerion angustifolium* | 16 | 14 | 8 | *Taraxacum officinale* | 1 | 3 | 7 |
| *Galium aparine* | 33 | 27 | 30 | *Angelica sylvestris* | 20 | 11 | 5 | *Viola hirta* | 1 | 2 | 6 |
| *Hyacinthoides non-scripta* | 21 | 20 | 21 | *Tamus communis* | 19 | 7 | 9 | *Convolvulus arvense* | | 1 | 7 |
| *Festuca gigantea* | 16 | 16 | 21 | *Galeobdolon luteum* | 15 | 10 | 8 | *Sanicula europaea* | | | 7 |
| *Viola riviniana* | 22 | 7 | 18 | *Lonicera periclymenum* | 18 | 11 | 4 | *Cirsium eriophorum* | | 1 | 4 |

areas derived from the surveys by C.W.D. Gibson, but also from their occurrence in the Dawkins plots, 79 of which are in ancient woodland, compared with 85 recent woodland plots.

Excluding trees and shrubs the Woods include records for 54 indicator species, but 16 of them show no differences in occurrence between ancient and recent woodland. If only presence–absence data for whole woodland blocks are considered, 20 species show a difference in occurrence between ancient and recent areas (for example *Anemone nemorosa, Euphorbia amygdaloides, Iris foetidissima, Milium effusum, Paris quadrifolia, Poa nemoralis*, and *Veronica montana*); a further 10 species (including *Allium ursinum, Epipactis purpurata, Lathraea squamaria*) may be acting as indicators, but were only recorded from one or two ancient blocks so it is difficult to draw any conclusions from their occurrence. The Dawkins plot data allowed further discrimination of ancient and recent woods based on species abundance. Six species (including *Carex sylvatica, Hyacinthoides non-scripta*, and *Lamiastrum galeobdolon*) that occurred in most blocks of woodland (whether ancient or recent) were more frequent in ancient than recent woodland. *Adoxa moschatellina* has so far been recorded (but at low abundance) only from recent woodland and *Potentilla sterilis* and *Tamus communis* also appear to be more associated with recent than ancient areas in Wytham.

These data illustrate that caution must be used in deducing woodland history from the occurrence of ancient woodland indicators, because in the right conditions these species may not be as immobile as sometimes suggested. Conditions for their spread are more favourable in Wytham than in many other places because of the interspersion of ancient and recent woodland patches. Some ancient woodland species are also found in old grassland. The fact that the recent woods at Wytham have generally developed on former common rather than on highly disturbed and nutrient-rich arable or improved grassland fields contributes to their unexpected richness in indicator species.

Plant lists of one sort or another are widely used in biodiversity assessments, in the evaluation of sites for nature conservation, and in exploring the likely origins of woods (e.g. Kirby 1993; Goodfellow and Peterken 1981; Woodland Trust 2007). Plants are easier to record than many other groups of organisms, but different ways of recording the ground flora—as 'complete' lists for woods, or from representative plot samples—can produce slightly different results. Each has its strengths: more species are picked up in the whole-wood lists, but a single bluebell plant or a complete carpet of them each count only as one 'record'; the plot lists include fewer species, but should provide more consistent recording of these species and more potential for quantitative analysis. Work in Wytham Woods, and sites in Scotland and Wales, has highlighted the potential sources of variation in plant lists arising from observer differences or the time of year (Kirby *et al.* 1986).

## 6.4 The woodland flora and stand dynamics

Radley (1979) described the early stages of woodland flora development under scrub at Wytham. Conditions for the woodland plants change as the shrubs and trees grow, only, in some instances, for them to change again as grassland has re-established. However, even in the older woodland light, temperature, and rainfall regimes, as experienced by the ground flora, all vary considerable over distances of a few metres depending on whether measurements are taken in a gap or, for example, under the deep shade of a mature beech tree.

In general, open plots tend to have richer ground floras than those under closed canopies; with more species under ash and oak than in plots under canopies of beech or sycamore (see also Morecroft *et al.* 1998). However this difference is partly offset by the additional shading effect of the understorey on the ground flora richness. In general the understorey in Wytham Woods tends to be much denser in stands dominated by ash, whereas beech and sycamore stands are more open below.

At any one point the conditions change over time as the trees above that point grow, expand their canopies, and eventually die, creating a gap once more. In the Dawkins

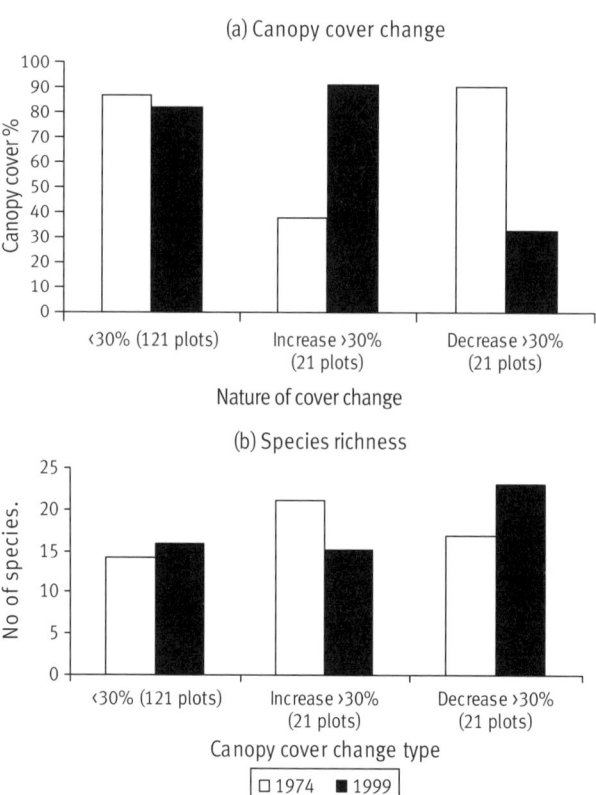

**Figure 6.2** Changes in canopy cover and ground flora richness for different groups of plots 1974–99. Only changes of at least 30 per cent were considered in the analysis because small changes could be simply the effects of recorder differences.

plots that showed a reduction in canopy cover between 1974 and 1999 there was an increase in species richness; those where the canopy was estimated to have increased showed a corresponding decrease in richness (Fig. 6.2). These changes are analogous to those that would have taken place when parts of the Woods were being managed as coppice. The richness of coppice ground floras is strongly related to this periodic opening up of the woodland canopy (Buckley 1992). Differences in light regime may also influence variation in distribution within a species—Wade *et al.* (1981) found that male shoots of *Mercurialis perennis* in Wytham are more frequent under canopy openings whereas female shoots are more frequent around tree bases.

Light reaching the forest floor encourages regeneration of all species from seed and promotes the flowering and spread of vernal species such as primrose that can survive in deep shade for many years as a few above-ground leaves. Other plants grow up from the buried soil seed bank (Brown and Oosterhuis 1981). In Wytham, mullein, *Verbascum* spp., (most evident on the limestone) and foxglove, *Digitalis purpurea*, (confined to a very few acid areas) are among those that tend to occur only in the few years of open conditions. Despite what would appear to be the natural rarity (without intensive management by people) of temporary openings in woods, these plants have specialist insects (for example, the foxglove pug, *Eupithecia puchellata*, and mullein moths *Cucullia verbasci*) that in turn have specific parasites feeding on them which also occur in Wytham.

## 6.5 Other changes in the woodland flora 1974–99

The plots where richness has increased are broadly balanced by those where it has decreased, so that overall the mean species richness in the Dawkins plots has changed little between 1974 and 1999. However, this masks some major shifts in the abundance of particular species. Osmaston (1959) noted that in most compartments bramble and bracken were abundant in the more open areas, with dog's mercury dominant under the denser shade. Nettle was locally common; bluebell and willow herb less often mentioned. The first four of these were among the most frequent species recorded in the Dawkins plots in 1974 and retained high frequency in the 1999 records (Fig. 6.3). Bluebell and willow herb were less common in the Dawkins plots, in line with Osmaston's comments, but whereas bluebell has retained its frequency the willow herbs (*Chamerion angustifolium* and *Epilobium montanum*) have both declined.

The most noticeable changes in frequency between 1974 and 1999 are the increase in grasses and sedges, particularly *Deschampsia cespitosa* and *Brachypodium sylvaticum*, to a lesser extent *Carex sylvatica* and *Poa trivialis*, and the decline in the herb *Circaea lutetiana*. These are all distinctive, easily identified species and most show little seasonal variation in detectability. *Circaea lutetiana* may sometimes be missed in surveys early in the year (Kirby *et al.* 1986), but the converse is true for *Poa trivialis*, which often dies back to dead thatch by late summer. The differences are therefore likely to be real.

Some of these species showed considerable changes in within-plot abundance as well over time. For the main six species highlighted by Osmaston (1959) a visual

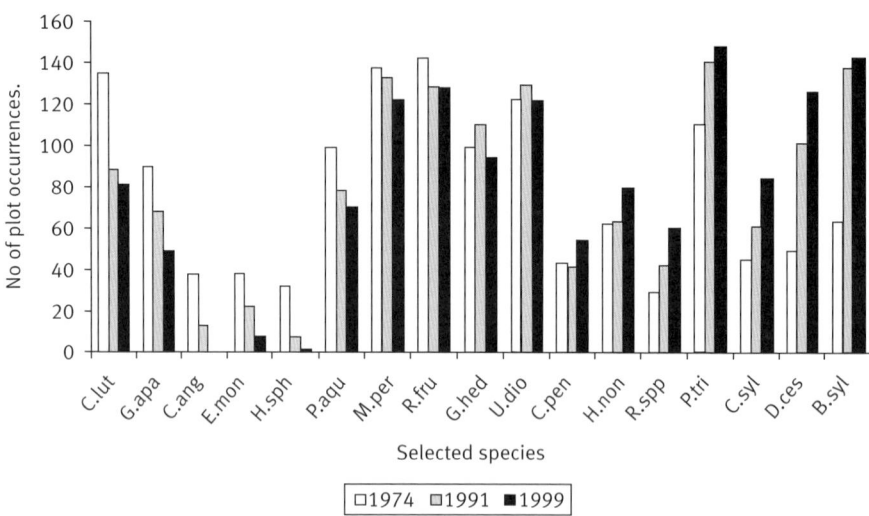

**Figure 6.3** Changes in frequency (out of 163 plots) for selected species 1974–99. Species abbreviations are: B.syl *Brachypodium sylvaticum*, C.pen *Carex pendula*, C.syl *Carex sylvatica*, C.ang *Chamaerion angustifolium*, C.lut *Circaea lutetiana*, D.ces *Deschampsia cespitosa*, E.mon *Epilobium montanum*, G.apa *Galium aparine*, G.hed *Glechoma hederacea*, H.non *Hyacinthoides non-scripta*, H.sph *Heracleum sphondylium*, M.per *Mercurialis perennis*, P.tri *Poa trivialis*, P.aqu *Pteridium aquilinum*, R.fru *Rubus fruticosus*, R.spp *Rumex* spp, U.dio *Urtica dioica*.

estimate of cover was made in 1974, 1991, and 1999 in the Dawkins plots. Bracken and willowherb declined and bluebell increased consistent with their changes in frequency. There was, however, a very marked decline in bramble, from covering about a third of the plots to less than five per cent. Smaller declines were shown for dog's mercury and nettle (Kirby 2004b) (Fig. 6.4).

For other species, changes in within-plot abundances are indicated by the number of circlet records (the 0.1 m² quadrats recorded across the diagonals of the plot, see Box 6.1). The circlet score for a species is given as 1 (i.e. present in the plot) plus the number of circlets (out of 13) in which it was recorded for that plot. Mean scores were calculated just for the plots in a given year in which the species occurred. The species shown in Fig. 6.4b illustrate different trends. The grasses *Brachypodium sylvaticum*, *Deschampsia cespitosa*, and *Poa trivialis* have become more abundant within the plots where they do occur as well as occurring in more plots (Fig. 6.3); *Circaea lutetiana* and *Galium aparine* have reduced abundance within plots and also occur in fewer plots; *Glechoma hederacea* has not increased in plot frequency overall (it was already widespread) but within plots seems to be increasing; while *Carex sylvatica*, which increased in the number of plots occupied remains scarce within them.

The flowers of the forest 83

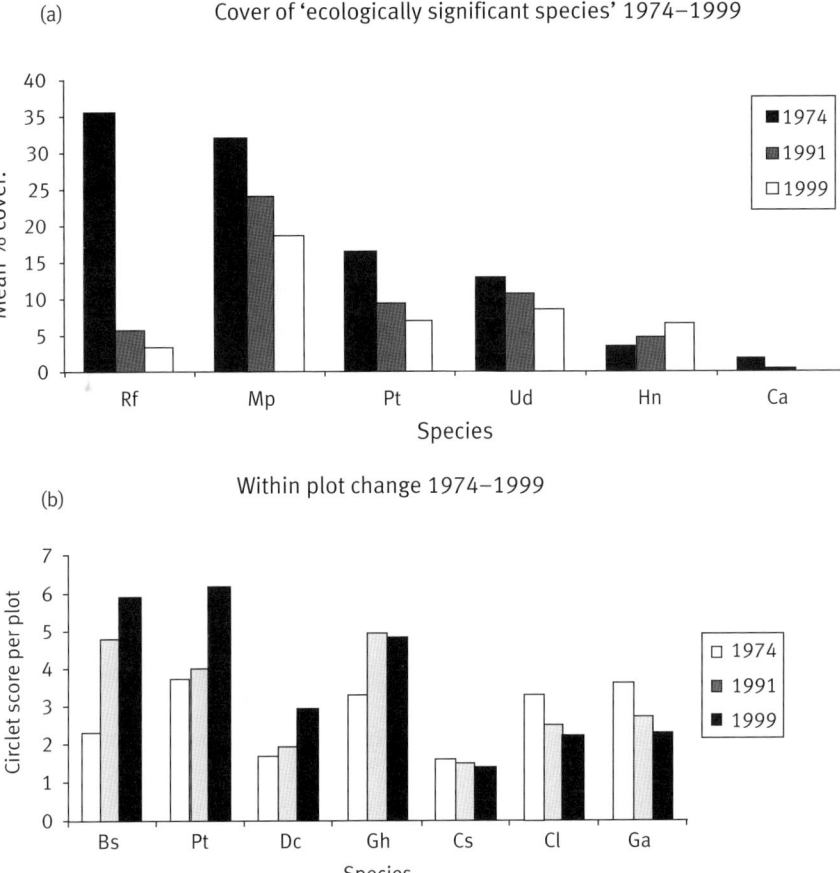

**Figure 6.4** (a) Mean percentage cover per plot in 1974, 1991, 1999 for *Rubus fruticosus* (Rf), *Mercurialis perennis* (Mp), *Pteridium aquilinum* (Pt), *Urtica dioica* (Ud), *Hyacinthoides non-scripta* (Hn) and *Chamerion angustifolium* (Ca) based on visual estimates for each plot. (b) Within plot cover change for *Brachypodium sylvaticum* (Bs), *Poa trivialis* (Pt), *Deschampsia cespitosa* (Dc), *Glechoma hederacea* (Gh), *Carex sylvatica* (Cs), *Circaea lutetiana* (Cl), *Galium aparine* (Ga). The score is based only on the plots in which a species was recorded in the year in question and equals (1+N) for each plot where N is number of 0.1 m² circlets (out of 13) in which a species was recorded for a plot.

## 6.6 Changes in Wytham Woods compared to trends elsewhere in the country

Across Britain as a whole (Preston *et al.* 2002a, b) plants of broadleaved woodland have not changed a great deal in their distributions compared to other habitats, based on 10 km square occurrences pre-1970 and in the subsequent 30 years. Braithwaite *et al.* (2006) looked at changes in the British flora (1987–2004) based on a sample of tetrads (2 x

2 km squares) and found some decline for broadleaved woodland species, but this was only marginally significant. There was a tendency for northern species to have declined and southern species to have increased, possibly a response to warmer winters; and for species of infertile soils in particular to have declined.

From the results of the Countryside Survey 2000, a stratified random sample of 1 km squares across Britain, Haines-Young *et al.* (2000) found evidence for increasing shade in the changes in broadleaved woodland flora recorded in Countryside Survey 2000 for England and Wales. They also found that plants of more fertile woodland conditions increased more than those of poorer soils between 1990 and 1998. However this latter effect may not be solely due to increased nutrient inputs, because their study included plots in new woodland created between 1990 and 1998, which had higher fertility scores than the older woodland plots. Smart *et al.* (2005) concluded that the eutrophication effect in the Countryside Survey results was less marked in woodland than in other habitats. The results from the 2007 Countryside Survey suggest that these trends have broadly continued with some decline in plant richness in broadleaved woodland.

Kirby *et al.* (2005) describe changes in woodland plants from 103 broadleaved woods (Box 6.2) based on surveys carried out in 1971 and 2001, in other words, a directly comparable period to the results from the Dawkins plots at Wytham Woods. To what extent, therefore, do the changes at Wytham Woods (Kirby and Thomas 2000; Kirby 2004b) reflect those in broadleaved woodland more generally, and do they have similar causes?

There has been a decline in the number of species found across the set of 163 plots, 173 in 1974, 143 in 1999 in Wytham Woods. However, unlike in the national woodland survey (Kirby *et al.* 2005), there has not been a decline in mean species richness of plots at Wytham, because some opening up of the canopy has taken place as well as stands becoming more shaded (Table 6.2). The changes are also not spread evenly across the Woods: Corney *et al.* (2008) found that whereas ancient and recent semi-natural areas tended to show some increases in richness after 1974, the plantation plots tended to decline in richness.

As in the national woodland survey, substantial changes in the frequency and abundances of many species have occurred. Using the same binomial test as was used in the national survey, 26 species showed a significant decline across the Wytham plots and 16 an increase in frequency. Some species that changed at Wytham (either up or down) showed no change in the national survey (for example, *Brachypodium sylvaticum*, *Galium aparine*) and, conversely, species that showed a change in the national survey did not necessarily change in Wytham. *Carex sylvatica* is increasing in Wytham, but decreased in the national survey. There are limits on using national trends to predict what will happen to a particular species at any individual site.

The same forces for change are, however, likely to be operating at Wytham as in the national survey, albeit at different rates. Thus there are some indications that smaller, stress-tolerant, or shade-tolerant species might be increasing most, although some such

**Box 6.2** Summary of ground flora changes in British broadleaved woodland 1971–2001 (Kirby *et al.* 2005)

1 There was an overall decline in species richness at plot and site levels, although with considerable variation in responses between sites (some showed increased richness) and with some species increasing as well.
2 Changes in species cover as well as frequency occurred, and both woodland specialists and non-woodland species were among the species showing decreases.
3 There was a shift to assemblages of plants that are more characteristic of shade; declines in species richness were associated with indicators of increases in the overall 'shadiness' of the woods.
4 There was some evidence for the effects of intensive agricultural activity surrounding the wood and increasing nitrogen deposition, as indicated by increases in nutrient-demanding species. The vegetation response to higher fertility might, however, be partially suppressed under shaded conditions.
5 Some species showed changes in frequency or abundance correlated with changing climatic variables.
6 There was an increase in signs of deer within woodland, and evidence of grazing increased particularly in the lowlands; there was a slight positive association between increased grazing and species richness, but otherwise no relationship between grazing levels, richness, or the cover of individual species was detected.

as *Mercurialis perennis* clearly declined. While the opening of plots, for example, through ride widening appears to have been enough to balance losses in species richness due to plot canopy closure elsewhere, the openings have tended to be small scale temporary clearances. There is less disturbance than during the major planting programmes of the 1950s; hence the decline in the typical species of clear-fells and recent plantings such as *Chamerion angustifolium* also seen in the national survey.

Wytham Woods' proximity to Oxford and major roads, and the occurrence of intensive farmland round parts of its boundary, might be expected to favour the increased growth of tall-growing competitive species as in some of the Countryside Survey data (Preston *et al.* 2002b; Haines-Young 2000). Between 1974 and 1992 there was evidence that the soils had become more acidic, although this trend has in part been reversed (see Chapter 2 Section 2.6) in line with national trends (Farmer 1995; Kirby *et al.* 2005, unpublished data) and that soil nitrogen levels had also increased. However, there is little evidence that this is having a significant impact on the flora within the Woods.

Nettles, often highlighted as a sign of recent enrichment, have been common through the Woods since at least the 1950s (Osmaston 1959) and have not increased since 1974. Species increasing did include some 'high nitrogen' plants such as *Poa*

**Table 6.2** Comparison of plot frequency changes for species in Wytham Woods with their change in site and plot frequency in the national woodland survey (Kirby et al. 2005). Increases and decreases were judged using a binomial test with 0.05 per cent probability used as the test for significance in both the national and Wytham comparisons.

| National site frequency change | Wytham no change | Wytham decrease | Wytham increase | National plot frequency change | Wytham no change | Wytham decrease | Wytham increase |
|---|---|---|---|---|---|---|---|
| No change | 82 | 13 | 13 | No change | 39 | 3 | 3 |
| Decrease | 21 | 8 | 0 | Decrease | 57 | 17 | 9 |
| Increase | 6 | 1 | 1 | Increase | 13 | 2 | 2 |
| No data | 53 | 4 | 2 | No data | 53 | 4 | 2 |

*trivialis*, but others decreased (for example, *Galium aparine*) (Fig. 6.4b). While Corney *et al.* (2008) showed that plots closer to woodland edges had higher nitrogen scores, this disappeared when the effect of higher light scores for the same plots was taken into account.

Tall, nutritious plants are also susceptible to grazing. The spread of woodland grasses and sedges (grazing tolerant) and decline in bramble (grazing susceptible) in particular point to increased grazing and browsing by deer as a major factor. Small exclosures around planted trees frequently contained bramble shoots that were bitten off where they grew through the fence. This change, though exacerbated by the fencing of the whole main Woods in 1989 that confined the deer population to the Woods, was already underway in the early 1980s (Horsfall and Kirby 1985). Parts of the Wytham Great Wood, which had continuous stands of bramble up to 2 m high in summer (Kirby 1976) had, by the mid-1980s, been reduced to island patches of bramble with grassy paths between.

In order to test whether deer are indeed the predominant factor causing the change in the ground flora, a series of 'exclosure' plots were established in 1997 to exclude deer from plots. There are three main deer exclosures in ancient woodland areas across the Woods (Morecroft *et al.* 2001). Each covers approx 0.3 ha and includes three 10 m x 10 m monitoring plots following the Environmental Change Network (ECN) design (Sykes and Lane, 1996; see Chapter 12 Section 12.2, Table 12.1). These plots have been regularly recorded, together with control plots elsewhere in the woods. Initially there was a slow response to exclosure, probably reflecting the low light levels under closed canopy forest. However, as time has gone by, the results have become clearer and by 1996 the exclosure plots had an obviously well developed ground vegetation with higher diversity (Fig. 6.5). The increase in species number was mainly a result of increasing numbers of herbaceous species.

## 6.7 Bramble as a key species

In the story of Wytham's woodland flora a case can be made for bramble as a key species. The intertwining stems contribute to the lower shrub layer (0.5–1.5 m) which is used for nesting and foraging cover by various small birds and mammals (Fuller 1995). In glades and wood edges the flowers and fruits are a valuable sources of food for invertebrates, birds, and small mammals (Gyan and Woodell 1987a, b; Snow and Snow 1988). It does compete with lower growing species, particularly as many of the leaves remain green overwinter; removal of bramble may lead to local increases in species richness (Kirby and Woodell 1998). This winter-greenness also contributes to its importance as a food source for deer, despite its prickliness (Bazeley *et al.* 1991; Gibson *et al.* 1993).

As deer numbers and their impacts are reduced, some increase in the bramble cover is expected, because it does still occur widely through the Woods and is also a common

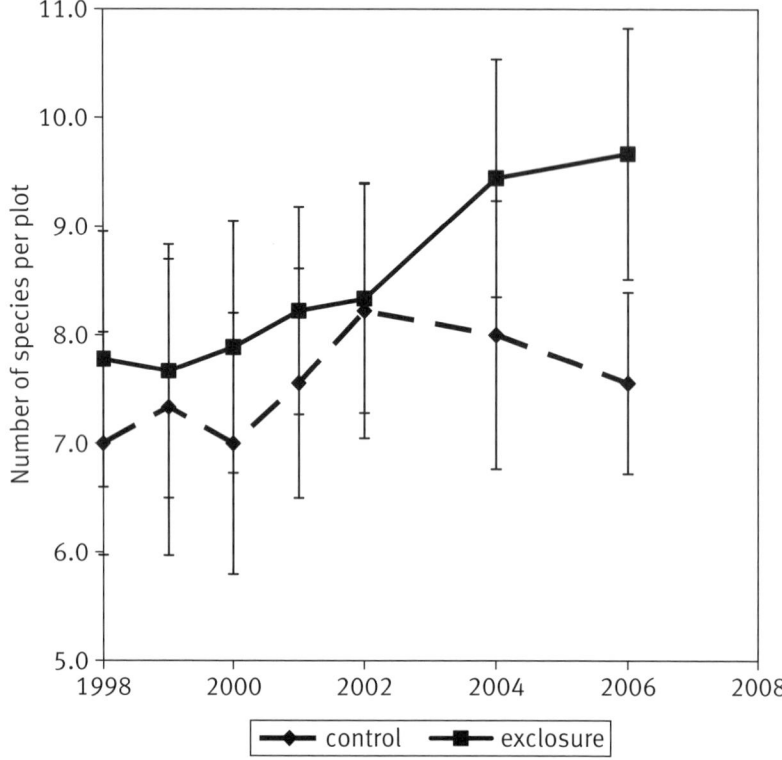

**Figure 6.5** Changes in species richness over time in control and fenced plots in Wytham, 1998–2006 (data from M.D.Morecroft).

component of woodland soil seed banks (Brown and Oosterhuis 1981). However, the rate of recovery will be affected by light conditions—spread and growth are much faster in more open conditions (Morecroft *et al.* 2001; Kirby 1976).

## 6.8 Conclusions

The broad patterns and changes in the flora across the Woods over the last 60 years are reasonably well understood. The importance of historical factors in plant species distributions has been confirmed, but also that even ancient woodland indicators can spread to new woodland if the conditions are right. Management through its effects on occurrence of gaps and glades has a strong influence on species abundance. Overlying these influences has been an increasing impact of deer on the distributions and abundance of particular species, in patterns seen elsewhere at Monks Wood (Cambridgeshire)

(Crampton *et al.* 1998; Cooke 2007) and at Bradfield Woods (Suffolk) (Gill and Fuller 2007).

There are signs that the deer management efforts are starting to be reflected in recent floral changes, but what the long-term outcomes will be remain uncertain. The effects of climate change on the woodland flora may start to appear. Hence it is important to continue with the permanent plot monitoring.

# 7
## The ecology of Upper Seeds—an old-field succession experiment
C.W.D. Gibson

## 7.1 Introduction: patterns in space and dynamics

At the beginning of the twenty-first century (Chapter 4), open habitats at Wytham form a network of patches and linear corridors amongst woodland and support some of the most diverse ecological communities in the area. Marley Fen, like some of the woodland patches, has an ancient origin and has probably shown little change over time in its structure. Other open habitats have passed in and out of arable use; at times they have been grazed or cut for hay; some have developed on former quarries (Chapter 3). Large areas that were open until the 1940s have since become covered with scrub or been converted to plantations; only a few areas have been restored to open ground since 2000. Virtually all the open habitats are therefore dynamic: on scales varying from centuries down to a few years and on varying spatial scales. This dynamism is one of the major determinants of biodiversity.

Despite or because of this dynamism, the open habitats at Wytham support 25 grassland and 12 arable plant species noted as rare in Oxfordshire, including nationally scarce species such as slender bedstraw (*Galium pumilum*) and the small mouse-ear (*Cerastium pumilum*). Survival of rare plants in these habitats has also been good compared to wetland species with a past distribution concentrated outside the Woods on the rest of the estate. At least 37 out of 54 (69 per cent) of Oxfordshire's rare grassland/arable species have survived, as opposed to only 11 out of 34 (32 per cent) wetland species.

This chapter discusses the way that vegetation and the invertebrate fauna of open habitats develop through an interplay of colonization, management, and site history (such as soil nutrient levels), focusing on a case study—the abandoned arable field at Upper Seeds. This shows how ecologists can tease apart factors that influence the success of different species; knowledge can in turn help conservation managers to create rich new grassland.

## 7.2 Patterns in time: the nature of change

After a major disturbance such as cultivation, community change in open habitats can be detected for a century or more (Gibson and Brown 1992; Gibson 1995), but major successional changes can be apparent in a few decades (Gibson and Brown 1985).

Elton (1966) and others documented the types and structures of open vegetation before and during the crash in rabbit populations when myxomatosis arrived in the 1950s. They saw the first effects of scrub invasion on the grassland and the spread of coarse tussock grasses such as tor grass (*Brachypodium pinnatum*) and bracken (*Pteridium aquilinum*). Plant diversity diminished, but the more varied structures allowed a greater variety of fauna to survive than in even, homogeneously-cropped short turf. Elton also helped to ensure the survival of much of the older open habitats that remain today, by securing reserves that escaped degradation from intensive agriculture or woodland plantations.

Other researchers have vastly extended knowledge about what has happened to the open habitats, but particularly the grasslands. Vegetation information was compiled by P.S. Lloyd (unpublished) until his untimely death and later by Woodell and Steel (1990). Others have studied invertebrates (e.g. Duffey 1962a, b; Whittaker 1969), small mammals (e.g. Chitty and Phipps 1966; Southern 1979) and soil invertebrates (e.g. Phillipson *et al.* 1979).

This robust observational base underpins and informs experimental studies about how the Woods work from the landscape scale down to that of individual populations. Wytham thus sits among other places in the world such as Harvard Forest (Massachusetts) (Foster and Aber 2004) where a long term synergy between observation and experiment allows critical questions in ecology to be answered.

The Upper Seeds experiments and structured observations were initiated by the author and colleagues in the early 1980s. Each experiment has allowed others to be generated and so, for example, the climate change experiments described in Chapter 12 are nested within the main system described here.

The initial focus was on the plants, which are easy to record and form the production base of virtually all terrestrial communities. However, plants only form a small proportion of the species in the ecological community and the observations and experiments in Upper Seeds have incorporated studies on the interactions between plants, invertebrates, and other ecosystem components.

## 7.3 The experimental system

The core experiment was designed to test the effects of different seasons and intensities of grazing on succession from land abandoned from arable cultivation on limestone. It can be seen as analogous to the many 'old field' studies of succession that have taken place in North America.

The Wytham studies are centred on the hilltop field known as Upper Seeds, and started in 1984 (Box 7.1, Fig. 7.1). The rationale was that most (even if only in prehistoric times) chalk and limestone grassland passed through one or more arable periods and hence, given time and appropriate management, it should be possible to restore or recreate such grassland from abandoned arable land on suitable soils. This work has acquired greater practical relevance since the development of the UK Biodiversity Action Plan

> **Box 7.1** An experimental approach to understanding the effects of different grazing treatments on limestone grassland restoration (full details in Gibson *et al.* 1987b).
>
> The experimental core has involved short periods of grazing by sheep in spring or autumn, with ungrazed controls. These three management treatments are replicated 6 times, giving a total of 18 paddocks (blocked in two grids) (Fig. 7.1b).
>
> Monitoring of the whole has been by permanent 1 m x 1 m quadrats with species lists recorded at least twice a year for each of twenty-five 20 x 20 cm cells in each quadrat.
>
> Surveys were made in 1987, 2000, and 2004–5 on a series of the other Wytham grassland patches of known history and management (including before and after new management that was in imposed after 2000) to provide an observational comparison with the results of the standard grazing regimes, but on areas with different histories and / or vegetation.
>
> Other studies that have taken place in the plots include:
>
> - Invertebrate monitoring of the insecticide-treated plots and their controls on Upper Seeds from 1985 to 1992 (Brown *et al.* 1988, Brown and Gange 1989).
> - Direct counts of spider webs in the same areas and in permanent quadrats on Bowling Alley and Sundays Hill (Gibson *et al.*, 1992a).
> - Direct counts of leaf-miners in permanent quadrats on Upper Seeds, Bowling Alley and Sundays Hill (1985–92) (Sterling *et al.* 1992).
> - Seed bank study of Upper Seeds (Woodell and Steel 1990).
> - Seedling recruitment studies (Fowler 1994).
> - Analysis of major soil nutrients in Upper Seeds in 1992 (Ollerton 1993).

(HMSO 1994): the action plan for chalk and limestone grassland includes targets for the re-establishment of at least 1000 ha of such habitats by 2010 (English Nature 1998).

Grazing is important if the grassland is not to develop into scrub and woodland. For the experiment, sheep were chosen because they were the commonest stock used on such grassland in the past and the easiest to manage at the scale of this experiment. Spring and autumn grazing were contrasted with ungrazed controls, because these are the two major periods of plant growth in unfertilized grasslands. This core experiment still continues.

Within the core experiment, smaller plots were established to determine the effects of below-ground and above-ground grazing by insects; two different pesticides were used to kill below-ground insects on some plots and above-ground ones on others. This sub-experiment ran from 1985 to 1992, by which time it was apparent (Brown *et al.* 1988; Brown and Gange 1989) that the effects in this system were relatively small compared to other early successional habitats, albeit statistically significant. However the scale of effects was very different to that of sheep on the system (Gibson *et al.* 1987a). A further

94 Wytham Woods

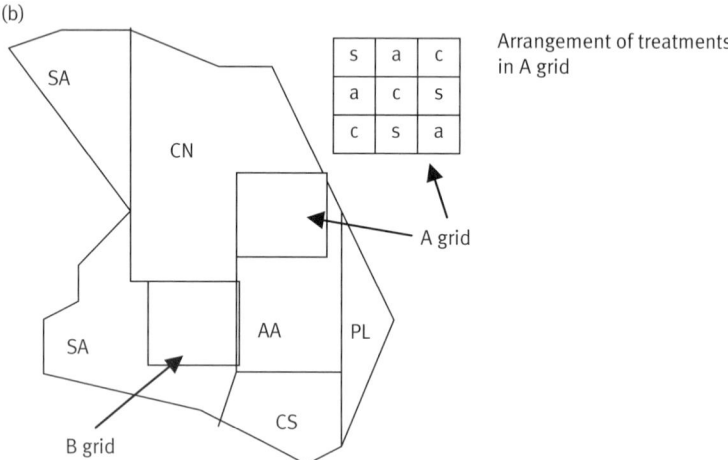

**Figure 7.1** (a) Oblique aerial photo of the Upper Seeds experimental system looking north; (b) layout of grazing treatments: SA heavy spring and autumn grazing, AA heavy autumn grazing, CN, CS deer-grazed only, PL autumn ploughed; paddock grid treatments C ungrazed control, A autumn-grazed, S spring-grazed.

sub-experiment, started in 2002, involved the removal of scrub from half of the control area, while leaving it free to develop on the other.

### 7.3.1 Putting the results into a wider context

More extensive areas of the same field have been grazed for longer periods in autumn or in both spring and autumn. These regimes can be regarded as more like the sorts of grazing regimes practised in normal agricultural management; they could not be used in the main experiment because the paddocks were not big enough to support the animals over such periods.

In 1988, the grazing trials were extended to compare the effects of the same grazing system (heavy grazing in both spring and autumn) on two other patches of old grassland in Wytham (Bowling Alley and Sundays Hill). These appear never to have been cultivated, so provide a contrast to Upper Seeds which had been an arable field in the early 1980s. The wider applicability of the results has also been tested against the results of monitoring of different types of past disturbance from 1989 at the Holies and Lardon Chase (both National Trust properties where the Thames cuts through the chalk hills at the Goring Gap) 35 km from Wytham; and from 1997 on Butterfly Conservation's 10 ha grassland restoration trials at Magdalene Hill Down near Winchester, 80 km from Wytham. Observations at these and other sites have helped to put the findings from Wytham into a wider context and have improved our understanding of the degree to which species-rich grassland can be restored from arable land (Gibson and Brown 1991a, b; Gibson 1995).

## 7.4 Colonization and species pools

Restoration of plant communities following a major disturbance such as cultivation depends on the right species being able to reach the restoration site. Previous studies have emphasized the slow colonization of new localities by species characteristic of 'ancient' habitats, particularly for woodland species (Rackham 2003). Such studies have usually relied on comparisons of the species found in habitat patches of known age to infer slow rates of spread: the oldest patches tend to have the greatest number of characteristic species for that habitat. This has, to some extent, been matched to species' biology: many species have large seeds and no very obvious means of long distance dispersal (for example, bluebell); or have adaptations such as elaiosomes (fleshy, protein-rich food patches) (for example, hairy violet) suited only to small dispersers (ants) which are assumed to move the seeds only over short distances. A clear demonstration of such slow spread in woodland is provided by Rackham's observations at Hayley Wood (Cambridgeshire) where species have been spreading into a patch of new woodland developed on an old field at rates of only a few metres a year (Rackham 2003).

This does not mean that colonization must always be slow. Even in Hayley Wood some individual plants did manage to establish well beyond the general colonization front. Indeed longer distance/faster spread and colonization must have happened in the past or many plants would not occur in Britain at all: there has simply not been enough time since the retreat of the glaciers for them to have moved by such small steps up from their southern European refuges. The findings from the Upper Seeds grassland experiment tend to reinforce the reality of history, namely that many plants can spread quite quickly if conditions are favourable.

The most likely seed sources for species that invaded the experimental treatments are other remnant patches of grassland that survive elsewhere in Wytham, for example the old quarry by the chalet. Plant lists were therefore made for these patches to estimate the available species pool and how far species might have had to travel to colonize the former arable land at Upper Seeds.

The sheer number of species that appeared in Upper Seeds was astonishing for an area of only about 10 ha, with 264 plant species being found at least once between 1983 and 2007. This is about 40 per cent of the flora of the whole 1240 ha estate (Gibson 1986). Nearly half of these were in the soil seed bank despite a period of arable use. This is a minimum estimate because seed bank studies only sample a small proportion of the soil and cannot provide a complete inventory. For species that were not in the soil seed bank, the information on where else they were recorded in the Woods provided an estimate of the minimum distance each species would have had to move to colonize the field.

For species not detected in the soil seed bank the nearest seed source was mostly within 150 m. However, some appeared to have spread from patches more than 200 m away in less than 10 years, and a few species also appeared that were wholly new records or re-appearances of species previously thought to be extinct on the estate (these are now termed 'overlooked' species). These observations may underestimate the ability of species to spread from different distances to the newly-created habitat patch, because some seed might have come from more distant patches, particularly if the more distant patch had a larger, more vigorous population, producing many more dispersible seeds.

The colonizing species include many common arable weeds and ubiquitous grassland plants, but also, less expectedly, species typical of neutral grassland and even some 'ancient woodland indicators' (Gibson and Brown 1991b). Of the late successional flora of chalk (Gibson 1995) and limestone grasslands available in the estate species pool, 65 per cent (17 out of 26) had appeared in Upper Seeds field by 1989. Five were present in the seed bank, six more had been recorded by 1985, with the other six moving in between 1985 and 1989. No more had appeared up to the time of writing in 2007.

It is unlikely that dispersal has ceased given that there are plenty more species that might colonize the field. What seems more probable is that as the grassland has developed over the last 20 years, the opportunities for new species to establish have declined. Whereas dispersal and seedling establishment were important in the first few years of the experiment, competition and survival through to adult plants may have now become the limiting factors on the appearance of new species.

Results from permanent quadrats provide another insight into the processes involved. The occurrence of species within each grid of 20 x 20 cm cells within the 1 m quadrat has been recorded over time and their increase in abundance within the quadrats can thus be followed. For example, a group of seven species typical of late-successional limestone grassland (*Trisetum flavescens, Galium verum, Primula veris, Viola hirta, Poa angustifolia, Lotus corniculatus*), had all established in the experimental plots by 1995. Initially, all were present in 10 cells or fewer across all grazing treatments, but thereafter they show an exponential increase in the number of cells occupied up to 2007. By contrast, another group of species (*Anacamptis pyramidalis, Centaurium erythraea, Linum catharticum, Carex flacca, Sanguisorba minor*) remain in relatively few cells, despite, in some cases, having been present since the start of the experiment.

At Wytham, therefore, succession from arable land to ancient grassland communities does not appear to be limited by the ability of the 'expected' species to arrive, but by their subsequent ability to establish and/or spread. These limits will presumably eventually be gradually broken down, but it might take one or even several centuries for the ancient grassland community to become fully established (Tansley and Adamson 1925, 1926; Gibson and Brown 1992).

## 7.5 Early and mid successional communities during an English old-field succession

In the first year of observation, Upper Seeds resembled a poor and weedy crop. Reminders of the cereal crop occurred, with wheat 'volunteers' accompanied by short-lived herbs (particularly small-flowered buttercup *Ranunculus parviflorus*, cut-leaved cranesbill *Geranium dissectum*, black medick *Medicago lupulina*, and field madder *Sherardia arvensis*) and the annual grass *Anisantha sterilis*, with numerous other species, to give 15–20 species per square metre. Wheat and the other annual ruderals, such as groundsel (*Senecio vulgaris*), soon disappeared. Perennial grasses established rapidly, with rough-stalked meadow grass (*Poa trivialis*) and creeping bent (*Agrostis stolonifera*) becoming dominant in all treatments. Over the next 15 years until the late 1990s distinct effects of grazing appeared. The following account is updated from Gibson and Brown (1992).

Ungrazed controls rapidly and permanently lost annual herb species (Fig. 7.2) as a common component. The only other consistent distinctive feature of the ungrazed

**Figure 7.2** Control paddock in the background contrasting with heavy autumn-grazed treatment in the foreground.

treatment was the rapid establishment and growth of woody species dominated by hawthorn (*Crataegus monogyna*) and wild rose (*Rosa canina* agg.), but with more than 20 other woody species recorded. Canopy closure to the extent of exclusion of ground vegetation was rare by 2000, but even so species richness remained steady and the least of any treatment until this time.

Autumn grazing favoured large-seeded herbs, for example common vetch (*Vicia sativa*), smooth tare (*V. tetrasperma*), and bluebell (*Hyacinthoides non-scriptus*) and annual grasses (*Anisantha sterilis* and *Bromus* spp., mainly *B. hordeaceus hordeaceus*). A few other perennial herbs, notably wild basil (*Clinopodium vulgare*), also did well, but overall there was a long term decline in herbs and total species richness. Woody plant invasion was slower than in the other treatments.

The spring-grazing treatment plots have generally a higher species richness compared to other treatments. However, some species favoured by autumn grazing were scarce or absent, as were some vernal species, such as cowslip (*Primula veris*) and hairy violet (*Viola hirta*). Presumably these were unable to recover from the early loss through grazing of their leaves and flowers, even after the sheep were removed.

The heavier grazing treatments in the field surrounding the formal experimental paddocks also differed in their effects according to season. When heavy grazing was confined to autumn it had little additional effect save to increase the dominance of grasses and large-seeded herbs seen in the autumn short-period grazing plots. The heavy spring and autumn treatment, in contrast, produced highly distinctive effects. Species richness varied over the years in an apparently cyclical manner, sometimes being the highest of all treatments and at other times no more than the average. Further investigation showed that these fluctuations were confined to annual species. Species-richness peaks were correlated with summer droughts in the year prior to observations. The drought acting in combination with the grazing led to the creation of micro-gaps (Watt and Gibson 1988) in the sward which could then be colonized by annual herbs. Excluding annual herbs, there was a gradual decline in richness towards the overall treatment means.

As well as its peaks in annuals the spring–autumn grazing treatment was distinctive in having a dense sward dominated, prior to 2000, by a mixture of creeping bent, rough stalked meadow grass, and Yorkshire fog (*Holcus lanatus*). This was usually accompanied by white clover (*Trifolium repens*), although this species, along with ryegrass (*Lolium perenne*), suffered periodic declines and resurgences. Robust biennials and short-lived perennials were also favoured, initially ragwort (*Senecio jacobaea*), but later wild parsnip (*Pastinaca sativa*) and especially woolly thistle (*Cirsium eriophorum*). Although the species have fluctuated in abundance, woolly thistle has been a regular and distinctive component for more than 15 years.

The other important characteristic of this heavy grazing is that it has been the only treatment that has suppressed false oat-grass (*Arrhenatherum elatius*). Its invasion and spread during the 1990s and subsequent decline has dominated the succession from former arable land in the other treatments.

## 7.6 Invasion and decline of *Arrhenatherum elatius* (false oat-grass)

False oat-grass is commonly regarded in Britain as an invasive nuisance, dominating less intensively managed areas, such as highway verges and abandoned quarries, or arable land where much more diverse vegetation would otherwise be desired. This seems to have been borne out by the early results from the Upper Seeds experiment.

Across all of the grazed treatments there was a slow decline in plant species richness per 1 m$^2$ quadrat to a low point around the turn of the millennium. The level was similar to that reached by the ungrazed controls nearly 15 years earlier.

False oat-grass forms individual tussocks and patches that range from a few centimetres to tens of centimetres across. The competitive effect of this species in excluding other plants, and hence reducing species richness, therefore becomes more apparent if the records for individual 20 x 20 cm cells are compared over time—an individual false-oat-grass plant may occupy most of a cell.

In spring-grazed (Fig. 7.3) paddocks false oat-grass first appeared in a quadrat in 1986, but by 1994 it had colonized half of all cells within spring-grazed quadrats. From the late 1990s until 2005 it was still almost ubiquitous, but more recently it has declined.

From 1986 to 1990, cells with false oat-grass were still rare and cells with and without the species showed little difference in species richness. After 1990, however, a significant difference in species richness developed; cells without false oat-grass consistently supported more species than those with it. It was apparent on the ground that false oat-grass was invading to the detriment of all other species during this period.

**Figure 7.3** Spring-grazed paddock in the background, showing clear browse-line effect on the few trees that have established; in the foreground is the more extensive heavy autumn-grazed treatment area.

**Figure 7.4** Autumn-grazed paddock showing some hawthorn scrub establishment amongst the grassland.

Since 2000 there has been resurgence in plant richness in all treatments including the controls with, for the first time, the heavily grazed (spring–autumn) treatment failing to be more diverse than the others. In all save the heavy spring and autumn-grazed treatments, the earlier decline was evidently associated with the spread of false oat-grass, which became a virtually ubiquitous member of the community.

In the spring-grazed paddock, for example, after 2000, the difference in species richness between cells with and without false oat-grass started to diminish and by 2004 it had disappeared. False oat-grass, although remaining in the sward, was no longer associated with reduced species richness. The pattern was similar for all the other grazing treatments (except the heavy spring and autumn grazing) although the speed of invasion of false oat-grass and its subsequent retreat has been slightly different. In ecological terms the invasion is over, or the invasive species has changed its character so it no longer behaves like one.

Other accounts of chalk and limestone grasslands refer to this changing abundance of false oat-grass, for example Hope-Simpson (1940), reporting on some of Tansley and Adamson's (1925) former arable plots on the Hampshire chalk. The observations of Leps and Stursa (1989) in Czech vineyard successions suggested a phase of false oat-grass dominance of between 50–100 years after last cultivation before more species-rich *Brometalia* grasslands took over.

Such general observations indicate that the false oat-grass dominance is a real phenomenon, but do not help to identify the cause. On Upper Seeds, because the changes happened within the context of a controlled experiment there might be pointers to what causes the change in the relationship between its abundance and plant species richness. This in turn might indicate how, elsewhere, species richness could be more quickly restored where this is one of the conservation management aims.

One cause of the invasion and decline appears to be ruled out. Grazing pressure by sheep on the different paddocks has been constant over time, so the different treatments define the range of grazing pressures within which the invasion and decline can occur. Only heavy grazing in both spring and autumn prevented the invasion of the grass.

False oat-grass is known to demand moderate to high levels of major nutrients (Grime *et al.* 2007). Could a single generation use up nutrients in micro-sites so they are too depleted for the grass to remain competitive? This would imply that one or more major nutrients ought to have been at elevated levels in the former arable land, compared to ancient grassland without false oat-grass from the beginning of the invasion period, and this nutrient should have later declined.

Ollerton (1993) measured levels of major nutrients across Upper Seeds in 1992, just as the false oat-grass invasion was accelerating. Nitrogen and potassium levels were universally low and similar to the nearby species-rich *Bromus erectus* grassland in Lower Seeds reserve. Only phosphorus levels appeared elevated. However, when the distribution of the areas with high phosphorus was looked at in more detail, it was found to be associated with the archaeological remains of Romano-British settlement (Chapter 3), whose influence extends into the heavy autumn grazing treatment. Most of the field, and all of his six samples (placed at regular intervals on a 50 m grid) that fell within the controlled grazing paddocks, had available phosphorus levels below 5 mg per 100 g, in other words as low as those in the Lower Seeds old grassland reserve. This tends to rule out the nutrient hypothesis as, even in 1992, elevated levels of phosphorus were restricted to only a small part of one of the treatments in which false oat-grass showed expansion and decline.

Another suggestion is that the competitiveness of late successional grassland species may change over time. Most, if not all, such plants have mycorrhizal fungal associations. Early colonists may come without suitable mycorrhizae and be susceptible to competition. Later, species may come 'armed' with their appropriate fungal associations and so be able to oust the false oat-grass. If this hypothesis were true, then late successional species should be able to colonize species-poor cells with false oat-grass post-2000 but not before. This was explored by following the change in distribution of *Viola hirta* in 20 x 20 cm cells, with and without false oat-grass and with different levels of species richness. Cells colonized by hairy violet without false oat-grass tended to be richer in other species than cells where it did not occur; and cells colonized by hairy violet with false oat-grass were also richer in other species than other cells with false oat-grass, but no hairy violet.

Similar patterns are shown by other late successional plants; they tend to make their first appearance in 20 x 20 cm cells that are already, for whatever reason, rich in other species, whether or not they contain false oat-grass. In essence, the late successional grassland species follow areas of already high species richness rather than invade species-poor areas. Where they invade false oat-grass areas, the grass has already declined sufficiently for overall richness to be high.

While different levels of sheep grazing appear to have little impact, could insect herbivory exert sufficient pressure on false oat-grass, as time goes by, to decrease its ability to keep out other plants? Above ground, the only insect herbivore to attain significant numbers is a bug (*Megaloceraea recticornis*). This is widespread on the Wytham grasslands and is a specialist on developing seeds and other protein-rich parts of false oat-grass (Wetton and Gibson 1987). The species was already present near the beginning of the invasion period (Brown *et al.* 1992) and has been common throughout. There has been no clear increase or outbreak pattern in this conspicuous and distinctive insect that could explain the decline in its food-plant. If below-ground insect herbivores were responsible for reducing the vigour of false oat-grass, then their reduction in the insecticide experiment should have shown up this effect. If anything, the opposite was the case, with more false oat-grass in controls compared with root insecticide-treated plots.

Frustratingly, the cause of the changing impact of false oat-grass remains to be found: perhaps there is a build up of some chemical in old clumps that inhibits its growth, perhaps there is some fungus that gradually weakens the plant. The experiment has, however, helped to rule out other more obvious causes.

## 7.7 How does the grassland on Upper Seeds compare with other grassland?

Ungrazed controls become dominated by scrub at the expense of all grassland plants. Heavy spring and autumn grazing appears to slow down or even prevent the calcareous grassland succession. This treatment appears stuck in a phase that is poor in plant species. Heavy continuous grazing may promote a high turnover of nutrients, and so maintain relatively high nutrient levels, and may also promote the establishment of nitrogen-fixing white clover. It would tend to favour highly competitive species and hence low plant species richness.

By contrast, the seasonal paddock grazing does appear to produce the correct conditions for succession towards plant communities typical of ancient grassland. Most of the grazed treatments now appear to be passing into a new successional phase, where species richness is increasing (Gibson and Brown 1992).

In Fig. 7.5, records for late successional species, as indicated by the number of 20 x 20 cm cells in which they occur, are shown for a number of sites. High and steady levels are found for these species at the Quarry at Wytham and ancient grassland at the Holies (Streatley) which are thought never to have been ploughed. Grassland at the Quarry is grazed only by deer and rabbits (uncontrolled) but has scrub removed by hand at two-yearly intervals. Grassland at the Holies is grazed under a SSSI management regime that is flexible where considered necessary to improve the grassland condition.

The Bowling Alley and Sundays Hill at Wytham are ancient grasslands, subject to the same heavy spring–autumn grazing regime used in Upper Seeds, but on deeper soils. They have lower scores for late successional species than the best ancient grassland at Wytham (the Quarry).

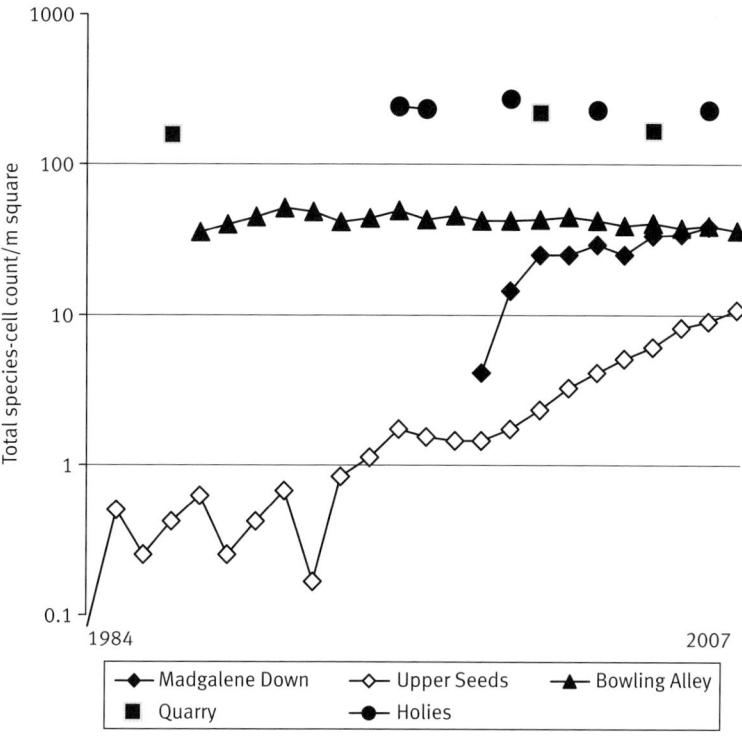

**Figure 7.5** Occurrence and spread of late successional species in grassland of different origins, 1984 to 2007.

The richness of late successional species is assessed as the sum of 20 x 20 cm cell-counts for all species per square metre quadrat. The sites are:

1 Ancient, unploughed shallow grassland—Quarry (Wytham) and Holies (Streatley); ancient unploughed, deeper soil grassland—Bowling Alley and Sundays Hill.
2 Restored grassland with additional seeding—Magdalene Down (Winchester).
3 Restored grassland, no seeding—Upper Seeds (Wytham).

Prior to 1993, the grazed paddocks in Upper Seeds showed random variation in cell scores associated with very low abundance of late successional species. Since 1993, there has been a fairly steady increase—speculative extrapolation suggests 60–80 years to reach the levels of the best ancient grassland, which is rapid compared to observations of grassland succession elsewhere.

These results were put into a wider perspective in relation to other calcareous grasslands in Southern England by Gibson and Brown (1991b). Similar records are shown for Magdalene Hill Down, a former arable site, into which seed harvested from chalk downland was sown, supplemented with a mixture of other chalk grassland species. Grazing management has been reactive and deliberately varied in order to achieve the most rapid results in terms of chalk grassland species. This 'acceleration' of natural succession clearly gave a rapid start, but the increase has since slowed down and reached no higher than the Bowling Alley and Sundays Hill ancient grassland levels.

## 7.8 Structure, plant taxonomy, and invertebrates

The overall experimental system on the former arable land has produced a complex structural mosaic, including tall scrub, low scrub amongst species-rich grassland, tall coarse grassland, short grassland, and bare ground. These cover the range of structural types originally described by Elton (1966) from his Wytham observations.

This combination of variations in vegetation composition and structure is important not just from a plant point of view, but also because they help determine the variety and composition of other components of the community. While little is known about the below-ground elements, dominated by fungi, bacteria, and other organisms, the above-ground invertebrates have been reasonably well-studied (see Chapter 8).

Ecological succession tends to be dominated by plant studies, but the Wytham grassland work provided an opportunity for parallel studies on the animal communities. The work built on the ideas started by Elton (1966) and since developed by Lawton (1983), Southwood *et al.* (1979), and others on the relative roles of habitat structure and plant taxonomy in affecting invertebrate assemblies, as discussed in more detail in Chapter 8.

The main Upper Seeds studies concentrated on a suite of animal groups chosen to have a spectrum of responses to taxonomy and structure. At one extreme, spiders were expected to be relatively indifferent to plant species composition, except inasmuch as this affects the structure of the vegetation. Leafhoppers (Auchenorrhyncha) mainly feed by sucking the sap of a wide range of grasses, and therefore might be expected to track plant chemistry and architecture rather than taxonomy. By contrast, herbivorous beetles (Coleoptera) and true bugs (Heteroptera) contain many species that are taxonomic specialists and are more likely to be influenced by plant species composition (Gibson *et al.* 1992b). Finally, leaf-miners were chosen as a group that reflects a feeding type rather than insect taxonomy, but which have the unique advantage that their feeding signs can be counted in direct relation to plant abundance, without the need to rely on indirect methods used to sample most mobile insects.

Each treatment and grassland age, not surprisingly, turned out to have conditions best suited to some species or other: that is, all components of the mosaic of open ground habitats at Wytham are contributing something different to the overall invertebrate diversity. Amongst the leafhoppers recorded in 1989, *Mocydia crocea* was truly ubiquitous; the hopper *Adarrus ocellaris* was everywhere on the early successional grassland developing from arable land but was absent from the old grasslands. *Adarrus multinotatus* was only present on the old grasslands, except for a single atypical occurrence (Brown *et al.* 1992).

The spittlebug, *Philaenus spumarius*, was commonest in the ungrazed and short-period grazed (autumn or spring) treatments, less common in the heavily autumn-grazed and wholly absent from the spring and autumn-grazed treatments, whether in the former arable land of Upper Seeds or in the old grassland of the Bowling Alley and Sundays Hill (SH). *Philaenus spumarius* also avoided the heavy grazed treatments, while *Macrosteles laevis* was most abundant in this treatment.

In general, both leafhoppers (Brown *et al.* 1992) and spiders (Gibson *et al.* 1992a, b) were more abundant and more diverse in species in ungrazed or lightly grazed treatments. Large web-spinning spiders generally chose high-standing robust plants, such as wild parsnip (*Pastinaca sativa*) although one group, the Clubionids, was a special case: they often make their retreats in rolled leaves in lower vegetation. Both live and dead plants were used.

The leaf-miner, *Phytomyza ranunculi,* was obviously only present where its food-plant (perennial *Ranunculus* species) occurred, but this was actually one of the very few species (Sterling *et al.* 1992) that showed a simple response to the abundance of a single food-plant. In many other cases, a single insect species has several potential food-plants and may show preferences amongst them as well as preferences for different degrees of structural complexity. The use of different species therefore changes over space and time.

One of the most widespread and common species of Dipteran (fly) miner, *Phytomyza nigra*, was recorded from 11 species of grass overall and from all the grasslands in the study, but the great majority of the mines were on only 3 species: creeping bent (*Agrostis stolonifera*), false oat-grass (*Arrhenatherum elatius*), and Yorkshire fog (*Holcus lanatus*).

The changes in miner numbers between 1986 and 1992 showed two peaks in overall abundance separated by a trough in 1989–90. Numbers on creeping bent steadily declined, while numbers on false oat-grass steadily increased with a relatively small fall in 1989–90. In contrast, the numbers on Yorkshire fog fluctuated violently, this grass being the main associate of the overall fluctuation in miner numbers.

Creeping bent was less favoured as a food-plant overall. There were relatively few mines even when this grass was at its commonest, while false oat-grass was used as soon as it appeared in the sward. Yorkshire fog, like the miner, fluctuated in abundance, but the fluctuations were out of phase with those of the leaf-miner. It is unlikely that the grass fluctuation was caused by the miner: the numbers involved were too small to have a great impact on the plant, and overall insect herbivory at Wytham both below and above ground had relatively little effect on succession (Brown and Gange 1989).

The leaf-miner populations thus appear to respond differently with respect to the different food-plants. It is not clear whether this is because of subtle differences in nutrient availability in the food, natural enemies of the miners, or an interaction between the two. Such dynamics are, moreover, occurring in a sward where all the grasses grow together, not in different areas or patches. When all the leaf-miner species are considered, there are directional changes in the community that cannot be explained solely by food-plant structure, plant species composition, or simple functions of the passage of time (Sterling *et al.* 1992).

## 7.9 Invertebrate movement between patches

The spiders, leaf-miners and leafhopper examples given above illustrate the complexity of understanding the presence or absence of a species at a patch of open ground. Another consideration is that specialist insects may take additional time to find and colonize new populations of their food-plant. It may also not be possible for a single patch of habitat to maintain the full range of specialist species all the time. Species may therefore need to move between patches. The size and distribution of patches, as perceived by the organisms concerned, therefore becomes important. Butterflies of open ground habitats at Wytham, which show different degrees of movements, illustrate this point.

At one ecological extreme the dark green fritillary (*Argynnis aglaia*) is a robust strong flying species that breeds in small patches of tall grass-herb communities and scrub edges where its larval food plant (*Viola hirta*) grows lush and tall. Adults range freely over the whole of the open habitats, feeding wherever suitable concentrations of the best nectar plants (often thistles and knapweeds) occur. The estate must be regarded as supporting a single population.

The scrub specialists, such as the black hairstreak (*Strymonidia pruni*), exist mainly in small colonies where their blackthorn food-plant has reached the correct stage. Usually adults appear sedentary, but a few individuals fly considerable distances in 'outbreak' years and give potential for finding new patches of suitable food plant. In 1960 and 1984, for example, individuals were found over 1 km from the nearest breeding area. Among grassland species, the recently rediscovered Duke of Burgundy (*Hamearis lucina*) may behave in a similar manner. It also illustrates the conservation value of extending the habitat mosaic through the Upper Seeds experiment. There was only a single record for the butterfly at Wytham between 1960 and 2003, since when it has spread to Upper Seeds and another of the grassland patches.

Among the least mobile species is the green hairstreak (*Callophrys rubi*), which at Wytham is a specialist feeder on rockrose (*Helianthemum nummularium*) and has, since 1970, been confined to the same small population areas without either local extinction or colonization of new patches.

Looking at a wider range of species, Table 7.1 shows the pattern of occurrence of specialist herbivorous invertebrates in relation to their food-plants. The species pool ('Pool' column) is the number of species that are reported in the literature as specialists on each plant concerned and have a geographical range in Britain that would allow occurrence at Wytham.

The 'Wytham' column shows the number of the pool species that have been recorded at Wytham, excluding old records without recent confirmation. The 'Upper Seeds' column shows the number of Wytham species that have colonized the former arable land experiments in the 25 years since the last crop was taken—just under a quarter of the species found at Wytham. The last column checks whether or not the food-plant has colonized the former arable land—for instance the milkwort specialist could not have colonized Upper Seeds because its food-plant is not there yet.

**Table 7.1** Specialist invertebrates of limestone grassland in Wytham Woods. The 'pool' is the number of specialist herbivorous invertebrates associated with a particular plant and whose range includes Wytham; other columns indicate whether they are present in Wytham and whether they have colonized Upper Seeds.

| Food-plant | No in Pool | Occurring in Wytham | Colonized Upper Seeds? | Food-plant present in Upper Seeds |
|---|---|---|---|---|
| Carline thistle *Carlina vulgaris* | 2 | 0 | 0 | No |
| Milkwort *Polygala vulgaris* | 2 | 1 | 0 | No |
| Sedge *Carex flacca* | 6 | 6 | 0 | Few |
| Rock rose *Helianthemum nummularium* | 15 | 10 | 0 | Few |
| Thymes *Thymus* spp. | 8 | 1 | 0 | Few |
| Wild liquorice *Astragalus glycyphyllos* | 2 | 1 | 1 | Yes |
| Tor grass *Brachypodium pinnatum* | 1 | 1 | 1 | Yes |
| Birdsfoot trefoil *Lotus corniculatus* | 12 | 4 | 1 | Yes |
| Cowslip *Primula veris* | 3 | 3 | 3 | Yes |
| Hairy violet *Viola hirta* | 3 | 2 | 1 | Yes |
| **Totals** | **54** | **29 (54%)** | **7 (24%)** | |

These results demonstrate both the range of species that can be supported and the importance of the dynamic nature of habitats: new open habitats can quickly (in ecological terms) provide additional population areas even for the most demanding species. It is an encouraging start towards full restoration of a rich limestone grassland community.

## 7.10 Conclusion

The Upper Seeds experiment has provided a striking demonstration of the potential for, and also some of the limitations in, trying to restore limestone grassland from an arable field. Re-establishment and re-colonization of elements of grassland communities (both plant and animal) are possible, provided the right continuity of management exists. For the plant components of communities, the best management for re-colonization is at surprisingly low intensity: heavy grazing produces communities that are of interest in their own right but still far removed, even after 25 years, from ancient grassland vegetation. This seems to have been borne out by the studies elsewhere.

The dynamics of the whole mosaic are important, with the restored habitats capable of producing new areas for some of the most specialist organisms within only 25 years of establishment. Each component of a mosaic of structure and species composition is important for a different component of the fauna, and so contributes to the diversity of the whole mosaic. This then will have implications for work more generally on chalk and limestone grassland re-establishment under the habitat action plans.

# 8
# Invertebrates
C. Hambler, G.R.W. Wint, and D.J. Rogers

## 8.1 Introduction

Research on invertebrates at Wytham has been very important in the development of ecological theory and in the application of ecological ideas to conservation and pest control. Few ideas in ecology have a single origin and there is a risk of over-stating the importance of Wytham, but this chapter indicates some of the fields in which research at Wytham has at the least been supportive of rapid changes in ecological understanding. In the first sections of this chapter we introduce the broad scope of ecology which Wytham has influenced, which include: the recognition of the importance of vegetation structure, dead wood and microclimate; detecting and understanding long-term population regulation; elucidating the relationship between diversity and stability in ecosystems; developing ecological energetics; and the development of population sampling methodology. We then give a more detailed case study of one of the best-known species—the winter moth (Section 8.16).

Wytham could not have provided the opportunities and insights it has if its habitats and invertebrates were less diverse or less well known. Its range of habitats include wet and dry areas, riverside, calcareous and neutral soils, a range of soil nutrient levels, and grassland which is heavily grazed, rabbit grazed, or lightly grazed. The woodland itself includes ancient and recent broadleaved deciduous areas and conifer stands and has had a range of management histories or 'treatments'. However, as Charles Elton remarked, the woodland at Wytham is a fairly 'ordinary' for central England, distinguished only by the sampling intensity it has received.

The importance of invertebrates in ecosystems was recognized by early pioneers of ecology, including Darwin—who had a strong interest in earthworms. Zoologists in the nineteenth century studied morphology and taxonomy and began to compile species lists for sites. Some specimen collection was occurring at Wytham even before 1822. With its large professional and amateur scientific community, the sites in the vicinity of Oxford, including Bagley and Wytham Woods, came to be relatively well-recorded for groups such as beetles, moths, and molluscs (Walker 1926; Elton 1966; Crowcroft 1991). Interest in the ecology of invertebrates developed partly out of an interest in pest control in forests and crops. To understand the historical importance of Wytham, we need to consider the role it played in the scientific careers of some of the most globally influential ecologists of the twentieth century.

## 8.2 Pioneers of ecology and ecological genetics

During the twentieth century, Oxford was uniquely placed to develop ecological theory and methodology. Oxford's very large community of zoological scientists were scattered in several departments and research groups (including the Department of Zoology and Comparative Anatomy, the Hope Department of Entomology, the Edward Grey Institute of Field Ornithology, the Bureau of Animal Population, and the Museum of Natural History). Researchers and teachers had the benefits of the second largest collection of invertebrates in Britain (and one of the largest in the world). The entomological collections in the Museum were curated from 1861 by the Hope Professor of Zoology (Entomology).

Early work combined the themes of description of species and lists of species with an attempt to understand patterns in the prevalence of species. Why might a species be more abundant in some places than others? Why might some species occur only in, or mainly in, some places? Which species and which 'abiotic' (physical and chemical) features of the environment were important or limiting? How could pests be controlled? The key ideas of regulation of animal numbers were being established by researchers in Oxford such as Charles Elton (1927), David Lack (1954), and 'Mick' Southern (1970, 1979), initially using animals that were relatively easy to count and monitor such as birds and mammals. Their ideas were exchanged with colleagues interested in invertebrates and the different insights and challenges these smaller animals presented, and invertebrate biologists could also call on colleagues in other departments—including many experts in plants, microbes, soils, and rocks.

Charles Elton was a founding figure in ecology and conservation biology (e.g. Crowcroft 1991; Southwood and Clarke 1999; Farnham 2007). He is arguably the ecological counterpart to Charles Darwin, formulating many of the central theories of the field, supporting them with data and presenting them in a readable form. His books and papers are still hugely influential and include *Animal Ecology* (1927), *Voles, Mice and Lemmings: Problems in Population Dynamics.* (1942), *The Ecology of Invasions by Animals and Plants* (1958), and *The Pattern of Animal Communities* (1966). Elton proposed the idea of a species' 'niche' as a *functional* unit—what he called its 'place in the biotic environment', including its 'relations to food and enemies'. 'When an ecologist says "there goes a badger" he should include in his thoughts some definite idea of the animal's place in the community to which it belongs, just as if he had said, "there goes the vicar"' (Elton, 1927). It was readily apparent to Elton that the niche could be delineated by very subtle minutiae of habitats and that detailed recording of many features of a site, both biotic and abiotic, would be needed to describe and explain the niche—and hence the presence and abundance of a species (Elton and Miller 1954). Invertebrates gave Elton many insights into the subtlety of a species' requirements.

The history of the development of Elton's ecological research group, the Bureau of Animal Population, has been reported in the book *Elton's Ecologists* (Crowcroft 1991) which refers to numerous publications on invertebrates. Dating from 1932, this research

**Plate 1.** The Oxford Skyline from Wytham Woods. (see also chapter 1)

**Plate 2.** Wytham Park and yellow ant reserve and the edge of the Woods. (Photograph by Ian Walsh). (see also chapter 4)

**Plate 3.** Wytham Sawmill. (see also chapter 4)

**Plate 4.** A yellow-topped marker for a 'Dawkins' plot. There are 163 such plots spread across the Woods; the measurement plot of 10 m x 10 m is located 14.1 m north-east of alternate points on the 100 m marked grid indicated by these posts. (see also chapter 5)

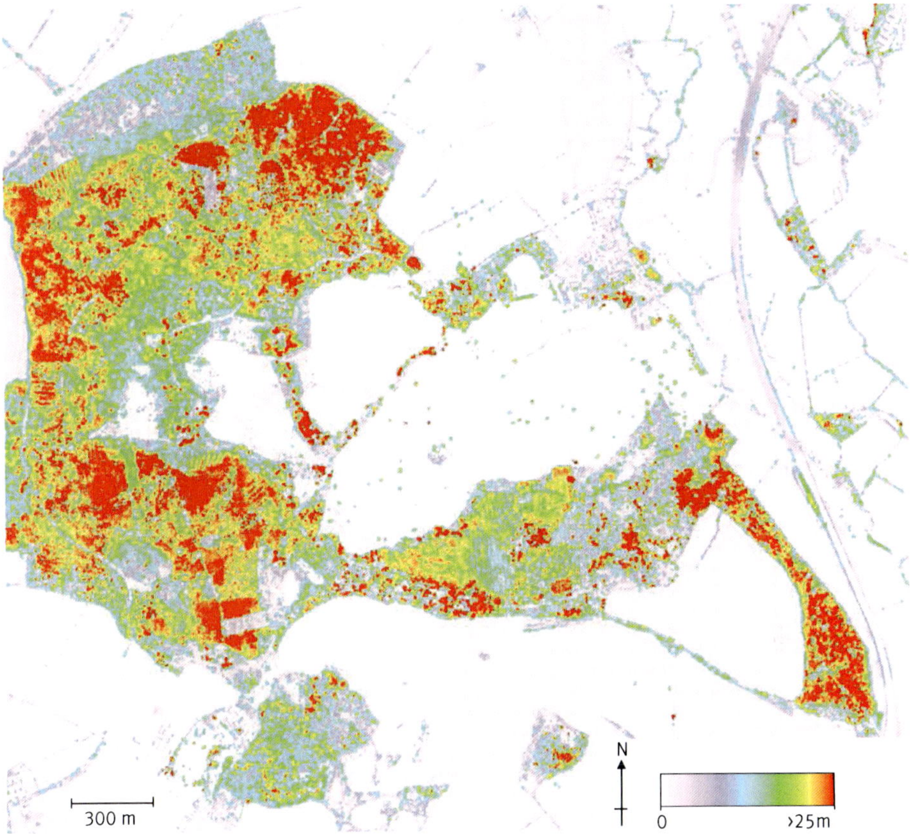

**Plate 5.** Digital Canopy Height Model produced from airborne LiDAR data acquired in summer 2005 by the NERC Airborne Research & Survey Facility. These LiDAR data were processed by Ross Hill (Bournemouth University) whilst based at CEH.

**Plate 6.** Bluebells in Marley Wood. (see also chapter 6)

**Plate 7.** A patch of herb Paris (*Paris quadrifolia*) in Rough Copse. This species is an indicator of ancient woodland. (see also chapter 6)

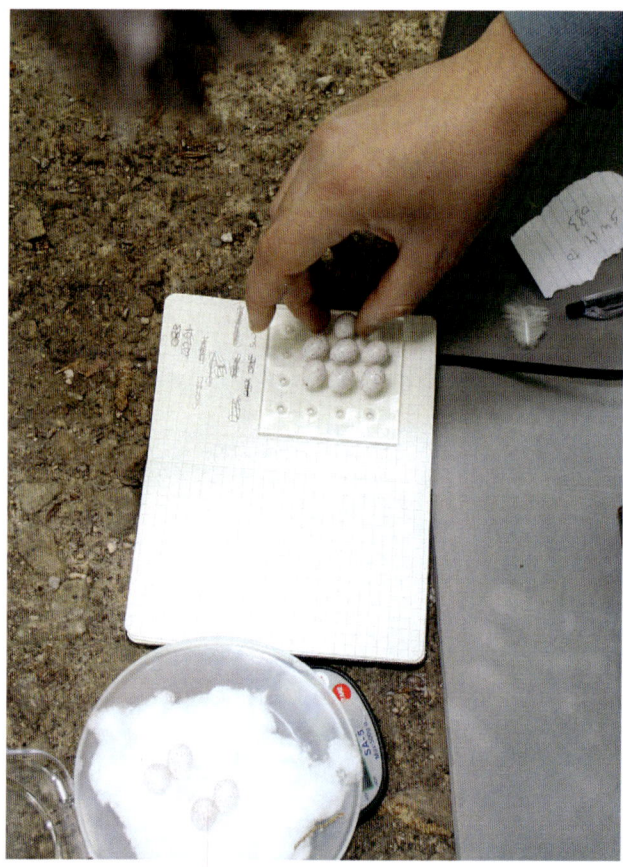

**Plate 8.** Weighing Blue Tit eggs. (see also chapter 9)

**Plate 9.** a) Blue Tit b) Great Tit c) young Tits - the long-term population studies of these two birds started in 1947, and have been a major contributor to Wytham's ecological reputation. (see also chapter 9)

**Plate 10.** Andrew Gosler examining a nestbox in Great Wood. (see also chapter 9)

**Plate 11.** A Sparrowhawk with young. This is a major predator of Blue and Great tits. (see also chapter 9)

**Plate 12.** Wytham badgers have been the subject of one of the most comprehensive long-term studies of medium-sized carnivores anywhere in the world. (Photograph reproduced by kind permission of Andy Rouse).

**Plate 13.** A bank vole. The impacts of predation by tawny owls and environmental factors on population densities of bank voles and wood mice was the topic of H.N. Southern's long-term research in Wytham Woods. (Photograph by kind permission of Chris Grady).

**Plate 14.** A veteran beech tree on the Singing Way. (Photograph by Fred Topliffe). (see also chapter 11)

# Edmund Brisco 'Henry' Ford (1901–1988)

E.B. Ford. Photo by courtesy of the librarian of the Hope Library, University of Oxford.

Henry Ford was an ecological geneticist. He was an experimental naturalist who wanted to test evolution in the field. As a schoolboy he became interested in Lepidoptera, the group of insects which includes butterflies and moths. Particularly interesting was the work done at an early age with his father on the effect of fluctuation in numbers on evolution in the marsh fritillary butterfly, and later his investigations into spot number in the meadow brown butterfly.

Ford read Zoology at Oxford, and was taught genetics by Julian Huxley who was a source of considerable inspiration to him. He originated the field of research known as ecological genetics. Ford had a long working relationship with the statistician, R.A. Fisher. His work on the wild populations of butterflies and moths was the first to show that the predictions on conflicting pressures within natural selection made by R.A. Fisher were correct. He was also the first to describe and define genetic polymorphism (the coexistence within a population of two or more discontinuous forms in some kind of balance) representing a delicate balance between conformity and diversity. He predicted that human blood group polymorphisms might be maintained in the population by providing some protection against disease. Six years after this prediction it was found to be so.

The book for which he is best known is *Ecological Genetics* (1964), which ran to four editions and was widely influential. He laid much of the groundwork for subsequent studies in this field. Amongst Ford's many publications, the most popular were the first book in the New Naturalist series, *Butterflies* (1945) and *Moths* (1954). He refined the technique of mark—release—recapture and this helped greatly to estimate the force of selection in wild populations. His investigation of the pigment of Lepidoptera was one of the most successful attempts at relating classification to chemistry.

In 1963 Ford became Professor of Ecological Genetics at the University of Oxford and was one of the first scientists to be elected a Fellow of All Souls College since the seventeenth century (National Archives 1989). Once there, he campaigned strenuously against the admission of female Fellows. He was considered decidedly eccentric.

Associated with Ford's scientific research was a great interest in conservation. In 1945 he became a member of the Special Committee appointed by the Ministry of Town and Country Planning to study the conservation of wildlife in Britain. The Committee's report resulted in the formation of the Nature Conservancy on whose governing body Ford served, from 1949 to 1959.

group included some of the founders of both field ecology and conservation biology, such as Eric Duffey. Many influential ecologists worked in or visited the Bureau, including famous visitors from the USA such as Eugene Odum, who used radioactive tracers to estimate the colony size of an ant nest on the Bowling Alley (Odum and Pontin 1961).

Parallel to the work of Elton and his associates, the field of ecological genetics was being founded by Edmund B. Ford. Ford worked extensively on butterflies and moths, examining ecological influences on the frequency of genes as revealed by differences in morphology in different locations and over time. Ford recognized the importance of local selection pressures and of gene flow through dispersal and some of this work was done at Wytham, (Ford 1971). Ford continued supervising undergraduate projects in Wytham up to the early 1980s.

Selection pressure through bird predation was illustrated for a different group of organisms in work on the evolution of banded snails (*Cepaea* species) by Arthur Cain and Philip Sheppard (1954). They compared shell colours and banding patterns at sites in Wytham and elsewhere and found that certain shells were more conspicuous to predators when in habitats of particular background colours (such as beechwood and grassland). This might eventually remove variety in the population, but sub-fossil shells showed the variety of banding patterns had been fairly stable for many years. Hence they concluded that the wide range of shell types (polymorphism) can only survive if the directional selection due to visual predators is counteracted by some other (unknown) selective advantage.

The Oxford biologist Bernard Kettlewell undertook one of the classic studies on evolution in action, demonstrating the increased frequency of dark (melanic) varieties of the peppered moth, *Biston betularia*, due to predation by birds of pale moths on polluted dark surfaces (Kettlewell, 1973). Whilst exploring the evolution of industrial melanism, Kettlewell reared various colour varieties (morphs) of scarlet tiger moths on comfrey at Wytham, including the yellow *bimaculata* morph obtained in Italy. Some of these spread far further than was anticipated and hybridized with native scarlet tiger moths. The resultant genetic contamination ('introgression') is still evident in very low frequency of *bimaculata* morphs which are found in moth traps in the Oxford area (Gibson, pers. comm.). Such unforeseen dispersal and hybridization may have relevance to the current debate on the spread of genetically modified organisms.

Early studies of invertebrates in Wytham also included work in the 1950s by Amyan Mafadyen on soil invertebrates of The Dell and collaborative work by Christian Overgaard-Nielsen on the energy balance of the pill millipede *Glomeris marginata* consuming grass litter. This was perhaps the first demonstration of external digestion by 'microbe stripping', in which invertebrates feed on the microbes attached to the surface of detritus whilst the plant residues pass largely unaltered through the gut (King and Hutchinson, 1983). John LeGay Brereton worked on factors controlling the population distribution and abundance of four species of woodlice in the wood, Stephen Sutton expanded work on woodlice movement and density, and Tom Huxley studied the diet of woodlice in detail.

Edward 'Bill' Fager worked on the invertebrate colonization and community organization of pieces of decaying wood, which became a major theme. Kitty Paviour-Smith (later Kitty Southern) studied the beetle communities of bracket fungi on birch, including the Australian invasive species *Cis bilamellatus*. Monte Lloyd worked on centipedes and other litter invertebrates and developed statistical descriptions of crowding. Roger Kitching examined the importance of the tree-hole fauna in woodland ecosystems. The invertebrates of dung, carrion, and animal nests were also studied in some detail (Elton 1966). Rory Putman (1978) showed that some of the less abundant species in the community were critically important in the mechanical breakdown of carrion—effectively a type of 'keystone' species. Through such research projects and a deliberate attempt at an inventory of some groups of invertebrates, a long list of species was being accumulated from Wytham.

## 8.3 The Wytham Ecological Survey

After the gift by Raymond ffennell of the woodland and grassland of the Wytham Estate to the University in 1942, survey and research work on its invertebrates became much more intense than at almost any other site in Britain. Elton recognized the need for base-line surveying of the habitats and species in the site and the value of monitoring subsequent changes. Hence, around 1950, he started the Wytham Ecological Survey, and with substantial help from Cliff Elbourn (1966) this became a massive card-index of sites matched by a collection of 'voucher' specimens to confirm the identification of species. Crucially, it was not only the abundances and locations of relatively exciting rarities that were recorded, but also the common species in some taxa such as harvestmen and woodlice (e.g. Brereton 1957).

The aims of the survey included understanding what Elton called the 'dynamic relations' between populations of species living in an area and finding out which species tended to occur together in particular habitats. Elton's discovery of similarities between the groups of grasshopper species found on grasslands at Wytham and Silwood Park in Berkshire contributed to the understanding of invertebrate communities (Elton 1966).

Through the activities of numerous researchers, including undergraduates, this survey and inventory of Wytham became unparalleled in the world. In correspondence to Professor M.V. Laurie of the Imperial Forestry Institute in Oxford, dated 15th February 1960, Elton says:

Briefly, over a hundred research biologists have worked there, the number is increasing, about a hundred papers have been or are being published and the Ecological Survey here has assembled more data about this square mile or two than exists for any spot on earth.

It was the knowledge of invertebrates that was particularly unusual, since conspicuous and species-poor taxa, such as plants, birds, and mammals, are often well recorded on sites. For example, in 1973 Wytham had the highest recorded number of species of spider on a reserve in Britain (Duffey 1974), including species known only from two or three sites in Britain. Only a few other terrestrial sites in Britain have received attention approaching that of Wytham, most notably the Institute of Terrestrial Ecology sites such as Monks Wood (Steel and Welch 1973) and other university study sites such as Woodwalton and Wicken Fens near Cambridge. The future of the Survey material was amongst Elton's greatest concerns later in his life (Crowcroft 1991). It is now housed in the University Museum of Natural History.

Because it has been so intense and employed a range of unusual and novel sampling methods, the Survey revealed several species new to Britain (such as wasps found by Marcus de Vere Graham and Lionel Cole, and the rare fly *Microsania vrydaghi* found by John Ismay which is attracted to bonfire smoke). Some species were discovered to be new to science—for example, Wytham is the 'Type Locality' for the wasp *Syntretus lyctaea*.

## 8.4 Invertebrates, habitat specialisms and landscape management

Elton's preparedness and ability to engage with invertebrate ecology helped him to one of his key insights, the importance of 'cover' (Elton 1939). Cover provides protection from predators, parasites, and the elements. Elton had recognized that invertebrates are strongly influenced by the physical structure of the plants and the physical and chemical ('abiotic') habitat around them. He recognized the importance of plant 'architecture' and, within woodlands, he became aware of the crucial role of dead wood. He concluded the landscape is divided into ecological units based on the structure of the habitat, irrespective of the species of green plants that are present, because most species are not specialist herbivores (Fager 1967; Strong *et al.* 1984). This led researchers to investigate further how the richness of invertebrates changes as the amount of plant structure and cover increase through the process of 'succession', whereby an abandoned grassland or new land surface may develop into a forest.

## 8.5 Studies of cover and succession

In the early 1950s, Eric Duffey studied the Wytham spiders for his doctorate at the University of London. He applied the experience in population processes developed on vertebrates in the Bureau of Animal Population in one of the first studies of invertebrates to use modern ecological methods. In line with the ethos of the Bureau to monitor population abundance in a reliable way, Duffey developed 'absolute' sampling methods which detect the density of animals per unit area. He investigated spider dispersal by 'ballooning' on silk threads (Duffey 1956) and the patterns of abundance and species richness of the group, especially on the limestone grasslands of Bowling Alley and Sunday's Hill within the Woods. In particular, Duffey developed Elton's theme of the

# Sir Richard ('Dick') Southwood (1931–2005)

Sir Richard Southwood. Photograph reproduced with permission of the Warden and Fellows of Merton College Oxford.

Sir Richard Southwood has been described as a towering figure among his generation of British zoologists, though he had at least four distinct careers: as research scientist, scientific administrator, Vice Chancellor of Oxford University, and advisor to governments on scientific aspects of policy (he chaired a working party on Mad Cow Disease (Bovine Spongiform Encephalopathy) and was largely responsible for the adoption of lead-free petrol in the UK). He had a natural talent for dealing with people, genuinely caring about the staff and students working under him. His infectious enthusiasm for insects in particular, and ecology in general, inspired new generations of biologists throughout his life.

After taking a first degree at Imperial College London, Southwood worked for his PhD. as an ARC research scholar at Rothamsted. He then spent a period from 1967 at Imperial College's Field Station at Silwood Park, becoming head of the Department of Zoology and Applied Entomology, and Director of Silwood Park. He developed Imperial's reputation into a world-leading centre both for teaching an integrated undergraduate Biology course and for ecological research. His own research at Imperial focused on insect communities and population dynamics. He also wrote the immensely successful and influential book *Ecological Methods* (1978), a handbook of experimental methodology and statistical design. This hugely influential work has been used by generations of ecologists and is now in its third edition (2000).

In 1979 Southwood became Linacre Professor of Zoology at Oxford University, and a fellow of Merton College. He spent the rest of his career at Oxford, as head of the Department of Zoology until 1989, as Vice Chancellor of the University between 1989 and 1993 and subsequently as Professor and Emeritus Professor of Zoology. He was responsible for attracting several very eminent scientists to the Department (which already had a record of outstanding scientists), including both Lord Robert May and Bill Hamilton (*The Times* 2005).

In 2003 he published *The Story of Life*, a book based on the first-year undergraduate lectures he gave at Oxford for 18 years, including during his time as Vice Chancellor. He is probably best known scientifically for an enormously influential paper, published in 1979 on *The relationships of plant and insect diversities in succession* showing how plant species richness per unit area is at a peak far earlier in succession than is the richness of invertebrates. Structurally complex habitats such as tall grass, scrub, and forest often support fewer plant species per unit area but more invertebrate individuals and species—as Duffey had reported at Wytham.

After Southwood arrived in Oxford, more use was made of the resources at Wytham for undergraduate teaching purposes.

importance of plant architecture and cover. Such work has been especially crucial to understanding the ecology of species and communities and in conservation.

In a pioneering study in which many of today's ecological understandings are already evident, Duffey (1962 a, b) compared spiders in three grassland types at Wytham. His work revealed that the habitats and locations in which there are the most invertebrate species are often not those with the most plant species, but instead are those with complex structure and abundant cover. After Duffey left Oxford he collaborated with Morris (1971, 2000) in further examination of the importance of cover and plant architecture to grassland invertebrates and its relevance to conservation management. Charlie Gibson developed the Upper Seeds project (Chapter 7) as a continuation of Elton's and Duffey's themes on the patterns that plants and invertebrates follow during succession, and this was further extended to investigate the significance of the soil fauna (Brown and Gange, 1989).

Southwood *et al.* (1979) published an enormously influential paper (based on work at Silwood Park) showing how plant species richness per unit area is at a peak far earlier in succession than is the richness of invertebrates, and he subsequently taught on undergraduate field trips at Wytham which explored this theme. Structurally complex habitats such as tall grass, scrub, and forest often support fewer plant species per unit area but more invertebrate individuals and species. This is a key fact in conservation since it makes it hard to please everybody on small nature reserves (Hambler 2004). As a succession from grassland to forest proceeds, invertebrate richness increases through time until a late vegetation stage in the succession. However, there is still debate as to whether (in Europe at least) richness peaks soon after the canopy closes, or whether it continues to rise until an ice-age removes the forest (Hambler 2004). Unfortunately, no undisturbed woods survive in Britain in which we can explore ancient forest invertebrate diversity, but the most natural forests surviving in Europe, such as Bialowieza forest in Poland, are thought to be very species-rich.

## 8.6 Understanding animal abundance and density

Work on factors causing population change continued in Wytham in the 1960s, with Chris Mathews studying the population dynamics of water shrimp (*Gammarus*) in the streams. The way in which mortality varied with population density ('density-dependence') was examined in several species, including bark beetles (Beaver 1967), elm bark beetles, moths (Varley and Gradwell 1968) and leafhoppers (Whittaker 1973).

Understanding density-dependence is vital to the application of ecology to pest control, conservation, fisheries and sustainable harvesting (Varley *et al.*, 1973; Hambler 2004). George Varley improved the analysis of mortality factors through the use of 'key factor analysis', which seeks to find the factor with the greatest contribution to population change; this may not be the factor which kills most individuals. Key factor analysis is widely used today (Jervis 2005; Speight *et al.* 2008). Work on beetles and winter moths in Wytham generated time-series long enough to advance our understanding of the density-dependent regulation of abundance (Varley 1951; Beaver 1966, 1967; Varley and Gradwell 1968; Hassell 1968; Den Boer and Reddingius, 1996). Population dynamics work on moths is being continued by Willy Wint (1983) and Lionel Cole. The history and importance of this work is outlined in a case study involving the winter moth at the end of this chapter.

In addition to studies of the dynamics of individual species, Wytham contributed to the study of the dynamics of groups of interlinked populations, a field known as 'metapopulation dynamics'. Our understanding of the persistence of clusters of populations through formation and extinction of populations has been greatly influenced by Illka Hanski (1980) for whose doctorate on the dung-feeding beetles in the Wytham area, models were developed of their distribution.

### 8.6.1 Competition, co-existence, and parasitism

Competition between species is a central issue in population ecology and has long been studied at Wytham. How can ecologically similar, competing species co-exist? This problem was considered in the 1950s by Mike Davies, working on predatory ground beetles *Notiophilus biguttatus* and *N. rufipes*. Although these species eat very similar foods, in some areas the complex structure of beech litter may be particularly beneficial in protecting *N. rufipes* from the parasitic wasp *Phaenoserphus vexator*, so the two beetle species may survive better in different habitats. Work on competition continued with studies by John Pontin (1961) on two species of ants—*Lasius niger* and the mound building *L. flavus*. Through transplantation of nests and other manipulation of ant density, he found *L. niger* suppresses the number of queens produced by *L. flavus*, but that *L. flavus* can somehow resist being completely displaced. He suggested that competition amongst colonies of each species was so intense that competition between species did not lead to exclusion of one of them.

Similarly, Gibson (1980) suggested that several species of bugs were able to co-exist on grasslands by having different timing of generations and slightly different preferred nitrogen levels and host plant ranges; there are thus enough 'refuges' for each species to prevent its complete exclusion through competition. The mechanism of competition was also examined by Robert Cowie and Steve Jones (1987) from University College London. They were interested in why the snail *Cepaea nemoralis* was being replaced by the similar *C. hortensis* on the Malborough Downs and why the two species tend not to overlap in range: *C. hortensis* is found in relatively cool regions in Britain and its occasional range-expansions might be driven by climate. They painted snails of both species with a dye which faded in the sun and kept them in cages in Wytham. They found *C. nemoralis* spent less time in the light than *C. hortensis*, but that this behaviour did not change when the other species was present. They concluded that *C. hortensis* was not displacing the other from its preferred micro-climatic light level and this helped to exclude one proposed mechanism for the decline of *C. hortensis*. Differential extinction and colonization rates of colonies might be more important, or competition might be for unidentified aspects of their niches. Such studies show how very difficult it can be to demonstrate competition between species in the wild.

The mortality caused by parasites is another key theme in ecology and evolution and early work in Wytham included studies of the water mites by Ian Efford (1965), including the life-history of the parasitic mites *Feltria romijni* which proved surprisingly abundant amongst its host midges. He recorded the highest water mite density known anywhere at that time (1368 mites per metre square), which is similar to those in terrestrial habitats. Falling leaves provided the energy source for the mites. Many young adult mites die in October, for unknown reasons, and this may be what regulates the mite population. Different mite species have different larval attachment sites on their hosts and when two species attach to the same host one may be displaced by competition to a less preferred site. The ratio of hosts to parasites is fairly constant at 4:1, despite great differences in host density in different habitats, but mites parasitize a higher percentage of hosts when hosts are scarce. The mites themselves have few predators and if they can survive through October they survive well as older adults.

The mechanism which underlies the impact of a parasitic fly on the population of its winter moth host was examined by Mike Hassell *et al.* (1998). This helped clarify the difference between a 'behavioural response' (revealed by increased *preference* for a newly abundant prey) and a 'numerical response' (in which *numbers* of parasites or predators increase in response to increased abundance of their prey). The fly focuses its hunt for prey on places where the prey are aggregated at high density. Behaviours of this sort can lead to a more rapid suppression of prey density than the numerical response. Such rapid negative feedback is potentially more of a force for population stability—just as a thermostat which responds rapidly to room temperature would regulate room temperature more effectively than one which takes hours to react!

The distribution of parasitic insects causing oak galls was examined by Richard Askew (1962). Graham Stone undertook studies of the evolution of gall-formation, considering galls as an 'extended phenotype' of the parasites (Dawkins 1982; Stone and Cooke 1998). The gall shape, colour, and scent are as much part of the phenotype as the body of the insects themselves and since they influence the survival and reproduction of the parasite, are under selection pressure to avoid predation, drought, and plant defences.

Many plants produce chemicals to defend themselves against herbivores, including insects. The subtle interaction between host plant defence chemicals and the herbivores was explored by Paul Feeny (1970), who went on to propose his very influential theory of 'Apparency' (Feeny 1976 and discussed further below). Parts of plants, or whole plants, that were easy to find in space and time would have to be particularly well-defended. In turn, the quality of food influences the parasites (including viruses): reduced plant quality may increase the number of viral generations per caterpillar and may increase ingestion of the virus on poorer quality host plants (Raymond and Hails 2007). Chris West (1985) suggested damage to leaves by leaf-mining moths induces plant defences which in turn impact on the survival and density of the moths.

**Figure 8.1** Sampling insects in oak canopies: Photographs: William Wint.
a) Using a long-armed pruner to find leaf-miners and galls.
b) A mist-blower carried on the back blows a short-lived pyrethroid insecticide into the canopy.
c) Standardized sampling trays collect the invertebrates for microscopic examination and identification.

The interactions between oaks and the many species which eat their leaves also underpinned a very big study by Southwood, Willy Wint, and colleagues of how and why different tree species support different numbers and types of insects. Southwood and his colleagues aimed to identify general features which might be important to the number of herbivore species on a tree and found that whether a tree is deciduous or evergreen may be more important than whether the tree is native to a region (Fig. 8.1). The work also underlined the complexity of these 'communities' and the degree to which the herbivores, parasites, predators, and even the insect 'tourists' (which arrive in small numbers) are all interdependent. The species occur in structured patterns determined by the ways in which the individual species interact.

Southwood's theories, developed in part from the sampling and identification of nearly 70,000 insects in 1500 species from several oak species in Wytham and nearby Bagley Wood, have helped to identify why some types of plant can be introduced into new areas and be susceptible to pests whilst others are less vulnerable. The work included seldom-studied but highly abundant and important groups such as Hymenoptera and many Diptera. The data for these studies took about 5 years to collect, but it was nearly 20 years before the final identifications were made in 2003, and another year to prepare the analysis for formal publication (Southwood et al. 2004). Ecological research is not for the impatient or those with short memories!

## 8.7 Stability of complex systems

Elton initiated one of the most important debates in ecology, the relationship between 'diversity' and 'stability' in ecosystems: are systems with many species more stable against change and invasion than simpler, less species-rich ones? Even the possibility that stability is increased by diversity is a potential argument for taking the precaution of conservation, in case biodiversity might drive the stability of 'ecosystem services' such as hydrology, pollination, or atmospheric gas balance (Hambler 2004).

Elton thought it very likely that diversity promoted stability and supported this argument using data on the community of tiny chalcid wasps associated with the oak galls at Wytham (Askew 1962). These wasps parasitize the other invertebrates which make the galls. Whilst some feed on only one host species, some parasites were more general feeders with several, or occasionally many, hosts. Elton (1966) talked of the food webs of ecosystems being populated by 'centres of action' and of communities having interlocked 'fields of force' analogous to those in physics. He suggested that the generalist chalcid species made important interactions between different gall communities and that these non-random links acted as 'girders' which stabilized the system; this is a remarkably insightful suggestion, which perhaps still has not been widely recognized in ecology (Southwood and Clarke 1999).

For many years, mathematical models by Robert ('Bob') M. May (1973) were taken as evidence that increased diversity in a network (such as an ecosystem) does not in general increase the network's stability. However, new models show that if connections are not random, but instead are 'disassortative' with particularly frequent interactions between nodes—like Elton's girders—the systems are in some ways more stable (May *et al.* 2008). Elton's insight, supported by meticulous work on invertebrates at Wytham, thus has relevance for stability in any complex network, from economies on the brink of recession to 'tipping points' in climate change. On a larger scale, this may be the basis for Gaian stability of the planet due to life (Hambler 1997, 2004).

## 8.8 Ecological energetics

The field of ecological energetics, the way energy flows through an ecosystem, was advanced considerably by the work of John Phillipson and colleagues through the 1970s and 1980s (Phillipson 1983; Phillipson *et al.* 1977, 1978, 1979, 1984; Wood and Lawton 1973). The ability to sample quantitatively and to identify so many of Wytham's invertebrates—including those of the soil—permitted population estimates and estimates of energy flow through different groups of organisms. The work on diet, distribution, and population density of detritivores such as worms and woodlice was a key foundation for such studies. Earthworms contributed some 4–5 per cent of 'soil respiration', some three times that of the tiny enchytraeid worms and 30 times that of the hugely abundant soil nematodes. The majority (often about 70 per cent) of the energy flow through a terrestrial ecosystem is through the detritivores—in contrast to the common assumption that the herbivores are the main route through which the sun's energy reached carnivores.

## 8.9 The origins of behavioural ecology

Working on invertebrates and birds, John Krebs and Nick Davies pioneered the field of 'behavioural ecology' (Krebs and Davies 1978). This examines the evolution of behaviour in an ecological context and includes studying how animals make 'optimal' decisions about food and mates given the costs and benefits of actions. Mike Hassell's work on behavioural responses of a parasitoid fly to host density (described earlier in this chapter) contributed to the understanding of energy-efficient 'optimal foraging'.

Nick Davies (1978) undertook work on territoriality in speckled wood butterflies. (*Pararge aegeria*). He showed that when the male butterflies compete to find a mate, they adopt a 'rule' that the first male to occupy a patch retains control over sunlit patches in woodland; these patches are the best places to find females. He moved butterflies in and out of patches to show occupancy was not simply a matter of fighting ability. An arbitrary rule such as 'the resident wins' can help reduce the risk of fights escalating to dangerous levels for both competitors—a bit like tossing a coin to decide an outcome.

In pens in pasture at Wytham Farm, Alex Kacelnik, John Krebs, Melissa Bateson, and others undertook studies of the way birds forage for invertebrates in an optimal way that minimizes wastage of energy and/or time (Bateson 2003). The patchy distribution and abundance of leatherjackets (the larvae of crane flies) in the soil was important to begin to identify the decision rules starlings use for giving up searching: for example, the giving-up time is a fixed proportion of the interval between food items. A spin-off commercial company was developed to help oil exploration, based on similar rules!

## 8.10 Farm wildlife

The farm on the Wytham Estate has permitted specialists in plants, birds, mammals, livestock, animal behaviour, crops, and invertebrates to get together easily to do joint work.

In one of the first studies of its kind, work on the potential benefits of beetles in reducing crop pests was combined with laboratory studies in developing the idea of 'functional responses' of predators—a combination of the behavioural and numerical responses to increased prey numbers. Jenny Cory (1984) and Martin Speight excluded beetles from areas of the farm using plastic sheeting and concluded that predatory beetles such as ground and rove beetles probably had limited potential for pest control in 'integrated pest management' (Speight *et al.* 2008).

Growing interest in 'sustainable' farming methods and in organic farming have raised the profile of 'beneficial arthropods', notwithstanding the results of Cory (1984). There is interest in using field margins or strips of land as 'beetle banks' to promote such invertebrates (MacLeod *et al.* 2004). Wytham was one of the first British sites for detailed, highly-replicated studies of how best to integrate farming and wildlife, particularly in the context of various European 'agri-environment' schemes: studies on field margins at the farm were extended by the Wildlife Conservation Research Unit (WildCRU), Oxford. In one of the largest studies of its type, some 16 management treatments were used on the margins of arable crops. The aim was to identify management techniques for crop field margins (and general 'set-aside' land) which benefited different types of invertebrate whilst at the same time not encouraging agricultural weeds (Smith *et al.* 1993). The focal groups were beneficial invertebrates (such as spiders and some beetles) and aesthetically attractive species such as butterflies. The effects on spiders could be compared to the effects of grazing at Upper Seeds (Baines *et al.* 1998). As with Upper Seeds, plant structure (particularly the height of the sward in the margins) was found to be more important than plant species richness in determining the abundance and richness of spiders.

Some beneficial arthropods, such as the money spider *Lepthyphantes tenuis* (which eats aphids), are more apparent in long or medium-length grass (Bell *et al.* 2002). However, other beneficial spiders which use agricultural land (such as some *Erigone* and *Oedothorax* species) were found to need short, warm grass—as they do in Upper Seeds (Gibson *et al.* 1992b). Different cutting dates or lengths favour different predators and field margins must therefore be managed with particular benefits or species in mind. Most of the species found on the farm were, not surprisingly, common—although some species considered nationally scarce (such as the spider *Syedra gracilis*) proved abundant and have probably been under-recorded nationally.

Many species of British butterfly can use field margins as a source of nectar or larval food. Even amongst the butterflies, whose larvae are almost all herbivorous, there are different habitats and thus ideal management requirements (Feber *et al.* 1996). Many species benefit from cutting margins in spring or autumn, but would suffer from summer mowing. The abundance of nectar sources is particularly important to the numbers of most species. In contrast, small tortoiseshell butterflies may benefit from mowing in summer, because when nettles regrow after cutting their leaves are rich in nitrogen and water, which in turn increases larval growth rates. Seeding new margins with a wildflower seed mix, reducing herbicide use, and permitting a more structurally varied margin and landscape helps establish a richer fauna more quickly (Feber *et al.* 1996).

The results from the field margins projects thus highlight the fact that an agricultural landscape must be varied if it is to meet the aspirations of a number of 'stakeholders', each with different interests in the countryside (Dennis *et al.* 2008; Feber *et al.* 1999, 2006; Gates *et al.* 1997; Haughton *et al.* 1999; Macdonald and Smith 1991; Macdonald *et al.* 2000; Smith and Macdonald *et al.* 1993, Smith *et al.* 1997; Vickery *et al.* 2004; Woodcock *et al.* 2007).

## 8.11 Wytham's invertebrates and British conservation management

Insights from Wytham have contributed to land-use change in Britain, particularly in the management of nature reserves and farmland.

Elton was a key figure in conservation science in Britain and was involved in the development of government policy, prioritization of sites for protection, and management techniques for protected areas (Crowcroft 1991; Farnham 2007). He was an early advocate of what is now known as 'restoration ecology'. In most parts of the world, restoration aims to use scientific evidence to produce relatively natural habitats (although in Britain, 'restoration' is also used by many conservationists to mean re-creation of traditional cultural landscapes and semi-natural habitats). A few years after the university acquired Wytham, Elton was vigorously advocating management to increase the amount of dead wood and to reduce the 'tidiness' of the Woods (Elton 1966; Paviour-Smith and Elbourne 1993). For example, in a letter in 1959 regarding proposed extensions to the scientific reserves in Wytham, he wrote:

There is also a very rich element in the invertebrate fauna dependent on the conditions found in old trees. Fallen timber in the open Park is, however, not so important as inside the Woods, because it tends to get dry and heat-sterilised when exposed to the sun.

The ecology of dead timber was amongst Elton's chief concerns in Wytham. In 1966 Elton wrote:

When one walks through the rather dull and tidy woodlands...that result from modern forestry practices, it is difficult to believe that dying and dead wood provides one of the two or three greatest resources for animal species in a natural forest and that if fallen timber and slightly decayed trees are removed the whole system is gravely impoverished of perhaps more than a fifth of its fauna.

He cited results from Wytham's Ecological Survey which show that Wytham supported roughly half the national total of about a thousand British species which use decaying wood.

Elton's view was informed by visits to active forestry and coppice sites, in which old timber and dead wood were either prevented from occurring by frequent harvesting, or were removed for supposed (and often erroneous) phytosanitary reasons (to try to reduce infestation by invertebrate pests or fungi). During the 1950s, management of Wytham became less like active forestry and more sympathetic to dead-wood fauna (Crowcroft 1991). The importance of dead wood in conservation has risen greatly in prominence following debates on the lack of it in coppice woodland (e.g. Sterling and Hambler 1988; Kirby *et al.* 1998). Work by Kitty Paviour-Smith and Elbourn (1993) on the dead-wood fauna of Wytham emphasized some special features of this hitherto relatively neglected habitat. Internationally, the importance of dead wood is also increasingly recognized in forest conservation, with biologists often citing Elton's work (e.g. Grove 2002; Wu *et al.* 2008). Dead-wood habitats and species are amongst the most threatened elements of biodiversity, not just in Britain but globally. Many British dead wood invertebrates were lost in pre-historic times and many are still at risk. Of 150 woodland invertebrates in the British Red Data Books, 65 per cent are threatened by removal of dead wood or old trees (Hambler and Speight 1995) and wood-dependent ('saproxylic') species are the most threatened community of invertebrates throughout Europe (Key 1993).

The ability to match invertebrate specimens to habitat details in the Wytham Survey permitted Mark Robinson (1991) to understand the habitat requirements not just of many beetles in today's landscapes, but of many beetle species he records in archaeological sites. The Wytham data has permitted characterization of the habitats in which the beetles were present thousands of years ago, complementing pollen evidence from archaeology which suggests a generally closed wildwood forest canopy. This has bearing on the highly controversial 'Vera hypothesis' (see Chapter 1.3) which suggests the forest was very open (see Chapter 1 and Moore 2005). Robinson's work thus has great relevance to the philosophy and targets of some modern forest reserve management: it tells us something of the 'baseline' conditions biologists may choose to aim towards in ecological restoration (Hambler 2004).

Whilst Elton's work (including perhaps his experience on overseas travel to more natural habitats) helped establish the need for conservation in Britain, the fieldwork and approach of Duffey (and his post-Oxford colleagues such as Mike J. Morris) was pivotal in moving conservation further onto a scientific footing, establishing the field of 'conservation biology' (Duffey 1966, 1975; Dempster 1977; Hambler 2004). Because the patterns of richness of plants and invertebrates differ in space and time, Duffey recognized that the then widely held belief amongst British conservationists, that looking after the plants will look after the animals, was erroneous. He highlighted the potential conflicts of management interests within British semi-natural habitats in a groundbreaking book *Nature Reserves and Wildlife* (Duffey 1974), in which he pointed out many species may have survived in Britain *despite* traditional land management, not because of it. Duffey and Watt (1971) edited the classic book *The Scientific Management of Animal and Plant Communities for Conservation*. Duffey subsequently became editor of the founding journal of conservation, *Biological Conservation*, one of the most influential in the field.

In part through Duffey's influence, spiders were one of only four invertebrate groups used by the Nature Conservancy Council in reviewing the list of National Nature Reserves and Sites of Special Scientific Interest (SSSIs) for Britain; within this Nature Conservation Review (Ratcliffe 1977), Duffey and Peter Merrett assessed the representation of spiders in the British reserve system, showing that at least one large invertebrate group was quite well represented in the network.

Elton's views were relevant to the selection and management of British nature reserves, which was initially dominated by botanists and vertebrate specialists. He stressed to his colleagues on the Scientific Policy Committee of the Nature Conservancy the importance of invertebrate survey (and the availability of invertebrate taxonomists) to the management of protected areas (Macfadyen 1992).

Strongly influenced by the writings of Duffey and Morris, and work on invertebrates at Wytham and nearby Little Wittenham and Brasenose Woods with Philip Sterling, Hambler and Speight (1996) published a highly controversial and widely-debated appeal to hasten a paradigm shift towards a scientific approach to conservation in Britain, in which the traditional forestry and agricultural practices were not automatically followed in nature reserves simply because they had been used for some time (Duffey 1974; Dempster 1977; Cooper 2000; Hobson and Bultitude 2004). Today, such scientific management is often (and less accurately) termed 'evidence-based' management, drawing parallels with similar changes in medicine from a dogmatic to a science base.

The relevance of studies of invertebrates and succession at Wytham to sustainable management and conservation is increasingly clear, particularly with the recognition that many habitats around the world are semi-natural (derived from traditional agricultural or forest management, Hambler 2004). Increasing British interest in re-creating floristically-rich grassland habitats (which had been lost through succession or agricultural 'improvement', such as use of fertilizer) led Charlie Gibson and Valerie Brown's team from Imperial College, London to fence about 20 experimental plots on Upper Seeds, which came out of cereal production in 1982—as described in Chapter 7 of this book and Gibson *et al.* (1987c).

The value of the Upper Seeds work on invertebrates and grazing is evident from its citation in papers on a wide range of management activities which change the structure or vegetation. There is relevance to the impacts of livestock and off-road vehicles in (for example) the Mojave Desert of California (Brooks 1999). One socially and politically contentious area in which the research is relevant is the impact of wildlife and pastoralist peoples grazing in East African savannah (Warui 2005). The relevance to livestock management in grassland is also evident in Australia (Churchill and Ludwig 2004) and Europe (Zulka *et al.* 1997; Woodcock *et al.* 2007). Knowing which invertebrates thrive in short, warm grass at Wytham has also helped understanding of the value of urban rooftop gardens (Kadas 2006). Understanding grazing is relevant to understanding the impacts of fire—another conservation management tool which reduces the structural complexity of vegetation in various regions and habitats (Swengel 2001), including Australian eucalyptus forest (Harris *et al.* 2003). Conservation planning for invertebrates in the USA is increasingly considering the requirements of invertebrates (Skerl 1999).

Wytham has also been involved in studies of the potential effects of climate change on plants and animals, as part of a remarkable project called TIGER (Terrestrial Initiative in Global Environmental Research). This is a 'NERC research programme designed to help meet the challenges brought about by changes in the global environment' (http://www.nwl.ac.uk/tiger). Rainfall and temperature were manipulated experimentally in replicated grassland plots at Upper Seeds, through use of rain-covers, hose-pipes, and underground heating elements (Chapter 12). Increased summer rainfall increased the abundance of leafhoppers (Masters *et al.* 1998). However, various soil-dwelling invertebrates responded in different ways to the experimental climates (Staley *et al.* 2007a, b), showing how hard it will be to predict the effects of climate change on biodiversity (including pests and rare species). The response of spiders and other taxa were investigated and some of this work is ongoing. Butterflies and some other invertebrates are also monitored on transects at the woodland edge and on the farm as part of the Environmental Change Network national monitoring programme on the influences of pollution, climate, and other changes (http://www.ecn.ac.uk/index.html and Chapter 12 of this book).

Public awareness of invertebrates and their importance has increased through wildlife photography and film. Oxford Scientific Films, which pioneered close-up photography of invertebrates, grew out of the Forestry and Zoology Departments, with founders including the spider-expert, John Cooke, and colleagues.

## 8.12 Invertebrate ecology in teaching

A major contribution that Wytham has made is through teaching and training large numbers of biologists. Further, many ecology teachers learned something of *how to teach* at Wytham: it has been a proving ground for numerous teachers of ecology. Invertebrates were included from the earliest days of teaching at Wytham. Elton and others taught a field ecology course, including quadrat samples of invertebrates of leaf litter, logs, and mark-recapture of grasshoppers, in the 1950s; some who took the course then developed similar courses elsewhere in the world (Crowcroft 1991).

Since the 1970s, undergraduates at Oxford have been introduced to the world of invertebrates in a number of ways, including the extraordinary year-long Animal Kingdom course in the Zoology degree, which ranged from protozoa to whales. Many students then took a popular option in entomology, taught by world-renowned figures such as George Varley and, from 1979, the head of Zoology, Dick Southwood. Students studying entomology visited Wytham as part of their fieldwork to see the diversity of insects and to help collect specimens for the collection of 50 insect species that comprised the practical component of this course option. Wytham's dead wood and other habitats sometimes looked locally devastated by their attentions! Indeed, the subsequent feeling of guilt over collection of lesser stag beetles by breaking open a rotting tree stump near the chalet, deeply influenced Hambler's interest in the conservation of such taxa and the habitat features of old trees. Numerous small projects, undertaken by undergraduates as part of their degrees, have added considerably to the body of knowledge about Wytham.

After Dick Southwood arrived in Oxford, more use was made of the resources at Wytham for undergraduate teaching purposes. A field course was moved to Upper Seeds and Malcolm Coe continued to oversee this course until the mid 1990s, after which it was redesigned and run by Dale Clayton and Hambler and based on the paper by Southwood *et al.* (1979). Students measured the changes in richness and abundance of plants and invertebrates through a succession from grassland to woodland. The invertebrates were studied in a variety of ways, with emphasis on energy flow, biomass, microclimate, and island biogeography. Various absolute sampling techniques were demonstrated, including vacuum samplers (Fig. 8.2) and knockdown of invertebrates by mist-blowing a pyrethroid insecticide into the canopy of scrub and forest (Fig. 8.1)—a technique Southwood had pioneered to give unbiased density estimates of canopy invertebrates from a known sample area. These field courses and others for various MSc courses, such as Forestry, influenced large numbers of biologists who subsequently

**Figure 8.2** An undergraduate samples insects on 'islands' of nettles using a Uni-vac vacuum sampler, on a field course at Upper Seeds. Photograph: Clive Hambler.

became professionals in fields such as conservation, wildlife film-making, and consultancy. In turn, experience from this work fed back into improvements in Southwood's definitive book *Ecological Methods* (which continues as *Southwood's Ecological Methods*, Henderson in prep.).

## 8.13 Improved sampling methodology and indicator groups

Field ecology is difficult, and would not progress without increasingly reliable methods. The ecological methods that were developed or refined at Wytham have had an impact around the world, which may be more lasting than many of the results of research itself. The 'gold-standard' in ecological methodology is an unbiased absolute sample of the density of organisms, and the desire to achieve this grew from Elton's interests in population regulation. This work has wide applied relevance: understanding sampling bias is vital in comparisons of sites as candidates for conservation, in testing the impacts of management, and in Environmental Impact Assessment (Hambler 2004).

Throughout its history, research on the ecology of invertebrates at Wytham has benefited hugely from the engineering, sampling, taxonomic, and photographic skills of Zoology and Museum support staff including Ken Marsland and Philip Taylor. They helped to develop some of the most important ecological sampling methods and devices (e.g. Kempson *et al.* 1963; Crowcroft 1991). Mark Williamson (1959) developed methods to extract molluscs from woodland leaf litter. Bill Fager (1968) developed improved heat extraction devices to sample invertebrates in logs ('modified Tullgren funnels'). He also developed standardized artificial 'logs' using boxes of sawdust, later modified by Peter Larkin to examine rotting oak boughs. Rory Putman developed specialized pitfall traps for the carrion community. The large numbers of samples from Wytham required an invertebrate extraction lab be set up in the Zoology department, in which numerous samples could be processed simultaneously in a standardized way for research and teaching.

Building on his work at Upper Seeds, Eric Duffey enormously improved the methods used to sample invertebrates in an unbiased way. It is important that features of the habitat do not lead to apparent increases in abundance or richness of species, simply because the sampling method works better in some habitats. He recognized that the abundance, species richness, and species composition caught in pitfall traps (where invertebrates fall into cups in the ground) would depend not only on which species were present and their abundance, but also on the sex, behaviour, and activity of the animals (Duffey 1962a). Factors such as weather, the density of the vegetation, and the surrounding habitats could influence the catch in this easily-used and thus popular technique. Pitfall trap results could therefore be highly misleading. Such 'relative' methods provide an 'activity-abundance' measure, in contrast to the 'absolute' sampling giving numbers per unit area which Duffey promoted (Southwood and Henderson 2000). Duffey compared results from a range of sampling methods, including visual searching of quadrats, pitfall trapping, and devices he designed to extract living animals from turf samples using lights.

Once he left Oxford, Duffey worked to develop the use of extraction sampling for invertebrates. A favoured method was the Dietric Vacuum sampler, the D-vac, a device used to sample crop pests. This method is still amongst the most valuable today and has been extensively used at Wytham and the farm (Gibson *et al.* 1992a, b; Brown *et al.* 1992; Macdonald and Stafford 1997; Baines *et al.* 1998). Notably, Duffey went on to publish one of the only direct comparisons between the species and abundances caught by a range or methods, showing many grassland invertebrates were not caught in a representative way in pitfall traps (Duffey 1975). However, if more individuals and species are caught in long grass—against the methodological bias—then the result may be real.

Since there are so many invertebrate species, and many are hard to sample or identify, there is much interest in finding groups of invertebrates which are responsive to environmental changes caused by management or climate. Groups which respond rapidly and clearly to an environmental change such as pollution, or which change in the same direction as many other groups which are harder to sample, are called 'indicators'. Indicators can be used in Environmental Impact Assessment, to monitor the success of management and to choose the best sites for nature reserves (Hambler 2004).

There is much debate over what makes a good indicator group, and studies on the effects of management at Wytham which consider more than one taxon have helped to show the problems of using only one group of organisms. For example, leafhoppers, leaf-miners, plants, and spiders all respond to management in different ways (Gibson *et al.* 1992a, b), with some groups such as spiders being strongly influenced by structure, whilst others are influenced by both plant species richness and the plant structure. A variety of sampling techniques have been found useful in recording such indicator groups in ways which are not biased by vegetation structure. When monitoring management treatments Gibson *et al.* (1992b) sampled spider-webs covered with dew, counted leaf-rolls, and complemented this using vacuum sampling to obtain a fuller range of species in quadrats. Sterling *et al.* (1992) found direct searching of quadrats for leaf mining flies and moths a valuable way to monitor the effects of management on wildlife.

For aquatic habitats, emergence traps developed by Denys Kempson were used by David Eccles for a thorough survey of aquatic insects in streams in Marley Wood (Crowcroft 1991). Roger Kitching developed novel sampling devices for species in semi-fluid substrates, such as water in tree holes, and went on to work on similar challenges—such as life in pitcher plants (Kitching 2000).

In contributing to refined sampling methods, Wytham's invertebrates have thus been disproportionately important in improving knowledge of ecology and management around the world.

## 8.14 Present and future work

Wytham's invertebrates continue to be studied and used in teaching. Undergraduates have worked there recently on issues such as island biogeography (using bushes as islands for spiders), wildlife corridors, and on snail distribution in relation to the calcium needed in birds' eggs.

The species recording at Wytham has revealed that a remarkably high proportion of Britain's invertebrates can be found there. Groups which have been relatively intensively surveyed include the lepidoptera (including, unusually, micro-moths), isopods, and spiders. In the better-recorded groups, it is often the case that Wytham includes around a third to a half of the known British species in a taxon. Many of these species are resident, rather than 'tourists'. This may indicate that the large British woods that survive can support a high fraction of a national invertebrate fauna from which the majority of very specialized forest species have already been filtered-out by landscape change. Indeed, as Table 8.1 shows, Wytham supports about half the woodland spider species of Britain and 70 per cent of those of the Vice County of Berkshire! It also supports some 60 per cent of the British chalk grassland specialist spiders. Similarly, Gibson (Chapter 7) found

**Table 8.1** Table illustrating the proportion of spider species at Wytham and at nearby Little Wittenham woods as a percentage of the available British and specialist habitat pools (i.e. species lists). Habitat specialisms as in Ratcliffe, 1977.

### REPRESENTATION OF SPIDERS FROM VARIOUS SPECIES POOLS

| POOL | WYTHAM | LITTLE WITTENHAM |
|---|---|---|
| **British total pool** (c. 650 species) | 230 species | 217 species |
| % of British list known at each wood | 35% | 33% |
| **British lowland pool** | | |
| = non coastal or montane (c. 555 species) | 40% | 40% |
| **British specialist pool** | | |
| Woodland (105 species) | 50% | 50% |
| Calcareous grass (29 species) | 60% | 40% |
| Wetland (76 species) | 20% | 20% |
| Acid heath (46 species) | 10% | 15% |
| Upland (61 species) | 3% | 0% |
| Montane (54 species) | 0% | 0% |
| Coastal (40 species) | 0% | 3% |
| **Berkshire specialist pool** | | |
| % of total Berkshire list (c. 350 species) | 70% | 65% |
| Woodland (c. 80 species) | 70% | 70% |
| Calcareous grass (c. 25 species) | 65% | 45% |
| Wetland (c. 35 species) | 40% | 45% |
| Acid heath (c. 30 species) | 20% | 20% |

A further 15 species are also known from the Wytham Farm, and 8 in buildings at Little Wittenham.

that over half of the pool of British chalk grassland herbivorous invertebrates which might realistically be able to reach Wytham do actually occur there.

This may actually not be that unusual. Little Wittenham, some 25 km from Wytham, has a similar range of habitats and a shorter period (three years) of intensive recording of spiders there has revealed almost as many species as Wytham (about 230). Of the combined list of 63 species of woodland spiders, some 70 per cent occur at both sites. Comparisons with other relatively well-known sites, such as the area within 10 km of Monks Wood in Huntingdonshire, have also revealed similarities in numbers and community composition, with the same proportion of their combined list of woodland specialists shared. It was initially intended that Wychwood Forest, a hunting estate on different soils and with a more continuous history of mature timber and dead wood, would be a companion site for Wytham to help in comparisons of determinants of community structure, but this has unfortunately never been pursued.

However, it is evident that even Wytham has not been thoroughly recorded for some major groups of invertebrates: only about 1000 species of beetle, 500 flies, 250 bugs, and 600 wasps, bees, and ants have so far been recorded—less than a quarter of the British species in these groups. Its wetlands have not been well-studied and several regionally common spiders have yet to be found there.

Only in the case of Lepidoptera has serious recording been continued in recent decades, in part due to a loss of invertebrate and field ecologists with taxonomic skills in the Zoology Department and the Museum and a decline in teaching in traditional entomology. Invertebrates are presently studied mainly in relation to their relevance as food for birds, such as the long term study on great tit population change (Perrins 1992; Buse *et al.* 1999) and for badgers (Hofer 1988; Da Silva *et al.* 1993). However, entomology is now being re-invigorated by the new Hope Professor, Charles Godfray, and colleagues, with a new undergraduate option.

The value of rare base-line and long-term data is increasingly evident as the political interest in climate change rises and invertebrates help reveal and explain trends which may be associated with climate change. When resources permit, the Wytham Ecological Survey will doubtless provide further valuable evidence of habitat and behavioural change. Computerization of this database would allow rapid examination of the habitat associations and communities, which Elton studied so carefully.

Since the 1980s, work on Wytham's invertebrates has been discontinuous, due to changes in staff, income streams, and priorities of heads of department in Oxford. However, it has been possible to maintain a remarkably long-term monitoring programme for species closely connected to the unique data sets on bird dynamics in the woodland. In particular, work on the abundance and timing of the availability of caterpillars on oak has revealed how crucial the timing is to fledgling survival (Perrins 1992 and Chapter 9). Long-term studies, such as the work on phenology and population dynamics of winter moth described later, will become increasingly valuable with time, but are often at risk in Britain and elsewhere through changes in research funding regimes.

## 8.15 Notable invertebrates in Wytham and their requirements

Elton (1966) noted that Wytham '...for many years has carried the reputation among entomologists of lacking many of the really "exciting" species of insects'. E.G.R. Waters commented in 1926 that:

In nearly all directions from Oxford are extensive woods which, though now circumscribed and difficult of access, are still the happiest hunting ground of the micro-lepidopterist. Wytham Woods, the most picturesque, are the least productive, though many of the usual woodland species occur there....

Although the entomological interest of the site derives largely from the exceptional amount of scientific work there has been, the site does, however, support some nationally notable species.

Through the influence of Elton, Duffey, Gibson, Speight, and others, the management of Wytham has been relatively and explicitly sympathetic to invertebrates for over 50 years. In addition to permitting the accumulation of dead timber, there are more 'non-intervention' areas than in many woodlands, where more natural forest can recover. Wytham supports species which may indicate forest habitat continuity, such as the spectacular leafhopper *Ledra aurita*—a lichen-feeder, but it lacks others such as the mountain bulin snail *Ena montana* known from other forests in the region such as Wychwood.

Having been extensively coppiced over about half its area and heavily managed elsewhere, Wytham is largely lacking in dead-wood specialist invertebrates more typical of wood pasture (Harding and Rose 1986). It has only a couple of beetle species strongly associated with old forest (Harding and Alexander, 1993). A few dead-wood species may perhaps have survived in, and might recolonize from, the nearby parkland estate. Wytham does not feature amongst some lists of European or British sites notable for dead-wood species (Harding and Rose 1986; Alexander 1988; Speight 1989), although over 40 species of less sensitive dead-wood beetles (and perhaps high recording effort) have put Wytham in the top 20 British sites for an 'Index of Ecological Continuity' (Chapter 5; Harding and Alexander 1993). More dead-wood species might recolonize naturally—although many have very poor dispersal abilities and some are flightless.

The Council of Europe Committee of Ministers recommended to governments that re-establishment of European dead-wood species be undertaken (Speight 1989), although there has been little progress in this regard. The size of Wytham and its recorded history may present opportunities for attempted re-introduction of true forest species which could not survive in smaller woods due to adverse effects penetrating from the edge (Hambler and Speight 1995; Ozanne *et al.* 1997; Hambler 2004). However, the small area of the woodland, by continental European standards, may not be sufficient for populations of forest specialists and such species may continue to disappear—a process known as 'relaxation'. Relaxation has been happening in Britain and Wytham since the fragmentation of the natural forest ('wildwood'), but the duration of the process is area-dependent and unclear (Speight 1989; Hambler and Speight 1995, 1996; Grove 2002). The exceptional recording of Wytham reveals that some species have become extinct in Wytham in historical times (Elton 1966). For example, the dead-wood specialist stag beetle *Platycerus caraboides* was recorded around 1820 by F. W. Hope and was extinct in Wytham and Britain by 1900; it may be a good candidate for re-introduction to Britain—at Wytham!

Europe's largest carrion beetle *Nicrophorus germanicus* was collected twice from game carcases at Wytham Park, before and during 1822; the species is thought to have become extinct in Britain during the nineteenth century. As with the more recent decline in Britain of the bone-skipper fly, extinction was presumably through loss of large carcasses as part of modern farm and forest hygiene, which makes it harder for carrion feeders to find enough habitat.

Mark Robinson (pers. comm.) suggests the fen at Wytham might reveal a wealth of interesting sub-fossil remains: in addition to pollen, preserved beetle wing-cases, snail-shells, and other parts of invertebrates might reveal much about the ecological changes in the area over thousands of years, perhaps including species that have become extinct. Such work on 'naturalness' might help set targets for conservation.

Active management through traditional methods (such as sheep-grazing on Bowling Alley and Sunday's Hill, coppicing, and scrub-management) have been increased recently, with a view particularly to promotion of flora and butterflies such as the black hairstreak (which needs long-rotation cutting of blackthorn) and the wood white. Some short calcareous grassland is being re-created from arable land, long-grass, or scrub and may protect short-grassland specialist invertebrates. These include the duke of burgundy fritillary and the moths *Cochylis flaviciliana, Coleophora silenella,* and *Cydia pallifrontana*. Such species will have declined since myxomatosis killed off most rabbits. The light-feathered rustic moth, *Agrotis cinerea* was last seen in 1956, and this nationally scarce species of chalk downland moth may now have died out; others are however recovering, as Chapter 7 describes. There may be species within the Woods and park that no longer occur there through changes in traditional management and successional conditions, or which occur at low abundance and might be promoted. Some management activity generates debate and controversy (e.g. Fuller and Warren 1993; Hambler and Speight 1995; Chapter 11 of this book), which itself can be helpful in teaching.

Many of Wytham's nationally uncommon species, such as the wetland spider *Trochosa spinipalpis*, the woodland spider *Clubiona caerulescens*, and the six-belted clearwing moth *Bembecia scopigera* have not been recorded in recent decades. In some cases, this is probably a result of misidentification or massively decreased recording effort. In other cases, it may be due to habitat and climate change. On the other hand, the purple emperor butterfly is now seen occasionally after a gap between 1986 and 2006 (Gibson, pers. comm.) perhaps in response to changed management or a warming climate. The nationally rare (British Red Data Book) spider *Tuberta maerens* was present in 1953 and is still present; this spider, as with some other invertebrates in Wytham, is being assessed as an indicator of ancient woodland in Europe. Wytham supports tens of the national Red Data Book invertebrates, including over 20 species of threatened fly.

Whilst the recording effort has been relatively good, very many invertebrates doubtless remain to be discovered at Wytham. Wytham presents many opportunities for future survey, conservation, teaching, and research. Many of Wytham's species lists are now out of date through taxonomic and environmental changes and greatly reduced sampling effort. It would be interesting to repeat the intensive survey of earlier years, to look for and try to explain changes. The following case study of one species, the winter moth, shows how detailed the recording at Wytham can be—yet how much one still needs to know to understand the abundance of even a single species.

## 8.16 The winter moth—an unlikely superstar

When selecting an animal to spend years working with to delve into some of the most important questions facing ecologists, one might think that a researcher would choose something impressive or entertaining—perhaps a lion or a meerkat—or at the very least something pretty and interesting. Not so. Imagine a small, very nondescript brown insect that flies in the dead of winter, (or indeed doesn't fly at all) and you have one of the superstar celebrities of the ecological world, the winter moth *Operophtera brumata*, Figs. 8.3 (a) and (b).

Why this species and why Wytham? To answer this one needs to find out about one of the early giants of the ecological firmament, George Varley—a keen entomologist who started his scientific career as a young post-doctoral student at Cambridge and then became Hope Professor of Entomology in Oxford at the (then) relatively tender age of 38 in 1948.

Varley wanted to understand the reasons for the changing abundance of insects through time. Why are insects pests in some years but very rare in others? What, if anything, prevents insect populations increasing without limits to their full potential? These questions have both a practical application (for example, for pest control) and a theoretical application in the general field of ecology where similar questions are asked about all sorts of other organisms. Varley assumed that animal numbers would be low if a lot died during development and would be high (or increasing) if fewer than average died during development. He therefore realized that one way to explain changes in numbers was to count insects at various points of their life cycle and then work out how much mortality had occurred between the different life stages in turn.

He decided to study a common animal occurring naturally within Wytham. The animal needed to be abundant for the simple reason that it is far easier to develop theories when you can sample anything in abundance and therefore are able to detect relatively low rates of mortality through sheer force of numbers. It also needed to be easy to sample, and as an entomologist he favoured insects.

Invertebrates 135

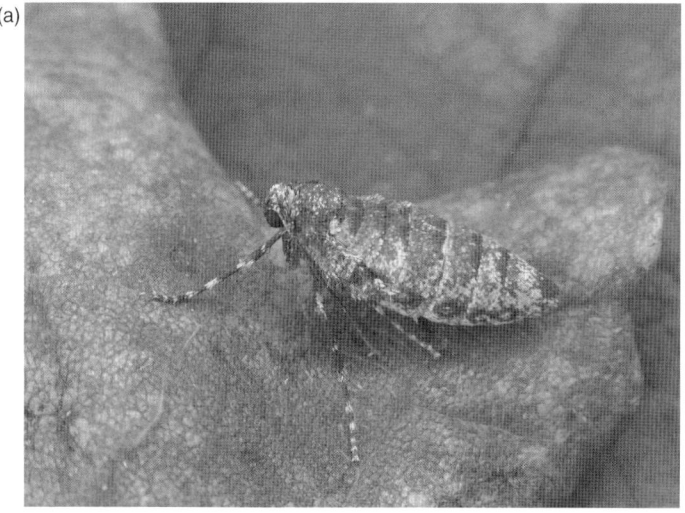

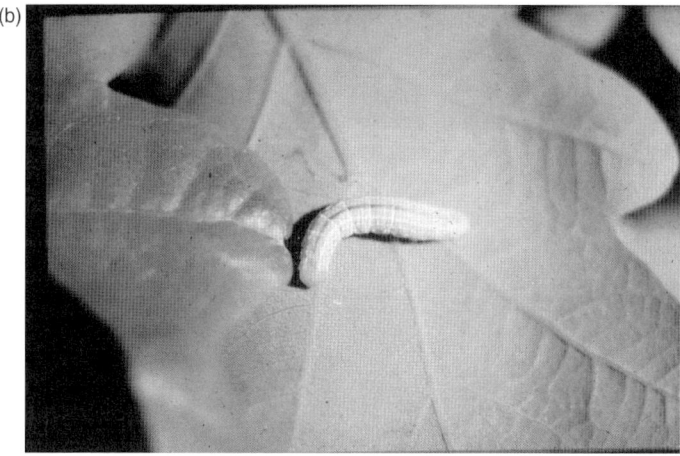

**Figure 8.3** a) The adult female winter moth is flightless and has only rudimentary wings. (Photo by Keith Tailby) (b) a winter moth caterpillar.

Two of the really common insects in Wytham Woods are the green tortrix and the winter moth, both feeding mainly on oak trees. The green tortrix is often the more abundant of the two, but both sexes of this species are winged and can fly. The female winter moth is large (when full of eggs) and flightless, making sampling of the female population relatively easy compared with the green tortrix. So Varley selected for his own detailed studies the winter moth. How much easier it would have been for generations of young

# George Copley Varley (1910–1983)

George Varley. Photo by courtesy of the librarian of the Hope Library, University of Oxford.

George Varley started his scientific career as a post-doctoral student at Cambridge. He became Hope Professor of Entomology in Oxford at the age of 38 in 1948. Though trained as an insect physiologist, he revolutionized the way the dynamics of populations are both studied and understood.

George Varley and his research assistant, George Gradwell, are probably best known for their study of a common animal occurring naturally within Wytham Woods: the winter moth (*Operophtera brumata*). They provided a strong basis for understanding the timing and abundance of the caterpillars. The tit-workers were particularly lucky that this was the case. As with many successful long term studies, this research provided the critical mass and intellectual stability to stimulate a series of spin-offs beyond its original objectives of developing the science of populations. As an example, the realization of the importance of the interaction between the caterpillars and their food plants prompted Varley to set various students to look at why the larvae did better on some leaves than others.

These studies on insect–plant interactions led to one of the first examples of a real understanding of the 'evolutionary arms race' that goes on between plants and herbivores. This suggested that big, obvious plants like trees devote more resources to defending themselves against their enemies, using high and costly concentrations of chemicals which reduced herbivore growth but usually did not kill.

From one small caterpillar and one man has developed a huge range of world class ecological research that has affected all our lives. This illustrates how having a fixed and undisturbed area for ecological studies readily accessible to research workers can lead to unforeseen discoveries and collaborations, and serves to emphasize just how valuable Wytham (and the winter moth!) have been to ecology.

His team's work led to the classic 1973 book *Insect Population Ecology*.

entomologists if it were the female green tortrix, with the relatively pronounceable scientific name of *Tortrix viridana,* that were flightless!

Varley and his research assistant George Gradwell were able to sample successive generations of a common insect in a relatively stable environment (the oak woodlands). He chose a part of Wytham Woods dominated by maturing oak trees that each appeared to support different densities of winter moth; this plot became Varley and Gradwell's 'oak patch'. From it came new methods to understand insect abundance.

## 8.17 The 'life table' of a species

It is late in the year by the time the adult winter moth hatches out from its resting stage (the chrysalis or pupa) in the soil—November to January. Female winter moths climb the nearest tree (often the one they themselves fed on as larvae) and, on their journey, are mated by the flying males. They continue to the tops of the trees and lay their eggs amongst moss and lichen on the oak twigs. The young winter moth larvae hatch in April, roughly when the oak leaf-buds start to open and at a time dependent on winter temperature (Chapter 9) and then feed on the leaves for the next six weeks or so, after which they drop to the ground on long silken threads. They then tunnel into the ground where they turn into pupae and remain there until the following November, when the cycle starts again.

The quantitative sampling that Varley and Gradwell undertook for the winter moth allowed them to build a catalogue of mortalities, in other words, an actuarial table like that used by insurance companies to calculate life insurance premiums for people, but for the moths. These came to be known as 'life tables' and had rarely been constructed for any creature other than humans, for which the risk of dying usually increases with age. The same is not necessarily true for small creatures such as insects; for them, the risk of death within a fixed future period of time depends on the risks they will encounter in that period, rather than on their age. For example, more young, wriggly caterpillars on a leaf might well be more likely to be killed (by spiders or birds that might eat them, or parasites that might attack them) per unit time than would older pupae, buried in the ground and protected by a hard exterior 'shell'. Also, crowding might increase the chances of starvation—an example of a population process that depends upon the abundance (or density) of the organisms concerned rather than just their age.

Varley and others therefore realized that producing an actuarial table for insects involved measuring not just the age of the insects, but all the many factors that could kill them in the next unit of time. Processes such as frost, floods, and drought kill a fixed percentage of a population more or less regardless of how many animals there are. The risk of death is a function of the temperature, or the excessive rainfall, and not the number of insects present. Such processes are described as 'density-independent' and cannot maintain populations in any stable way; an animal with a fertility of 100 offspring per individual and suffering, on average, 97 per cent density-independent mortality, would nevertheless increase 3-fold per generation; if the mortality were, on average, 99.9 per cent then the population would fall and would continue to fall until extinction. Only in the very unlikely event that the mortality allowed exactly one offspring per parent to survive would population be stable, and this is so unlikely that we can dismiss it as a real possibility.

Other processes, such as competition between animals for a limited amount of food, or some forms of parasitism and predation, act more severely at higher population densities of the animals concerned; the more there are, the higher the *per capita* risk of dying. Such density-dependent processes can stabilize populations because they cause a higher mortality when the population is above average and a lower mortality when it is below average: the population tends to move up and down around its long-term average value.

The debate that ecologists were having from the 1930s onwards concerned the relative importance of density-independent and density-dependent processes in bringing about the observed stability of animal and plant populations. One group, championed by Australian ecologists, felt that populations could be regulated by density-independent processes alone. Another group, championed by, among others, George Varley and another young ecologist in Oxford, the ornithologist David Lack (see Chapter 9), included a brilliant lone maverick Australian, A.J. Nicholson. They argued that only density-dependent processes are able to regulate populations within the limits we observe.

Partly because of George Varley's work, we now know the latter group was right and that density-dependence is the *sine qua non* of population regulation, but this was not so obvious at the time. In fact, at one International Congress of Ecology in the 1960s, and after a full airing of all sides of this debate, a group of world-famous ecologists got together, one suspects in the bar in the evening, and signed a document they could all subscribe to that was headed 'Ecology is what I do and you don't'. For many years, this signed document adorned the tea rooms in George Varley's Hope Department, an inspiration (and also maybe a warning?) to the Hope's doctoral students of the battles that lay ahead for them.

The quantitative measurements that George Varley wanted to make of the winter moth would allow him to investigate the relative importance of density-dependent and density-independent processes in the natural regulation of the winter moth in Wytham, and thus contribute to the global debate that was raging at the time.

The questions for Varley were, however, initially very practical ones. How many oak trees should be studied? How could the winter moth females be sampled as they crawled up the oak trees? How could the caterpillars be sampled six months later when they dropped to the ground? Do very young winter moth larvae re-distribute themselves between the oak trees in early spring (Varley knew that these young caterpillars occasionally spin out short threads of silk which, in windy conditions, are enough to carry them on the wind, spider-like, to other trees)? Which other species should be measured on the oak trees (when Varley began his study it was really not clear whether other species might be involved as potential competitors, or which were parasites, pathogens, or predators of winter moths). How could survival of the pupae in the soil be measured?

In the end, Varley and Gradwell decide to measure the populations three times a year. The first occasion was when the adult female winter moths climbed the trees at the end of the year. They were sampled by small lobster-pot style traps (Fig. 8.4), mouths open to the ground, that were nailed to the oak trees at about chest height. The Hope Department Librarian, Audrey Smith, saved up her used tights each year, to make and repair the lobster-pot traps for the winter moth study!

These traps were arranged so as to sample one quarter of the number of females ascending each study tree. By counting the females in winter and by dissecting samples

**Figure 8.4** 'Lobster pot' trap for catching female winter moths as they ascend trees in winter.

of these females to see how many eggs each on average contained, Varley and Gradwell could work out how many eggs were carried to the top of each sampled tree by the remaining three quarters of the females that were not caught in the traps.

The second sample was when the mature larval population dropped down from the oak trees to pupate on the ground about six months later. Two metal trays, $0.5 \times 0.5$ m were put under each oak tree's canopy, on a slight slope, and partially filled with detergent solution to kill the caterpillars (which naturally walk downhill). They could later be counted and dissected to see if they contained any parasites. Thus each metal tray (called a 'water trap') sampled a nominal 0.5 × 0.5 m column of canopy for all the caterpillars that had been produced by it that year; it was a simple matter to multiply up by the canopy area to discover the total number of larvae produced by each tree.

The same metal trays were used towards the end of autumn and right through until the following spring, but turned upside down in a different position beneath the canopy to sample the insects that emerged from the soil, to account for all the caterpillars that had spun down earlier in the year. As the insects emerged from the soil they were attracted to the light coming through small holes in the corners of the trays (called 'emergence traps'), where they were trapped in tubes and could be sampled.

Varley and Gradwell counted and studied the winter moth population in great detail from 1950 to 1968; the census was resurrected and continued in part by others intermittently thereafter, but the main body of analysis was done on the data from the continuously sampled period.

Because Varley was not sure what species other than the winter moth itself might be important in the winter moth's dynamics, he decided to examine and do approximate counts of other species of potential importance, judged on the basis of their own abundance and their likely proclivity to attack or eat winter moths. Given that the winter moth itself is one of the most abundant species on oak, any abundant natural enemy occurring on oak (such as a predator or parasite) might possibly be involved with the winter moth. One such species is a small active beetle called *Dromius quadrimaculatus* that is found on oak bark at most times of the year, even in autumn. Varley correctly guessed that *Dromius* is a predator, and showed by simple laboratory experiments that it eats winter moth eggs; so *Dromius* was added to the list of suspects and counted for some of the sample period.

The sampling of the winter moth showed quite rapidly that there were several key species of natural enemy attacking the winter moth at different stages of its life cycle. *Cyzenis albicans* Fig. 8.5 a) is a hairy blow-fly-sized brown, nondescript fly, the larvae of which live as parasites within winter moth larvae. The adult *Cyzenis* emerge from the soil early in spring (before the winter moth larvae hatch). They presumably mate and then mature their eggs in readiness for the winter moth larvae. When winter moth larvae feed, they damage the oak leaves which ooze sugary sap from the chewed leaf edges. *Cyzenis* females fly between the oak trees in search of such sugary sap and lay their own very tiny eggs where they encounter a lot of it. With luck (from *Cyzenis*' point of view) these eggs are later eaten by the winter moth caterpillars; they hatch out as small larvae in their gut and tunnel through their tissues, eventually ending up in the salivary glands (Fig. 8.5b) where they slowly mature. Of course, these *Cyzenis* larvae are carried wherever the winter moth larvae go, and eventually end up within the winter moth pupae buried in the soil. At this stage the parasites 'take control' and prevent the winter moth adults from developing within the pupae; instead the parasites then begin to complete development themselves and form their own pupae within the host's pupal skin. The adult *Cyzenis* hatch from these pupae not in November, like the hosts, but in the following spring, just in time for the next generation of host larvae. Only one *Cyzenis* is eventually produced

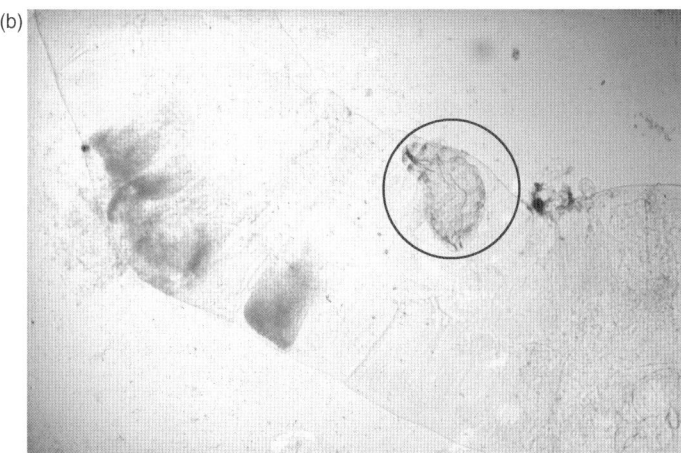

**Figure 8.5** a) *Cyzenis albicans*, a parasite of winter moth larvae. (b) Larva of *Cyzenis albicans* within the salivary gland of the host caterpillar.

from each winter moth, so a great battle goes on inside each host that eats more than one *Cyzenis* egg; the parasitized larvae fight each other and there is only ever one survivor that completes development.

Winter moth pupae buried in the soil between about May/June and November may fall victim to a number of soil predators, but key among these are two beetles, one called *Feronia madida*, a Carabid (a group that contains many predatory, soil-dwelling species) and the other *Philonthus decorus*, a Staphylinid (in the same family as the Devil's Coach horse). The problem with predation amongst insects is that you rarely see it happening in the field; you simply record that there are fewer prey around later in their life cycles than were present earlier. In the case of the winter moth it was clear that something was happening to the pupae within the soil because not all the larvae that spin down to the ground each year could be accounted for by the number of winter moths emerging later the same year plus the number of *Cyzenis* emerging early the following year. By designing some very neat experiments, Varley, Gradwell, and some of their D. Phil. students identified *Feronia* and *Philonthus* as the likely suspects causing these winter moth losses in the soil each year. *Feronia* and *Philonthus* are active throughout the year, so winter moths provide only a fraction of their diets—more in years when winter moths are more abundant. Such high food availability in turn seems to boost the populations of these two predators, so that the populations of the winter moth and the beetles are loosely coupled.

The third and final important cause of mortality through parasitism or predation was identified when Varley and Gradwell looked at the insects that emerged in the emergence traps. Among them was an ichneumon wasp called *Cratichneumon culex*, about which little was known at the time. *Cratichneumon* was confirmed as a parasite of the winter moth and its numbers monitored over the sample years. It is active in late summer and searches for the winter moth pupae in the soil, so it is a pupal rather than larval parasite.

Other causes of winter moth mortality were identified during the study (including a microsporidian disease of caterpillars), but none was very significant in determining the winter moth's numbers over time.

Once several (insect) generations' worth of data had been collected, Varley and Gradwell began their analyses. They quantified the losses for each generation and produced their life tables for the winter moth. By examining which mortalities contributed most to changes in the total generation mortality (this was done visually by plotting all results on the same graph) they were able to identify which sub-mortality was the 'key' to changes in population size over time. This 'key' mortality is called the 'key factor' and the whole operation is called 'graphical key factor analysis'.

Natural enemies were responsible for the death of a number of winter moths but the analysis revealed that the 'key factor' in winter moth dynamics is not a natural enemy at all, but is in fact the loss of individuals between the egg stage (the over-wintering stage) and the time when the mature larvae spin down to the soil and are sampled by the water traps. This so-called 'winter disappearance' is extremely variable from one year to the next and is the major contributory cause to changing winter moth numbers through time.

But what is the cause of winter disappearance? Many elegant studies by a succession of students showed that winter disappearance arises through the varying synchrony of the hatching of the winter moth larvae from their eggs and the leaf flushing of the oak leaves from their buds in spring. Because the signals used by the caterpillar and the tree are rather different, in some years the caterpillars hatch before the buds have burst and many caterpillars die of starvation. In fact they will migrate from their natal tree to try to find food elsewhere, but this is a very dangerous process, resulting in high losses. In other years the caterpillars hatch well after the buds have burst. The leaves are then a bit too mature and full of defensive tannins to provide good food for the larvae and prevent many caterpillars from developing at all. So, high winter moth losses occur in years when caterpillar hatch does not coincide with bud-burst.

Winter disappearance is a density-independent process as it operates irrespective of the density of the winter moth. By definition, therefore, winter disappearance cannot regulate winter moth populations. So what does stop the winter moth increasing without limit, or disappearing completely from Wytham? By relating the severity of each sub-mortality to the density of winter moths on which that sub-mortality acted, Varley and Gradwell found that pupal predation was greatest in years of high winter moth abundance. Pupal predation is therefore the main density-dependent mortality in the winter moth system. So active and efficient are the predators in the soil at Wytham that they reduce the importance of the larval parasite *Cyzenis*. The beetles eat both the healthy host pupae and those pupae in which *Cyzenis* has taken over. In effect the beetles regulate *Cyzenis* at the same time as they regulate the winter moth and thus render *Cyzenis* itself rather ineffectual at regulating the winter moth.

Interestingly there are fewer guilds of soil predators in Canada, where the winter moth became an important introduced pest in apple orchards. *Cyzenis* was introduced from Europe (along with a pupal parasite that does not occur in England) in an attempt to control the winter moth and confounded expectations based on the Wytham experience by causing a very high level of mortality on the winter moth, thereby reducing its populations to much less damaging levels. Thus *Cyzenis* controls the winter moth in Canada, but not in Wytham!

Varley and Gradwell tested their understanding of the winter moth system by constructing mathematical models for the winter moth population over time. When all the observed life table mortalities were inserted into this model, it reproduced the changing numbers of winter moth, *Cyzenis*, and *Cratichneumon* over time in a very convincing way. This implies that Varley and Gradwell had an extremely good understanding of this system through their sampling programme and the resulting key factor analysis. It may sound today like a trivial exercise, but at the time it was heroic and ground-breaking. One sceptic was heard to mutter loudly at an international conference at which these results were first presented 'How can you expect to learn anything by sampling from only five oak trees?' Well, by sampling from these five trees we have learned more about the dynamics of winter moth than we know about any other insect. However, equally, or even more importantly, the techniques that Varley and Gradwell developed in Wytham, later went on to see use in the analysis of population data from species as varied as cockles and wildebeest, and provided equal amounts of insight in each case.

## 8.18 From moths to generalities

As with many successful long term studies, Varley's long term winter moth research provided the critical mass and intellectual stability to stimulate a series of spin-offs beyond its original objectives of developing the science of populations. As an example, the realization of the importance of the interaction between the caterpillars and their food-plants prompted Varley to set various students to look at why the larvae did better on some leaves than others. This involved feeding caterpillars with different aged leaves, analysing the leaf contents, and measuring what went in and what came out (thus condemning the poor students to years of weighing caterpillar faeces).

What emerged from these rather bizarre activities was some of the earliest demonstrations that plants actively defended themselves against herbivores. They produced chemicals or toxins, or physical defences like toughness or hairiness that limited the caterpillars' growth, the size of the adult moths into which they eventually developed, and so reduced the number of offspring they produced.

These studies on insect–plant interactions led to one of the first examples of a real understanding of the 'evolutionary arms race' that goes on between plants and herbivores, as described by the Apparency Theory. This suggested that big, obvious plants like trees devoted more resources to defending themselves against their enemies, using high and costly concentrations of chemicals which reduced herbivore growth, but usually did not kill them. Familiar examples are the tannins and polyphenols which make tea and wine delicious and are beloved of those trying to rid themselves of free radicals. On the other hand, less obvious plants like short-lived or small annuals keep defence costs down by producing only small quantities of toxins which kill most herbivores, but can be detoxified by specialist herbivores. Many of the active ingredients in herbal medicines and plant-derived drugs fall into this category. Gardeners might also think of the infamous cabbage white butterfly which can detoxify the cabbage's defenses and thrive where cabbage is artificially apparent.

Varley and Gradwell's legacy also contributed to many later and ongoing studies in Wytham. The understanding of what affected winter moth and caterpillar populations in general became an important component of one of the other long term Wytham Classics, looking at blue and great tit numbers, as described in Chapter 9 of this book.

From one small caterpillar and one man has developed a huge range of world class ecological research which has affected all our lives.

## 8.19 Acknowledgements

We thank Lionel Cole, George McGavin, Steve Gregory, Keith Kirby, Mark Robinson, Martin Speight, and Philip Sterling for comments and information and David Green and Keith Tailby for help with Keith's photo of the winter moth. The late Charlie Gibson gave his usual perceptive help, and this chapter is dedicated to his memory.

# 9
# Birds
C.M. Perrins and A.G. Gosler

*This chapter is dedicated to the memory of our friend and colleague Robin (Mac) McCleery, who helped many of us to understand the tit populations in Wytham, and whose sudden loss in 2008 is sorely felt.*

## 9.1 Introduction

Birds have featured strongly in the ecological studies undertaken at Wytham since the University acquired the estate in 1942. In particular, the long-term population studies of tits, started in 1947, have been a major contributor to Wytham's ecological reputation. Most of these studies have been carried out by members of the Edward Grey Institute of Field Ornithology (EGI), part of the Department of Zoology and named after Viscount Grey of Fallodon, a former Chancellor of the University and Foreign Secretary 1905–16. These researches have always been aimed at addressing significant issues in field biology, including the evolution and ecology of reproductive parameters, such as the timing of breeding in relation to food availability, or the number of eggs (Plate 8) that a female lays in a clutch. Although some of these studies were initially purely academic, they have become increasingly important for understanding how birds respond to environmental variation such as climate change.

## 9.2 Bird species in Wytham

The extensive collection of scientific papers that have come from the tit studies at Wytham might lead one to suppose that the tits, and perhaps a few of the flock 'associates' (long-tailed tits, nuthatches etc.) and predators (sparrowhawk) that crop up in tit studies from time to time, were the only birds at Wytham. However, whilst not documented in as minute detail as the tits, Gosler (1990) recorded 153 species for Wytham, consisting of 78 'regular' species (63 breeding, of which 50 were resident), and 75 'irregular' species; the latter included a number of formerly-regular breeding species (including grey heron, tree pipit, redstart, and nightingale), and an eclectic roll of vagrants, many of which were seen flying up the Thames past Wytham. This list includes gannet, razorbill, marsh harrier, and goosander. Since the 1990 report, a further two species have been added bringing the total to 156: a kittiwake seen flying over the wood in March 1995, and a pair of mandarin ducks near the chalet in the summer of 2002.

Until about 1980, the Wytham bird community was essentially what one would expect to find in any large, chiefly broadleaved, wood in central England, with a high preponderance of oak and a well-developed understorey, that is, tolerably high densities of tits, thrushes, finches, breeding warblers, nightingale, and other common woodland passerines, three species of woodpecker, woodcock etc. The riverside vegetation of the Thames at Wytham provided nest-sites for reed and sedge warblers and even marsh warblers have been recorded there (1976). Kingfishers patrol this stretch of river, even coming into the Woods on occasion, and on the adjacent flood meadows curlew, lapwing, redshank, and snipe bred. Wytham's ponds provided habitat for little grebes, and were visited on occasion by ducks such as mallard and teal. Up until 1963, a heronry had flourished on Hither Clay Hill for at least 80 years. It was deserted, as was a smaller one in Broad Oak as a result of the extremely cold winter, coupled in the case of the Hither Clay colony, by some disturbance caused by woodland operations in the run-up to the breeding season.

## 9.3 Changes in species

Prior to University ownership, records of several species are known only through the species having been shot by the keepers, whose zeal for destroying predators appears to have been extensive. Hence gannet (1838), rough-legged buzzard (1920), razorbill (1926), and eight long-eared owls (1924) were all recorded by the Oxford Ornithological Society as having fallen to the keepers' guns.

More recent changes in bird populations in Wytham mirror those in the broader countryside, but others are intimately bound up with changes in the woodland itself. The planting of Radbrooke Common (1946–53) resulted in the loss of a heathland bird community (nightjar, woodlark, tree pipit, grasshopper warbler, and stonechat) that has not yet been replaced by a mature woodland bird community, such as can be found on the north side of the hill, or in Marley Wood. Woodlarks have not been recorded in Wytham since 1971, though tree pipits continued to breed on Rough Common and at the top of Lower Seeds Field (the scrub reserve by Brogden's Belt) until 1987. One or two pairs of grasshopper warblers still breed in some years in scrub and plantation edges but, while stonechats are sometimes recorded on Upper Seeds field, they have not been recorded breeding for many years.

The other two changes are inter-related. The cessation of the forestry practice of thinning maturing trees, and the rapid increase in deer populations, have both contributed to a loss of understorey. The closing of the canopy has reduced the light to the forest floor and so reduced plant understorey growth; in the case of the deer, they ate it! The near-total destruction of the shrub layer, in which many species nest, and creation of a distinct browse-line (Fig. 5.7) at about a metre in height through most of the woodland estate, has severely reduced the populations of a number of species including warblers and thrushes, and eradicated the nightingale as a breeding species (Gosler 1990; Perrins and Overall 2001).

Significant improvements have recently been made to both the main Wytham ponds, and these have benefited birds. Little grebes have returned to the woodland pond and bred in the five years 2003–07. The re-excavated pond in the park also now attracts wildfowl, and teal has been noted there again after a long absence. The Thames riverside habitat provides nest-sites for kingfishers, reed and sedge warblers, and the bubbling call of the curlew out on the meadowlands is still a welcome surprise to researchers visiting the tit nestboxes in the northernmost 'rounds' of the Wytham Tit Study. Using Gosler's (1990) criteria, the current 'regular' list stands at 75 (that is, a net loss of just 3), but a simple list hides important changes which have occurred in the relative abundance of species. Whilst most common resident species (tits, thrushes, finches) continue to thrive, there have been substantial declines in summer migrants. Most warblers have become relatively scarce (at least compared with the early 1980s), and several formerly-regular breeders have disappeared altogether. The list of losses includes both those noted by Gosler (1990) as declining or disappearing in the 1980s, namely the tree pipit, redstart, and nightingale, but also turtle dove and spotted flycatcher, which are now very rare in Wytham. Whilst decline in the numbers of warblers, and the demise of the nightingale population in Wytham, have been related specifically to habitat degradation caused by overgrazing by deer (Gosler 1990; Perrins and Overall 2001), which is itself a widespread problem in lowland England, the declines in both turtle dove and spotted flycatcher numbers reflect wider problems whose causes are not fully understood. Nationally, the turtle dove and spotted flycatcher have declined by about 80 per cent and 70 per cent respectively on farmland in the 30 years since 1975.

Whilst these losses are of critical concern for the conservation of these species, they also represent a significant aesthetic loss to the human visitor. Just 30 years ago, Wytham was one of the most reliable sites near to Oxford where people could hear the rich haunting song of the nightingale, the last voice singing in the evening echoing through the woodland, enriched yet further with the glorious sight and smell of bluebells. The soft purring of turtle doves on Rough Common on a June afternoon is another memorable sound now lost to us.

## 9.3.1 Willow tits

The first losses of formerly-regular resident breeding species have also occurred in the last 15 years. In 1990, Gosler described the willow tit as a regular resident breeder, but noted that it had never been abundant in Wytham and that its population was restricted to the '…damper more diverse ancient semi-natural woodland areas around the foot of Wytham and Seacourt Hills'. He named as 'traditional' areas: Bean and Marley Woods, Common Piece, Ten Acre and Nealings Copses, and Further Clay Hills. The species appears to be extinct in Wytham now, with no reliable records since 2002. Until its decline, the willow tit was not studied intensively in Britain, because it only occurred at very low densities, and also because it did not use nest-boxes. By chance however, the first study of it in the UK was undertaken in Wytham. There, Foster and Godfrey (1950) showed that there was usually only one pair, sometimes two, in a study area of some 27 ha of woodland. The Wytham decline is consistent with the widespread national decline in the UK (80 per cent since 1970). The reasons for its decline in Wytham are unclear. Because it excavates its own nest cavity in dead wood, this species is vulnerable to several possible pressures: decline in the standing 'crop' of dead wood (perhaps due to high deer numbers), competition from an increasing blue tit population, increasing predation from great spotted woodpeckers who raid their nests. Other studies done by the RSPB indicate that the drying out of woodland may also be a contributory factor. None of these can be ruled out as factors in the decline of the species, whether in Wytham or more generally in the UK.

### 9.3.2 Other losses

The willow tit is not alone in being a lost former resident breeder. Gosler's 1990 review also listed corn bunting and tree sparrows as regular breeders on Wytham farmland. It is possible that even by that date they had ceased to be regular breeding species, although a roost of 27 corn buntings reported near Swinford in 1984 suggested that this species might by 1990 still have been present. Today, corn bunting is almost certainly lost as a breeding species from Wytham and, while tree sparrows have bred in a sparrow nestbox on Wytham Farm in recent years, the species is today very scarce (Section 9.6).

While species such as the nightingale, tree pipit, and willow tit have clearly gone, a few species persist at such low densities that we are uncertain of their status. For example, in 2007, lesser spotted woodpecker was seen drumming at one site, and at the nest at another. It is unlikely that there were ever more than five or six pairs in Wytham anyway, but has it declined? Similarly, while neither species has been recorded for more than five years, the status of firecrest and hawfinch is difficult to state with confidence. Both species are enigmatic and can be very difficult to locate in the breeding season (in fact there are no breeding records of firecrest from Wytham).

### 9.3.3 Gains

The most notable gain is the return and subsequent breeding of the common buzzard. Buzzards probably did not breed in Wytham in the twentieth century (there are several pre-war records of them having been shot), and they were rare at other times of year. However, they became noted more frequently in the Woods through the 1990s and the first nest was found in 2003. The return of this species has little to do with any management changes in Wytham. Rather, it mirrors the recovery across southern

England in the 1990s that occurred in the wake of the red kite reintroductions in the Chilterns in the early 1990s. At this time, significant efforts were made to protect all raptors from persecution in the introduction area. To date, red kites themselves have not nested in Wytham, although the habitat is suitable. It is therefore high on the expectation list for the next few years. Another persecuted species, the raven, is now increasing across southern England, has been seen in Wytham and may one day breed there.

Another welcome recovery in Wytham has been that of the barn owl, largely lost during the 1970s and 1980s. In 1990, Gosler noted that there had been only one record since 1963 and that had been on Rough Common in 1979. Since 2000, estate workers and researchers have spotted barn owls frequently hunting over Rough Common, Upper and Lower Seeds, and farmland to the north of the sawmill; barn owl feathers were used to line a blue tit nest in 2003. It would be surprising if the species didn't also hunt over Wytham Park.

## 9.4 The Wytham Tit Study

### 9.4.1 Introduction

The long-running studies of tits, chiefly of great tit (Plate 9) and blue tit in Wytham is the bird study for which Wytham is best known. This work was started by David Lack shortly after he became Director of the EGI. One of his earlier studies, carried out while he was a school-master at Dartington Hall, Devon, had been of European robins (Lack 1943). As with all good studies, this one raised lots of new questions which needed to be answered. However, as anyone who has ever tried to hunt for robin nests in woodland will readily attest, they are extremely difficult to find.

After being appointed Director (in September 1945) Lack visited the Netherlands where he met H.N. Kluijver (later spelled Kluyver) and visited the area where he studied great tits. Lack was immediately impressed by how easy, compared with robins, it was to study a bird that nested in nestboxes, the birds find 'our' nests, rather than us having to find theirs! In fact, while Kluijver may rightly be considered the father of all the current European tit studies, he had himself taken over a still-earlier study. H. Wolda, a forester, erected boxes in Oranje Nassau's Oord in 1912 and kept records of the tits in them. The boxes were put up by foresters who hoped that by increasing the numbers of tits, they could reduce the numbers of defoliating caterpillars in the forests and so improve the growth of trees. Curiously, the seemingly important issue of whether increasing tit densities reduces pest abundance is still not fully resolved. Despite maintaining a long run of records, Wolda himself published little from his study; it was left to Kluijver to put the tit studies onto the world stage. Using nestboxes, it is relatively easy to make observations on numerous pairs of birds such as tits and pied flycatchers, which is why these species have now been intensively studied in so many countries. However, set against

# David Lambert Lack (1910–1973)

David Lack in about 1964. Photo by courtesy of Andrew Lack.

David Lack was the leading British ornithologist of his time, though he also achieved great success as an evolutionary biologist, ecologist, and population biologist. Like other field-oriented biologists of his generation, he found university courses unstimulating (he read Zoology at Cambridge). He became a school teacher at Dartington Hall where he was able to get time off to travel, do field work, and make professional contacts. It was while there that he made his ground-breaking study of robins. The *Life of the Robin* was published in 1943, but 10 years later re-published as a Penguin book, a measure of its popularity.

In 1937 he went on an expedition to the Galapagos and in 1947 published his classic study *Darwin's Finches*. Lack became increasingly interested in the relation of Darwinian natural selection to the manner of population regulation, concluding that density dependent forms of control dominated. His ideas on speciation, ecological isolation, group selection, migration, and the evolution of reproductive strategies are best summarized in his two most influential books, *The Natural Regulation of Animal Numbers* (1954), and *Ecological Adaptations for Breeding in Birds* (1968).

He was appointed Director of the Edward Grey Institute of Field Ornithology at Oxford University in 1945. He planned to extend his study of robins, since, as with all good work, his earlier studies had raised many new questions which needed to be answered. However, soon after his appointment he visited the Netherlands where he met H.N. Kluyver and visited the area where he studied great tits. Lack was immediately impressed by how easy, compared with robins, it was to study a bird that nested in nestboxes. Thus began the long-running studies of tits, chiefly of great tit and blue tit in Wytham, described in this chapter. This is the bird study for which the Woods are best known, being continued to the present day. The Dutch and Oxford studies rapidly became famous and spawned many other studies of these species across Europe.

The fact of climate change has given the study a greater significance than the largely academic interest that it might once have had, since the Wytham tit data, which stretch back long before climate change became recognized as an issue, offer some of the best insights available from any wild population anywhere in the world into what is happening to natural systems.

the immense practical benefit of using nestboxes, some caution is required: nestboxes are not exactly the same as natural cavities, particularly in relation to parasite loads and risks from predation; the latter can influence the population's dynamics subtly, and this must always be remembered when using the results of nestbox studies to address ecological questions in the wider world. In practice, however, this is not as great an issue as it might seem, as will be shown below.

So it was that in the winter of 1946–47, a new student at the EGI, John Gibb, put out 100 nesting boxes in and around Marley Wood, part of Wytham Woods. Apart from some small changes (the numbers of boxes were increased to 200 three years later and concentrated in Marley), Gibb studied the birds in great detail in this one area of Wytham, and became the first EGI student to obtain his D. Phil from a study of the breeding biology of the tits (Gibb 1950). When Gibb later moved to the Brecks in East Anglia to study tits in conifer forests, the Wytham tit records were maintained by Denis Owen until 1957 when Chris Perrins joined the EGI. David Lack and Chris agreed that for his project he should study the factors affecting clutch-size, so developing an earlier idea that Lack had had about the evolution of clutch-size in birds (Lack 1947/48). In order to do that, Perrins needed more nests than were available in Marley. So, in the winter of 1957–58, he put up boxes in two areas of the Great Wood, and in the following years extended these to other areas of the wood. By 1963, the EGI had just over 1000 nestboxes throughout almost all areas of Wytham, a pattern that has essentially remained constant (apart from replacement of boxes where trees have fallen down) until almost the present time. For purely practical reasons, the nestboxes in different parts of Wytham are named in 'rounds' (for example, Marley, Singing Way, Common Piece etc.) of which there are 9 in all that have now been monitored for nearly 50 years.

An important aspect of nestboxes that might bias data collected using them (as alluded to above) is that, compared with holes in tree trunks, they are conspicuous and easy to find. If allowed to, weasels raid nestboxes and take the broods, plus the female if they find her there. The weasel predation is interesting ecologically in its own right. For example, they raided the boxes more frequently in years when there were more tits nesting in the boxes, and also when small mammals were scarce: both observations that one might expect from an efficient hunter trying to maximize its chances of getting prey (Krebs 1970; Dunn 1977). However, it also became obvious that weasels rapidly learned that nestboxes were both easy to find and worth visiting, and so they would raid all the boxes in their territory. The birds nesting in the nestboxes suffered higher losses than they would have done in natural tree-cavities and weasel predation of tits was very heavy in the late 1960s. At that time some members of the EGI were looking at heritable characteristics of great tit breeding. One of the studies was the first to show heritability of clutch-size in a wild bird Perrins and Jones (1974). The weasel predation made this study almost impossible, and so in the early 1970s a switch was made to using the German-made hanging boxes that are still used today (Fig. 9.1), since these are largely weasel-proof. These boxes are made of cement and sawdust (woodcrete), are very long lasting, and they have the extra benefit of saving on the replacement of large numbers of wooden boxes that were wrecked each winter by great spotted woodpeckers.

Over the years, additional nestboxes have come and gone. For example, John Krebs added nestboxes in small copses on the Wytham estate to the south of the main woodland block (notably Higgins Copse) in the late 1960s in order to study the role of song and territoriality in determining the dynamics of great tit populations. These were maintained only into the early 1970s. These were replaced in the late 1980s together with more

**Figure 9.1** A weasel-proof hanging nest box of a type used in Wytham since the early 1970s. It is made of cement and sawdust (woodcrete), is very long lasting, and saves on the replacement of wooden boxes that are destroyed each winter by Great Spotted Woodpeckers.

than 150 others in outlying copses, hedges, and gardens by Roger Riddington and Andy Gosler, who were interested in movements of the tits into and out of the wood, and in comparing breeding success in these different habitats (Riddington and Gosler 1995; Verhulst *et al.* 1997). At the time of writing, some of these 'external' boxes in copses are being maintained by Teddy Wilkin for further studies of movements in relation to population genetics, habitat variation, and breeding success.

Partly so as not to interrupt the long-term run of population data for the great tit in Wytham, Ben Sheldon (the Director of the EGI since 2003) focused his research on the blue tit when he joined the EGI in the late 1990s. However, severe competition from great tits (which usually win in competition for nest-sites with the smaller blue tit) in some years meant that the number of blue tit pairs available for study might not have been adequate for all the research he had planned. So in 2002, he added 200 nestboxes with smaller holes (so excluding great tits) to the stock of 1016 nestboxes that were accessible to all tit species.

### 9.4.2 Populations and survival

Records of the numbers of breeding pairs of great and blue tits in Marley Wood (see Plate 10) have now been maintained for 60 years, probably a longer run of data for a single place than anywhere else (Fig. 9.2), and for more than 40 years for the rest of Wytham. There is no very marked trend in the great tit population in Marley over this time, although there have been slightly more pairs in the last few years than the average over the rest of the study. In contrast, the blue tit population rose steadily through the earlier part of the study, but seems to have levelled off since about 1980. There can be marked year-to-year changes in numbers, with the populations sometimes even doubling or halving in successive years. In demographic terms, these changes come about largely as a result of marked variations in the survival of the young birds; in the case of the great tit, as few as 3.5 per cent and as many as 22.2 per cent of the fledglings from one year have survived to breed in the boxes the following year, a seven-fold variation. Survival of the adults varies in parallel with the survival of the chicks, but is not so marked.

A number of factors affect these survival rates. Perhaps surprisingly, the severity of the winter does not seem very important, though extreme cold, coupled with snow cover that makes feeding on the ground impossible, can affect survival, especially for great tits which feed a lot on the ground. Plentiful food is important and one food—beech seed—seems key to how well the birds will survive the winter. The beech is one of the trees that 'masts', that is, it has years in which many of the trees have large quantities of seeds and others when no trees produce seeds (Fig. 9.3). Good mast years apparently occur in years following a hotter than average July (Matthews 1963). In Wytham, in addition to synchronizing with beech trees elsewhere in the UK and even continental Europe, areas of beeches seem to have sub-cycles within this such that, for example, Blenheim beeches might mast well in years when Brogden's Belt beeches do not and vice versa. It is not known how the trees synchronize in this way. However there is no doubt that synchrony in masting is beneficial to the trees themselves, since by swamping seed predators, such as the tits and small mammals, the trees are more likely to have some seeds left to germinate. Perhaps even the non-masting years are also important to the trees: by 'ganging-up' to have no-crop years, the trees may lower the number of potential predators.

The presence or absence of beech mast is the most conspicuous environmental factor with which changes in tit numbers are associated since, broadly speaking, tit

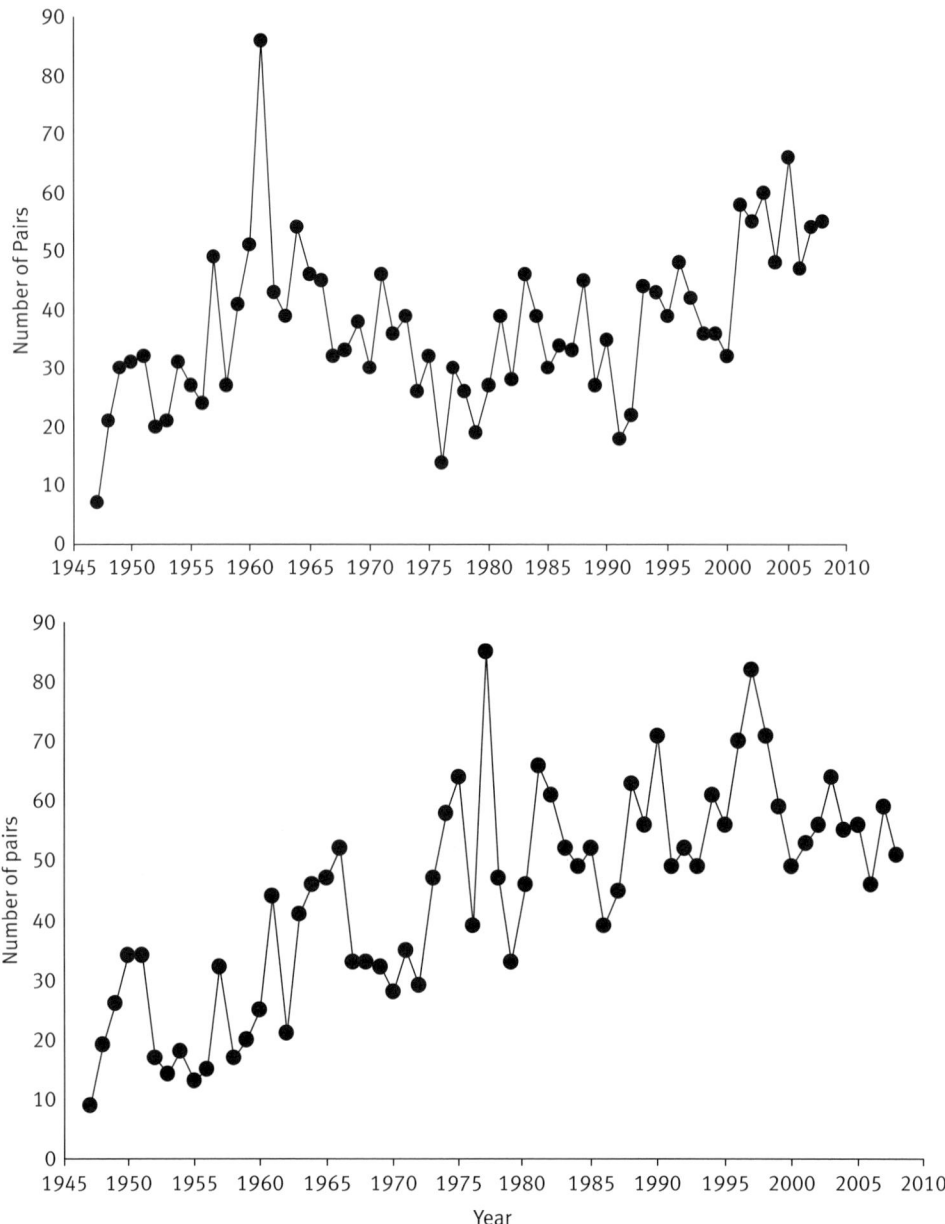

**Figure 9.2** Great tit (top) and blue tit (below) population trends Marley Wood, 1947–2007.

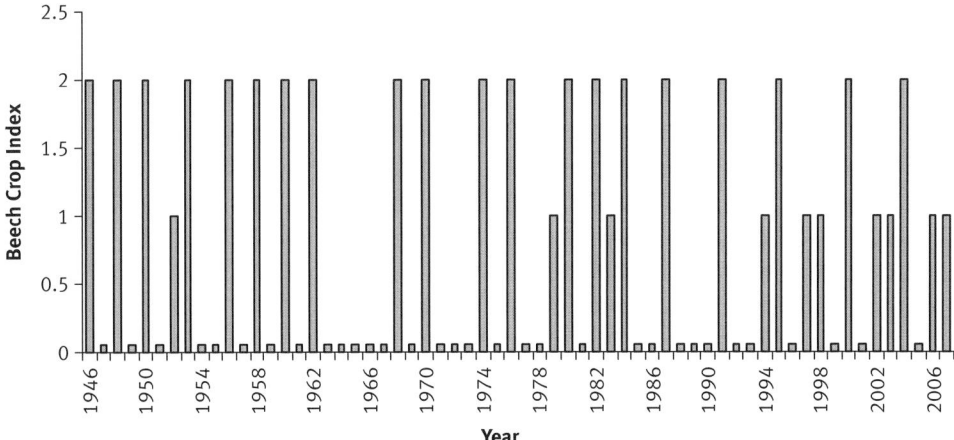

**Figure 9.3** Occurrence of beech mast. Years are scored as zero (none), 1 (intermediate), and 2 (good crop). In winters with a good crop the tits survive better than in those when there is no crop; good crops do not occur in successive years. Since 1946 there have been 20 years with plentiful seed and 32 years with none.

populations are higher after winters with a beech crop, and lower after ones without one. However, although these changes in tit abundance are associated closely with differences in survival of adults, and more so of juveniles, in the latter case at least, most of the mortality associated with these population changes has already occurred in late summer, well before the tits move onto mast as a food source in the winter.

Juvenile survival through the summer is closely related to the food supply available to their parents when provisioning the brood at the nest. When food is plentiful, the chicks leave the nest heavier and more of them survive to the following season. Hence population changes from year to year are driven to some extent by the abundance of prey in the breeding season.

Within the fledglings, those that were heavier when they left the nest are more likely to survive than those which were lighter in weight. This seems to be not just because they have more reserves; larger birds tend to become dominant in the summer flocks of fledglings and this dominance may enable them to get the best food. However, the individual's position in the dominance hierarchy is also affected by the date at which they leave the nest. Being earlier enables them to become established in the flocks before the late-comers arrive; the later-fledged young must compete with these older, more experienced juveniles. So it seems sometimes to be better to be a light, but early, chick rather than a later but heavier one.

In the early part of the twenty-first century, personality traits have been receiving much attention from researchers abroad, especially in the Netherlands and the US. In the EGI, Ben Sheldon, John Quinn and students, have been assessing great tit personalities. Simplifying greatly, there appear to be two sorts of personality and these are apparent in the exploratory behaviour of the birds when placed temporarily in an aviary: 'fast' birds, tend to be bolder, more dominant, more competitive, take more risks, and disperse further, while 'slow' birds are shyer individuals that tend to show the opposite kinds of behaviour to all those shown by fast birds (Quinn *et al.* 2009). Clearly these two kinds of individuals are likely to be favoured differently under selection in different years, which differ in food availability and bird (competitor, predator, and mate) densities, and there is recent evidence that these differences may be linked to specific genetic differences between the birds and with the types of territory that they acquire.

### 9.4.3 The breeding season

Many aspects of the breeding biology of both blue and great tits are closely associated with the summer food supply. The tits have large broods and need a plentiful supply of food if they are to raise their young successfully. The main prey for the chicks is caterpillars from the oaks and in particular winter moth (*Operophtera brumata*) and green tortrix (*Tortrix viridana*). In Wytham the early tit-workers were very lucky that these were being studied by George Varley and George Gradwell in the Hope Department of Entomology (Chapter 8, Section 8.16), one of the big advantages of working in a University with strong groups of field-workers. They provided a strong basis for understanding the timing and abundance of the caterpillars and, since then, similar valuable data on the caterpillars have been collected for many years by Lionel Cole.

The abundance of these caterpillars, and the period during which they are available, varies markedly from year to year and their abundance is important to the tits. The numbers available for the parent tits to raise their chicks vary in three ways. First, their numbers simply vary; there may be many or few of them in any particular year. Second, because there are early springs and late ones, the date at which the caterpillars become available for the birds also varies; they have been present as much as a month earlier in some springs than in others. Lastly, because they are cold-blooded, the caterpillars tend to grow much faster in warm than in cool springs. In a very hot spring, the winter moth larvae may complete their development in as little as three weeks, in a cold spring they may take seven and so be available as prey for the tits for much longer.

This year-to-year variation in the timing of prey availability and its abundance has a strong influence on two key breeding parameters of the tit populations, namely the 'lay-date' (date on which the first egg of the clutch is laid) and clutch-size (the number of eggs laid in a single clutch by each female). Whilst there is some evidence that both of these characteristics has some genetic component (that is, females that lay early, or lay large clutches are, more likely than random themselves, to have come from an early or a large brood, respectively), there are also strong environmental determinants of both parameters (and especially of lay-date).

The normal clutch of great tits varies from about 5 to 11, while that of blue tits is about 6 to 12; larger clutches—up to 16 in blue tits—occur in many years. In order for the parent birds to be able to provision such large broods it is crucial that they have their young in the nest when the caterpillars are most abundant. Even then, the parents are often unable to completely satisfy the brood. One outcome of this is that the larger the brood the more mouths the food has to be divided between, and so the young tend to be less well nourished in larger than in smaller broods. Young from large broods leave the nest lighter in weight than those in smaller broods and have a lower chance of survival. The most successful broods are those in which the largest number of young survive to become breeders.

Clutch-size declines through the breeding season; this can be seen in terms of the female tit trying to increase the fledging weights (that is, quality) of her brood so that they will survive better after fledging. There may be many reasons why a particular female lays late in any year, but the ultimate limitation is that she must have sufficient food herself to form a clutch, which is demanding in terms of energy and nutrients (especially protein and calcium), and she must try to predict when the peak caterpillar availability will occur and lay her eggs in time to have them hatch near to this time. Both these constraints may differ between territories, for example because of differences in the densities of oak trees, and this sets up much of the variation in timing that contributes to chick survival. This may also help explain why the birds lay smaller clutches later in the season; a reduction in clutch size allows the laying female to accelerate the hatch-date of her brood. She can only lay one each day and so laying one egg fewer also saves a day. The incubation period (which typically starts on the day the last egg is laid) is relatively inflexible at 12–14 days (typically 13) and must also be accounted for in her 'prediction'.

This ability of female great tits to predict the peak in prey availability is remarkable, and how they do this has still to be explained. Note that their prediction is not only of the timing (phenology) of caterpillar emergence, but also of their abundance, since on average, the tits lay larger clutches in years in which there *will* be more caterpillars. It is theoretically possible that both aspects might be predicted partly through a response to the spring weather itself using a simple rule-of-thumb such as 'lay early if it's warm', and partly by monitoring the availability of small, recently-hatched, caterpillar instars, or even lepidopteran eggs on twigs, but nobody knows for sure that this is how they do it. That they do not have perfect predictive ability is well demonstrated by data from recent years in which the birds have mis-timed laying because of unusually warm or cold periods of spring weather.

As a result of climate change, the springs are getting warmer earlier. This has led to advancement in the breeding season of about 2–3 weeks since 1980 (McCleery and Perrins 1998), but broadly the birds have shown themselves able to track the changes in temperature (Charmantier *et al.* 2008). One other effect of climate change is that the weather is less predictable, that is, more chaotic, than it used to be. This means that in an evolving system (the birds' breeding parameters) recent past events, which contributed through natural selection to the genetic variation underlying variation in the traits of laying-date and clutch-size, may not be a good guide to future conditions. The population's gene pool is always running to catch up with environmental conditions, and if the climate does not stabilize, it will never get there.

The evolution of this finely-tuned system has been a focus for research over many years, but climate change has given it a greater significance than the largely academic interest that it might once have had. The Wytham tit data, which stretch back long before climate change became recognized as an issue, offer some of the best insights available from any wild population anywhere in the world into what is happening to natural systems.

### 9.4.4 The sparrowhawk

A major predator of the tits is the sparrowhawk (Plate 11), whose main habits are described in Section 9.6. However, some aspects of these studies relate directly to the tit population study and so are covered here. People who feed birds in their gardens are sometimes incensed by the way in which small birds are taken from bird tables by these hawks, which have frequently been accused of reducing bird populations. For some years members of the EGI erected pylon hides at the hawks' nests to see how many tits they brought in to their nestlings. At times the hide was placed *against* the side of the hawk's nest and by using 'lazy-tongs' to 'borrow' the prey from the hawks it was possible to read the rings and so find which bird had been taken. This allowed a much more detailed understanding of the effect the hawk was having on the fledgling tits; at times they might be removing as many as 30 per cent of the fledglings in the six weeks or so that the young hawks were dependent on their parents. Of course, the hawks preyed on tits the whole year round; so what effect were they having on the population?

By chance there is a good way of addressing this problem. In the late 1950s, certain seed-dressings, such as DDT and Dieldrin, were being used to control insect damage. These were also being picked up by seed-eating birds such as finches and, as was later discovered, being accumulated in birds of prey to levels which interfered with their ability to breed, or even killed them. The sparrowhawk was one of the most seriously affected birds and the populations in many areas of the UK collapsed. They dropped from six or more pairs in Wytham to none after 1959; they were virtually extinct as a breeding species in Oxfordshire. These losses of birds of prey (peregrine was another species seriously affected), highlighted the problem and led to severe restrictions being placed on the use of many of these substances. As a result, the populations recovered and, by the early 1970s, sparrowhawks were breeding again in Wytham. One outcome of this was a sort of natural experiment: the long-term study of the tits started when sparrowhawks were present, went on through a period of ten or more years when they were absent and continued after they had returned. The results were quite surprising. The tit populations did not increase while the sparrowhawks were absent (Fig. 9.4a). The one very high year (1961) arose from a very heavy beech crop combined with an unusually mild winter. Since the hawks take parent tits while they are busy searching for prey for their young, one might have expected that adult survival would have been higher in the absence of the hawks, but this was not so (Fig. 9.4b). Faced with these rather counter-intuitive findings, some people have thought that Wytham might be an island of good habitat where

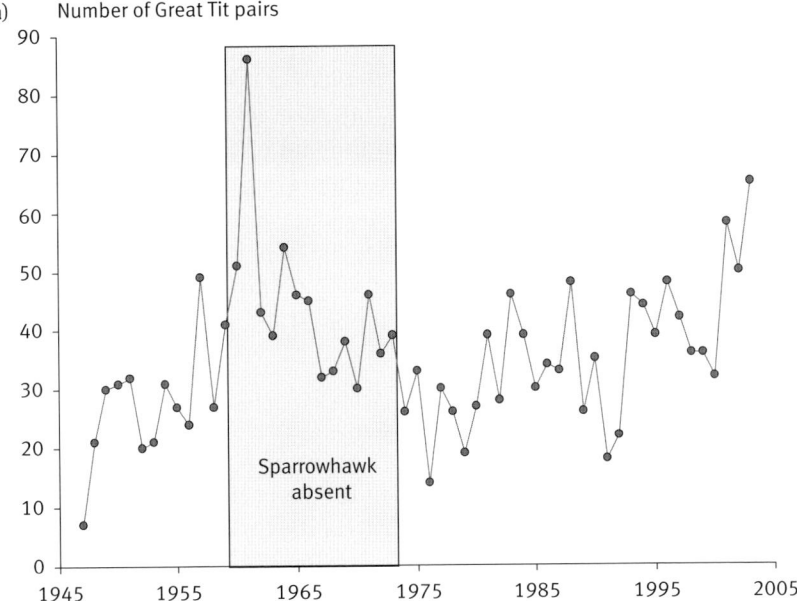

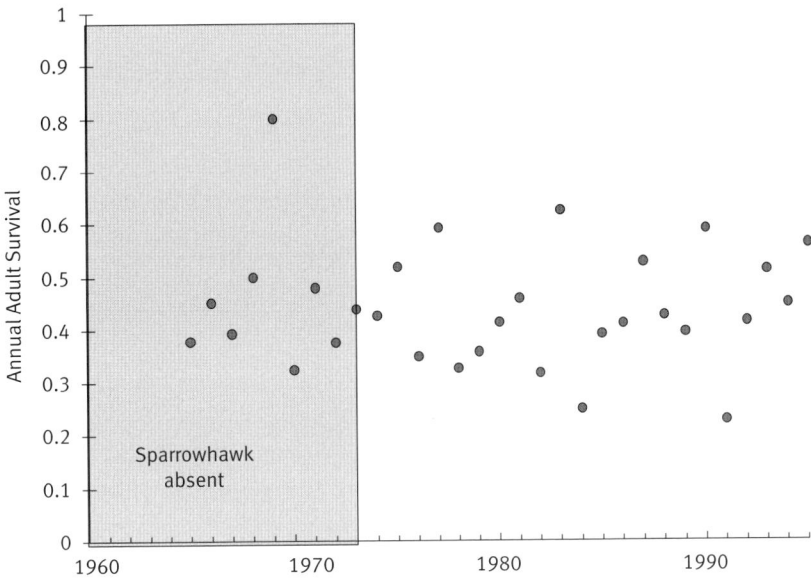

**Figure 9.4** (a) Number of pairs of great tits in Marley Wood 1947–2005 in relation to the presence and absence of sparrowhawks and (b) adult survival rates of great tits with and without sparrowhawks.

birds would prefer to breed and so they would continue to move in there (so that numbers would not change) while the hawks were lowering the numbers elsewhere. However, this is not true because the proportion of the breeders in Wytham that were born outside the wood did not increase when the hawks were present.

The explanation for this is that although the hawks take a lot of tits, there are even more tits there. The average pair of tits may, say, raise seven young in the spring; there are then nine tits (including the parents) at the end of the season. If the hawks take about one third of the fledglings, roughly two, there are still seven left. The hawk predation, in addition to other mortality factors, would have to lower this number to below two by the following spring, if it were to reduce the tit population. The hawks may alter the cause of death and timing of deaths, but they do not control the overall numbers that die. This is controlled by competition amongst the juvenile tits; we say that this mortality is density dependent.

### 9.4.5 Weights

While the sparrowhawks have not reduced tit numbers significantly, their presence in the woods has a very large influence on the lives of the tits in a completely different way. Small birds such as tits have a daily weight cycle; they lose weight overnight and put it on again during the day. This daily weight gain is in the form of fat which enables them to survive the next night. In mid-winter, they have just about eight hours to store sufficient fat to survive the sixteen hours of darkness. Hence they live 'close to the edge', needing to do this every day or face starvation in the night. It would seem sensible for them, if they could store more fat, to do so; in this way they might be able to survive more easily through a very cold winter night as they would carry a bit of extra fat to draw on.

However, studies of winter fattening in the great tits as part of the ongoing winter ringing programme of the tit study, showed that when conditions became easier (for example, when there was more food available or ambient temperature was higher), the birds carried *less* fat, not more (Gosler 1996). Furthermore, the more dominant individuals (for example, adult males), with greater access to resources, actually carried less fat than did the subordinate individuals (for example, first-winter females). This suggested that the tits treated fat reserves as insurance against starvation, but that, like insurance, it came at a cost and that the cost might be the risk of predation from sparrowhawks (Gosler *et al.* 1995).

The hypothesis was simple: those heavier, small birds would be less manoeuvrable and so more easily caught by sparrowhawks. So, the tit has to balance the need to avoid starvation by maintaining adequate fat reserves, while at the same time guarding against the risk of predation as a consequence of being too heavy. It does so in a number of circumstances, but the most striking is that the birds are lighter (that is, have less fat) in years when beech-mast is available as food, than they are in non-beech years. In beech years it is easier for them to find food so they have less need to store food. Further evidence to support the hypothesis comes from the period when the sparrowhawks were absent. In the absence of this threat of predation, the tits were heavier (Fig. 9.5): the balance of starvation risk and predation risk had shifted towards avoiding starvation by laying down more fat.

Further experiments using plastic sparrowhawks flying along washing lines, to increase the birds' perception of the risk at a feeding station, have shown that it really is

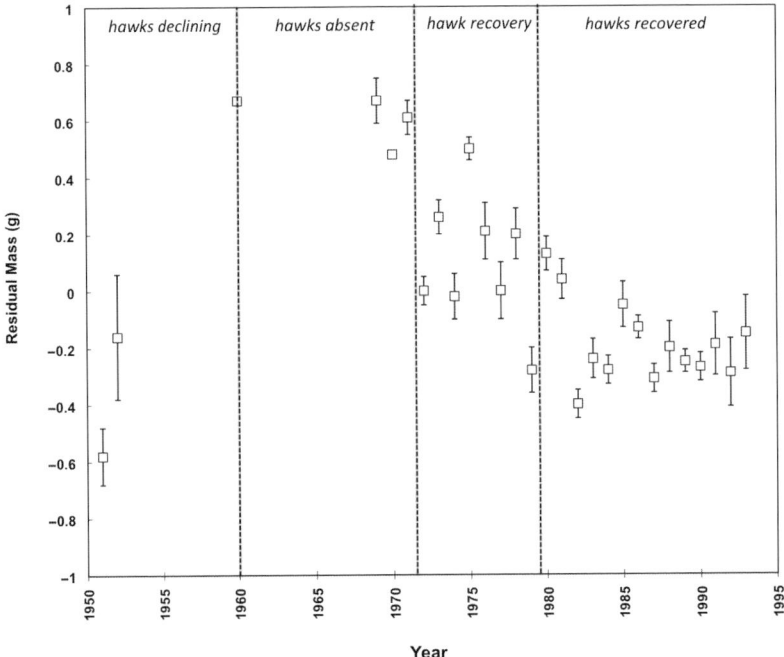

**Figure 9.5** Weights of great tits with and without sparrowhawks (redrawn after Gosler *et al.* 1995).

their perception of risk, and not the physical exertion of being chased about by hawks, that results in their lowered fat reserves (Gentle and Gosler 2001; Macleod *et al.* 2005).

### 9.4.6 Great tit song

The great tit's song, described as 'teacher-teacher', or less flatteringly as 'like a squeaky bicycle pump' is one of the most distinctive spring sounds in the Woods. With almost all the woodland occupied by territorial pairs, it is hard to escape this song as the male loudly proclaims ownership. This song was much studied by John Krebs and his students. The birds sing from early in the New Year onwards until they are raising young, there is also a brief resurgence of song in autumn after the birds have moulted. They sing most loudly just after dawn; by playing tapes of a particular individual's song to selected great tits they showed that the birds recognized their neighbours and, after a while, react less vigorously to their songs than they do to songs of strangers. After initial disputes, established territory holders come to accept their neighbours, but still react strongly to an interloper who may be trying to take over their territory.

Song analysis showed that there are many elements to a song and each bird has a repertoire drawn from this range; an individual may have as many as nine of these. There is some evidence that the better birds (as judged by their breeding success) have a slightly larger repertoire size (McGregor *et al.* 1981).

When singing against a neighbour, a bird may use a range of these elements, often switching to match those that the rival is singing. The different elements have different acoustic properties, such as how far they carry, and for any element these differ between habitats, so that, for example, great tits singing in dense coniferous forest use different elements from those in more open parkland (Hunter and Krebs 1979).

The male has not only to defend his territory, but also his paternity. Although one regards the great tit as monogamous, and indeed it is extremely rare to find more than a single female and a single male at each nest, studies of the DNA of chicks in the nest have shown the position to be more complex. In 18 per cent of the nests one or more of the young has been fathered by a different male, usually one in an adjacent territory (Chapman pers.comm.). This comes about because, during the fertile period—which starts about four or five days before she starts to lay her clutch—the female will accept matings with these males; indeed it sometimes seems as if she actively seeks out copulations with more dominant males in adjacent territories. Not surprisingly, therefore, males guard their mates during this fertile period in order to maximize the number of the young that they father. This involves remaining closely with her throughout the day and seeing her to her roost (often in the nest in which she will lay her eggs) at dusk and being present before she leaves in the morning (Mace 1987).

### 9.4.7 From field and to aviary

Another very common feature of woodland in winter is the tit flocks; usually these are composed of great, blue, coal, and long-tailed tits, often with additional species such as nuthatch and treecreeper, and perhaps goldcrest. Many students have studied flocking behaviour in the field, but so many things are going on at once that it is difficult to make precise observations. One solution has been to take small groups into the aviary where they can be watched in greater detail, at close range, in order to understand the behaviour of the birds in these flocks.

There are a number of advantages of flocking over foraging on one's own. These include an increased likelihood of spotting a predator—many eyes increase the chance. It is also often advantageous for finding food. If one bird finds a new type of food and starts to feed on it, the other members of the flock will rapidly start searching in similar places. It is not equally advantageous to all birds to be in a flock from the point of view of finding food; it depends on what sort of food it is. The birds in a flock have a dominance hierarchy or peck-order in which older males are at the top and the youngest females at the bottom; young males and older females are intermediate. When they are feeding on small items, which each individual can rapidly consume, then all birds get to eat everything they find. However, when the items are larger, such as a beech nut which takes a tit some time to open and consume, then the subordinate birds can lose out. Before they have finished opening the seed they may be supplanted by a dominant and so lose the meal.

The feeding flocks move at varying speeds through the wood. Many detailed studies have been made of birds in aviaries to understand what influences the way in which they forage. The birds are supplied with prey at different densities (they have to search for them because they are hidden from sight), which affects the rate at which they can continue to find food as the amount remaining diminishes. In the field, when searching for beechmast, the tits change their preference as the winter progresses and the profitability of foraging under a particular tree changes as the seed supply declines. Early in the winter they prefer to feed under trees with good cover nearby to avoid sparrowhawks. Later in the winter, however, they must forage at more risky sites (Walther and Gosler 2001). From these studies in lab and field it has been possible to learn a lot about how the tits decide whether to keep searching in a patch or to move on.

### 9.4.8 Storing seeds

Seeds are an important part of the diet of the tits outside the breeding season, as described for the great tits and the beechmast. One way of making the most of more food than one can eat at the time is to store it. Great and blue tits do not store food, but coal, marsh, and willow tits do. Most of the work in Wytham has been concentrated on the marsh tit because it is a prolific hoarder. The birds show a remarkable ability to store food and remember where it is.

This work has also contributed more widely to our understanding of the brains of birds and how they work. Observations made by the bird through one eye are stored mainly in the half of the brain on the other side of the head. The information is stored in the hippocampus and the hippocampus is relatively larger in species which store food than in species, such as the great and blue tits, which do not (Krebs *et al.* 1996).

## 9.5 The future—where next?

It is the minutiae of the birds' lives, which studies such as these continue to uncover, that have made the Wytham Tit Study so productive. However, such detailed investigations would have been very much weaker had they not been conducted within the framework of a long-term population study. This is because for the vast majority of great tits in Wytham, each bird is known: data are available on when and where they were hatched, who their parents were, and very often their parents before them; what they weighed at fledging, and a range of other details on body size and condition at various stages through their lives.

In recent years, ever more sophisticated approaches have been applied, for example in relation to population genetics, than could previously have been hoped; for example, the sex of nestlings is now routinely determined by DNA analysis of tiny blood samples taken from a small feather plucked from the birds. Other researchers have returned to look at some questions related to understanding of the effects of large-scale environmental changes on birds. For example, Teddy Wilkin (Wilkin *et al.* 2006) found that the actual density of oaks in a bird's territory was critically important to a pair's breeding success. Further, the distance between the pair's nest and the edge of the wood was also important. This edge effect has important implications in a world in which woodlands are often fragmented by other habitats. Significantly for Wytham, there are signs that deer browsing (Chapter 5, Section 5.6) may affect the breeding success of Wytham great tits, and this seems to be related as much to the quality of the oak tree canopy (where most food for nestlings is found) as it is to the understorey.

People sometimes ask: when will the work on the great tit be completed? 'Surely' they say 'there can't be anything more you can learn after all these years'. However, as theory develops, new questions arise. The new questions are, themselves, intimately linked with both the development of technology that makes new studies possible, and to environmental changes that expose the Wytham great tit population (and others) to new challenges. For example, the phenomenon of inbreeding depression (that is, the poorer reproductive success of closely-related individuals breeding with each other) was first studied in the Wytham great tit population some 30 years ago (Bulmer 1973). Then, the identification of close relatives was based on ringing records: chicks ringed in the same brood were assumed to be siblings, and the adults feeding them were assumed to be their genetic parents. Whilst this is very reliable in the great tit, females do undertake extra-pair copulations and so not all the chicks in their nest are necessarily offspring of the male who helps to raise them. The issue has recently be investigated again by Marta Szulkin, using both pedigree (Szulkin *et al.* 2007) and DNA evidence to determine the degree of relatedness. She demonstrated that inbreeding has substantial consequences in terms of breeding failure, indicating that there will be strong selection against inbreeding and in favour of mechanisms that favour the avoidance of kin.

DNA (in this case the detection of the DNA of parasites in the blood) has also been important in Matt Wood's work on the incidence of avian malaria in the great tit and blue tit populations. His studies, which have shown malaria to affect many of the birds nesting near to the River Thames, constitute some of the first investigations of disease in this population (Wood *et al.* 2007).

Two further studies indicate how developments in technology and theory are being used at the time of writing (2008) to further the understanding of the tits' ecology. Post-fledging and subsequent movements of the tits could previously only be studied by retrapping them frequently to read their rings. Teddy Wilkin is now fitting tiny PIT (Passive Integrated Transponder) chips to plastic leg rings on all the great tits. Both the adults and chicks in the nest received these in 2007. These chips can be detected and read at special feeders set up in, and moved around, the wood every few days. The data are then downloaded periodically, to give a picture of 'who' is visiting where for food. So far this study has largely confirmed what was believed to be the case about the distances that the birds travelled in the first six months of life, but that in itself is very valuable, because it has a bearing on almost every aspect of the species' ecology.

The second study is being conducted by Andy Gosler, and concerns the function of the reddish pigment spots on great tits' (otherwise white) eggs. The distribution and intensity of these spots is related to the thickness of the shell, but variation in both characteristics is related to the availability of calcium to the female during egg formation in April (Gosler *et al.* 2005). Thanks to the varied geology of Wytham, calcium availability varies massively between sites on clay in the Thames Vale, and on the limestone of Wytham Hill. The tits obtain their calcium by eating small snails, whose densities reflect that difference in calcium availability. Detailed records of eggshell pigmentation collected over 20 years in Wytham, indicate that the pigment pattern, and by implication, the shell thickness, has changed. It appears that reduction in eggshell thickness (so far 7 per cent since 1988) is related to calcium leaching of Wytham's soils due to exposure to acid rain. Hence, even now, the great tit study can alert us to aspects of environmental problems that might otherwise have gone undetected. Because of the Wytham studies, the great tit has sometimes been described as the ecologist's lab rat. Perhaps the miner's canary might be a better analogy.

## 9.6 Other Bird studies in Wytham

### 9.6.1 Owls

With such a large piece of woodland on Oxford University's doorstep, it is hardly surprising that Wytham has been the focus of many other studies of woodland birds.

The second longest bird study in Wytham is the one on the tawny owl, studied by H.N. (Mick) Southern from 1947 to 1959) (Southern 1970) and then by two students Graham Hirons (Hirons 1976) and Bridget Appleby (Appleby 1996). After the very hard winter of 1947, when many owls succumbed to the cold, there were only 17 pairs. Thereafter, the number rose fairly steadily, reaching 30 pairs in 1955 (equivalent to roughly 18 ha each) and remaining remarkably stable from then until the end of the study (30 and 32 pairs in 1958 and 1959 respectively). Later, Hirons (1976), recorded populations of 30 and 31 pairs in 1973 and 1974 and Appleby again recorded closely similar numbers in 1993–95.

Southern's study uncovered a lot of information about the owls' main prey; this was done by collecting the regurgitated pellets of the owls which contain the bones of their prey. They fed mainly on two species of small rodents, the wood mouse and the bank vole. By putting small rings on the mice and voles and retrieving the rings from the regurgitated pellets, Southern showed that each pair remained strictly within its own territory. So important were these two prey to the owls that their breeding was closely related to the fluctuations in numbers of these two species (Fig. 10.1); in years with good numbers of voles or mice, the breeding success of the owls was high, many pairs laying three eggs and raising a high proportion of their chicks. In contrast, in years with few rodents the owls fared badly; indeed in 1958 when the rodents were very scarce, no owls attempted to breed, not a single egg was laid. The owls also take many earthworms, much harder to recognize in their pellets, and very few tits.

Hirons (1976) showed that some of the territories were occupied by three birds, two females and a single male. This was not noticed before because the two females could only be distinguished on the basis of their calls and this required careful recording and comparison of their voices.

The main reason why the owl population remains so stable is that, unless there is a very hard winter and the adult birds are unable to find sufficient food, adult survival is quite high (about 80 per cent survive from one breeding season to the next). In September, the new juveniles (the fledglings are cared for by their parents throughout the summer) vigorously vie for any 'spaces' in the territories, rapidly filling any available vacancies. If they are unable to obtain one, the young birds may wander further afield and, if still unsuccessful, eventually perish.

### 9.6.2 Sparrowhawks

Geer (1979) and Gray (1987) made detailed studies of the sparrowhawks in Wytham. Despite the fact that the hawks were known to be major predators of small birds, it came as a surprise to discover quite how many they were taking. Geer (1979) estimated that, during the three months of the hawk nesting cycle, the birds took roughly 15 per cent of the adults and 20–35 per cent of the fledglings. In order to be successful, the hawks had to breed early enough so that they had their chicks in the nest when the young tits had just fledged, since these were much more vulnerable in the first week after fledging than later. Tits were particularly important prey when the hawk chicks were small, because at this time the female guards them continuously and so the male has to do all the hunting. Being much smaller (about 150 g), the male largely hunts small birds such as tits and finches. The female (about 250 g) can provide larger meals, such as thrushes and starlings, for the bigger chicks, but she can only start hunting when the chicks are large enough to be safely left on their own.

Gray (1987) studied sparrowhawks outside the breeding season. They hunted mainly along edges, either of the woodland itself or between habitat types within the wood. For great tits he estimated that 36–64 per cent of the adults and 37–40 per cent of the juveniles were taken by the sparrowhawks (and about 80 per cent of all adult and 45 per cent of all juvenile mortality was caused by sparrowhawks; comparable figures for blue tits were 25–31 per cent of adults and 30–31 per cent of juveniles, equivalent to about 40 per cent of all adult and 35 per cent of all juvenile mortality). However, the picture is complicated because in winters without a beech crop (Subsection 9.4.2) the tits may be quite short of food by the end of winter. If the winter tit population is reduced by sparrowhawk predation, the remaining tits may well have a higher survival rate towards the end of winter than they would have had in the absence of the hawks. Overall, the hawks have little effect on the number of pairs of tits in summer.

### 9.6.3 Rooks and jackdaws

One of the earliest studies in Wytham was that made by Jim Lockie (Lockie 1954) on the rooks and jackdaws. This partly stemmed from the studies made during the war by the

Oxford University Research in Economic Ornithology (from which the EGI developed) on rooks and wood pigeons. At that time there was a premium on maximizing the production of food and an urgent need to know the extent of the economic damage wrought by birds, and other animals, that raided the nation's supplies. Lockie looked in detail at the nesting biology and the food brought to nestlings of rooks and jackdaws. April was the best month for the rooks to breed because it is milder than March and the drier ground of May made earthworms more difficult to find while the longer grass made it harder to forage for leatherjackets. Jackdaws bred later, taking food from the grassland surface and caterpillars from the woodland trees.

### 9.6.4 Finches

In 1961, Ian Newton started a study of the finches in Wytham; in this he made important contributions to the understanding of the diets of the different species and of how their body and bill structures influenced their foods and feeding habits. In addition, he studied the moult of several species, and how it related to the timing of breeding. Although in the end his thesis focused on the bullfinch (Newton 1964), his studies of other species culminated some years later in his New Naturalist *Finches* (Newton 1972). At the end of the three years of his D. Phil., Ian stayed on for another three years, concentrating on bullfinches. For those too young to remember, at that time the bullfinch was much more numerous than today, and a serious pest of fruit crops, especially plum and pear, taking large quantities of the swelling flower buds in late winter and early spring, and thus reducing the subsequent crops. In the main fruit-growing areas of England, such as Kent and Worcestershire, bullfinches were being culled in large numbers, but this was not enough to prevent serious damage. Newton showed that the amount of bud-feeding by bullfinches each year was dependent on how much ash seed was available to them in woodland; when there was little seed, the birds started feeding on buds much earlier in the winter, and caused much more damage than in years with good ash crops. Over a period of years in the 1960s, ash cropped biennially, so severe orchard damage occurred mainly in the intervening years when ash seeds were scarce or unavailable. Even when they were eating the swelling buds in mid winter, some bullfinches were actually starving; at that time they were simply unable to get enough energy from the buds to survive, despite the damage they were causing.

Since that time, the bullfinch has undergone one of the most dramatic declines of any English bird, going from horticultural pest to redlisted (57 per cent decline) in just 30 years (Gregory *et al.* 2002). In one two-month period in the autumn of 1965, Newton caught more than 250 different bullfinches in Wytham, limited only by the number of mist-nets he could operate single-handedly. Yet by 1998 when another student, Fiona Proffitt, started a study funded by the RSPB to look at reasons for the reduction in bullfinch numbers, she had to move elsewhere to study them; they were just too rare in Wytham (Proffitt 2002). The reasons for the decline are still not wholly clear, but in Wytham the massive reduction in bramble and other scrub through woodland maturation and deer browsing (Chapter 5.6) greatly reduced the habitat and food sources available to the species.

### 9.6.5 Blackcaps, warblers, and chiffchaffs

Two other species which have undergone marked declines are the blackcap and the garden warbler; in this case part of the reason for the decline is also thought to be the loss of the Wytham understorey (Chapter 5, Section 5.6). These two rather similar-sized species were studied in Wytham in the late 1970s when they were still abundant, by Ernest Garcia. The two are well-known to birdwatchers, partly because many people find them difficult to tell apart by their songs.

In Wytham, Garcia (1981) found that the foraging behaviours of blackcaps and garden warblers were very similar, though the latter showed a slight preference for lower and denser vegetation. The two species were strongly aggressive to one and other and maintained inter-specific territories, each species defending its territories against intruders of both species. While such behaviour is known in other parts of the world, this is the only British pair of species known to behave in this manner.

The blackcap was the more aggressive and tended to be dominant over garden warblers. By using playback of the two species' songs, Garcia showed that both species reacted to the songs of both, again with blackcap doing so rather more strongly. Also, by removing some of the blackcaps from his study plot, Garcia was able to demonstrate that the garden warblers would move into, and successfully occupy, habitat from which they had been excluded; once they had become established, they would successfully defend their territories against both garden warblers and blackcaps. It is therefore possible that under some conditions, an increase in blackcaps could be a significant factor in determining the size of garden warbler populations.

Another warbler which has declined greatly with the loss of understorey is the chiffchaff. This was studied in the late 1990s by Marcos Rodrigues. By that time, chiffchaffs were already scarce in most of the Woods and Marcos concentrated his study around the main track up to the chalet from the sawmill where, perhaps because the passage of many people kept the deer away, there were still quite extensive patches of brambles.

In many species where a single pair occupies a territory, it is known that this does not prevent males from trying to enter an adjacent territory and copulate with the female; by so doing the male leaves more descendents to ensuing generations. Marcos found not only that the males guarded (by staying very close to) their mates in just that short fertile period prior to and during laying, but also that they were more likely to intrude into an adjacent territory if the resident male was temporarily absent.

### 9.6.6 Wrens

One small bird which does not seem to have suffered greatly from the loss of the understorey, despite the fact that it lives low down in woodland, is the wren (Perrins and Overall

2001). As with many other species of wren, the winter wren (*Troglodytes troglodytes*) is polygynous. This bird has been the subject of two major studies in Wytham, the first by Peter Garson 1975–77 and later by Jo Burn (1993–95). Peter Garson described how the males are polygynous, building a nest and guiding the female to it and then going on to build more nests and display to other females. It seems possible that the females choose which mate to breed with on the basis of his nest-building abilities, though the best males may defend better territories as well as building more or better nests. There is a high predation rate on nests and the presence of other nests enables the female to start to breed again almost immediately after she loses a nest. It may pay the second female to breed in the territory of a bigamous male, instead of mating with a male without a mate. If the bigamous male possesses a territory with a better food supply, she may be better able to raise the brood on her own than she would be if she had a mate who would help, but who had a territory with a poorer food supply.

Burn (1996) went on to show that females might choose a male that was the one with the first available nest when they were ready to start laying. However, they did not necessarily mate with that male, sometimes choosing to mate with another, often larger, male in an adjacent territory and then using the nest of the other male (and often getting his help in raising the young).

### 9.6.7 Long-tailed tit

Another small bird that has not suffered from the loss of the understorey is the long-tailed tit, but in this case, unlike the wren, it does not use the understorey much. Despite its name, the long-tailed tit is not a close relative of the other tits, being in a separate family—the Aegithalidae. Its habits are also very different. Outside the breeding season, the birds live in small parties, usually in the range of 8–12 individuals. These parties are largely composed of a brood from the previous season, plus their parents, and sometimes other adults. These parties have been studied in Wytham by Gaston (1973) and Glen and Perrins (1988). They defend quite large group territories against other families and dispute the boundaries vigorously.

Long-tailed tits are very tiny birds, weighing about 6–8 g. Hence they are very sensitive to severe cold in winter. During long spells of cold many of them may die. At the end of the extremely prolonged and cold winter of 1962–63, it seemed as if the species was extinct in Wytham. Fortunately, they bred well for several years thereafter and rapidly regained their earlier numbers. Long-tailed tits have one trick which they use to help them survive the cold. The party roosts together and, when it is very cold, they huddle close to each other so reducing the heat loss during the night.

As spring approaches, the parties break up into pairs. Interestingly, the pairs are not formed from two members of the same winter party. The males from the party take up areas within the winter flock territory. But they find themselves a mate from outside the flock. The likely explanation for this is that since the flock is composed largely of a single family, if the males took mates from within it they would be mating with their own sisters.

Also unlike the true tits, the long-tailed tits do not nest in holes in trees, but rather build a beautifully intricate nest of feathers, mosses, and lichens (Lack and Lack 1958). Although often hard to see, many are found by predators and robbed. If this happens early in the season, the pair may build another nest and have a second breeding attempt. But if it happens later, the birds may abandon any attempt to raise their own young and go and help at another nest. Intriguingly, this seems usually to be a nest of a relative (there is an evolutionary advantage in increasing the nesting success of your relatives because they share the same genes). To do this is relatively straightforward for the males, since the adjacent nests are ones belonging to a brother. But for the female this is not the case. To find her relatives she has to go to another territory, further away. And although the evidence is not strong (it is hard to get), it seems as if this is exactly what the females do, they go back to their winter territory to find a pair (which includes a brother) to help. It is these helpers who make up most of the 'extra' adults in the winter flocks (Glen and Perrins 1988).

### 9.6.8 Blackbirds

Blackbirds are common in Wytham, but have not been studied there as intensively as one might have expected. They are shy and many of their nests are taken by predators. By far the most detailed study of blackbirds was made by David Snow (1958), but this was primarily made in the Oxford Botanic Garden. However, he did make a number of comparisons with observations in Wytham Woods and woods elsewhere. Blackbirds breed much more successfully when the ground is moist and so worms are more easily available. Their long nesting season is circumscribed by cold weather in spring (too many frosts) and hot weather in summer (ground has dried out). Compared with gardens, the birds in Wytham have larger clutches and the clutch-size is greatest in May when the woodland blackbirds can switch from worms to cash in on the abundant caterpillars (Chapter 8, Section 8.16). Because of the richer food supply, nestlings in Wytham grew better than those in the Botanic Garden.

More recently, a study was made comparing the breeding success of the birds in different parts of the Wytham Estate, including both the farm and the woodland (Hatchwell *et al.* 1996a and b). The birds were more successful in the woodland than on the farm, but their success in the woodland tended to depend on the density of the vegetation; in the most vegetated parts of the wood the birds nested at higher densities and were more successful. The major cause of nest failure was predation, and successful nests were better hidden than those that failed (Chamberlain *et al.* 1995).

### 9.6.9 Farmland studies

There have been a number of studies of birds on the University Farm and on some of the other farmland on the Wytham Estate. These are not covered here in detail, because they are outside the Wood. Two farmland species, starling and house sparrow, are amongst those British birds which have undergone the most marked declines in recent years, and the Wytham populations have shown similar reductions. The swallow has also declined, but not so markedly. The starling once used the Woods at just one brief time of year; it used to come, in considerable flocks containing the new fledglings, to feast on the caterpillars on the oaks, and then depart again back to the fields.

David Seel made studies of house sparrows around the Woods, at the University Farm and in Wytham Village and at Hill End Camp (Seel 1968). He also studied the tree sparrows that nested in the tit nestboxes within the Woods. Although this is another species which has undergone a dramatic decline, 95 per cent in 25 years), serious enough to have got onto the RSPB's red list (Gregory *et al.* 2002), it is not always remembered that it was formerly uncommon in Wytham. In the first 15 years or so of the Wytham tit study it was not recorded breeding in the nestboxes. Then, during the very mild winter of 1960–61, some tree sparrows were caught in the Woods, feeding with the tits on the rich crop of beech mast. From the spring of 1961, they used the boxes; they were largely grouped in 'colonies' in Bean Wood, Marley Plantation, near the river in the Great Wood, and around Three Pines on the road up to the chalet. The OOS reports record that there were around 30 pairs by 1962 and as many as 70 occupied boxes by 1963. They remained plentiful for a few years and then, almost as suddenly as they arrived, they faded away again.

## 9.7 Conclusion

To date, some 40 students have completed theses which involve some aspect of the study of tits; the majority of these studies were made in Wytham, but others have been lab-based behavioural studies. A further 20 have studied other birds in Wytham or on the University Farm; about half of these were studies of starlings, jackdaws, or house sparrows, again stressing the convenience to field-workers of studying hole-nesting species. Doubtless in the future there will be many more!

# 10

# The Mammals of Wytham Woods

C.D. Buesching, J.R. Clarke, S.A. Ellwood, C. King, C. Newman, and D.W. Macdonald

For many visitors to Wytham, sightings of elusive mammals are amongst their most treasured encounters. Shy, cryptic, and largely nocturnal, the nature of mammals makes them some of the hardest species to observe and study. Researching their natural history not only requires the long hours of patience typical of field biology, but also a reliance on indirect techniques and increasingly intricate technology. Furthermore, as larger mammals can live for many years, long-term studies are often required to observe trends through different cohorts and generations. As a microcosm of the lowland English countryside, Wytham Woods are close to unique as a site for such studies. Indeed, the suite of inter-related research projects undertaken in the woods, in many cases stretching back for decades, provides a cutting edge model for future long-term, inter-disciplinary studies of biological processes and communities. This is the focus of the work by the Wildlife Conservation Research Unit (WildCRU) under the directorship of David Macdonald.

In general, the smaller the mammal, the more abundant the population and the more dynamic and responsive is its ecology to environmental factors. As primary consumers on the first rung of the food chain, creatures such as wood mice and voles were the first subjects of mammal research in Wytham, and are also the starting point for this chapter. Later came studies of larger mammals, and major investigations of predators, to which we will turn as this story of Wytham's mammals unfolds.

## 10.1 Wytham's small mammals

Charles Elton (1924) provided the initial motivation for research into the ecology of small rodents and their population dynamics in Wytham Woods. From 1943 onwards, initially under H.N. (Mick) Southern (the founder of Britain's Mammal Society), this research developed into one of the longest continuous small mammal studies in the world.

Elton was amongst the first to apply quantitative methods to the understanding of mammalian population dynamics (Elton 1942). He analysed and interpreted the plagues of small rodents that occur in many parts of the world, the fluctuations in the size of populations of voles, lemmings, and wood mice (*Apodemus sylvaticus*) and, by ingeniously making use of the fur-trading records of the Hudson Bay Company, he was able to

# Henry Neville 'Mick' Southern (1908–1986)

Mick Southern in 1953.
Photo by courtesy of
Mrs K. Southern.

Mick Southern took his first degree at Oxford in 1931, in Classics. He was by then already deeply interested in natural history and ecology. His talents as a bird photographer were already evident and he published a book on the subject, in 1932, based on photographs taken largely while he was an undergraduate (*Close-Ups of Birds*). It was not surprising that four years after graduating he was up at Oxford again reading for a degree in Zoology. During this period and in the remaining pre-war years, he was actively publishing papers on birds.

Amongst his many interests, perhaps the main two at this time were the arrival dates of migrants in relation to spring weather and the problems of polymorphism in birds; his first paper on the frequency of the bridled form of common guillemots was published in 1939. After graduating for the second time in 1938 he joined Charles Elton in the Bureau of Animal Population (then in the Department of Zoology and Comparative Anatomy). His study was of rabbits, using new techniques for catching and marking them. Then came the Second World War, when the Bureau switched to work on the control of pests of our limited food supplies: rats, mice, and rabbits. He was also secretary of the executive sub-committee of the Oxford Committee for Ornithology, a body which helped to oversee parallel work on potential bird pests such as rook, wood pigeon, and house sparrow.

After the Second World War, Southern was appointed Senior Research Officer in the newly formed Department of Zoological Field Studies (comprising the Bureau of Animal Population and Edward Grey Institute). He then started the work for which he is best known: a 13-year study in Wytham Woods of the numbers and turnover of a population of tawny owls and of its two main species of prey, the bank vole and the wood mouse. It remains one of the most valuable and careful investigations of a predator and its prey ever carried out (Perrins 1987). This was the second-longest study of a bird population in the Woods, and that of the small mammals which continue to be 'index-trapped', thus making it one of the longest such studies in the world.

The Bureau's war work was published in 1954, at the Clarendon Press, Oxford, in three volumes: *Control of Rats and Mice*. The first two volumes on *Rats* were edited by D.H. Chitty and Volume 3, on *House Mice*, edited by H.N. Southern.

A founder member and, later Chairman, of Britain's Mammal Society, Southern edited the first edition of *The Handbook of British Mammals* (1964) and the next (1977) with G.B. Corbet. He had an important impact, too, as editor of the British Trust for Ornithology's journal *Bird Study* from its inception in 1954 to 1960 and the British Ecological Society's *Journal of Animal Ecology* from 1968 to 1975. He was President of the BES from 1968 to 1970.

understand the demographics of fur-bearing mammals in parts of Canada. Importantly, Elton drew attention to the fluctuating nature of some mammalian populations, and concluded his remarkable book (Elton 1942) with the hope that the study of animal population dynamics would lead to '...understanding, not for power alone, but on account of its own wildness and interest and beauty, the unstable fabric of the living cosmos'.

Elton had a holistic approach to ecology, illustrated by his pioneering investigation on wood mice in Bagley Wood near Oxford (Elton *et al.* 1931), undertaken long before Wytham Woods became the University's site for ecological research. This study estimated the population density of wood mice, investigated their fecundity and their parasites, and paved the way for all subsequent small mammal research in Wytham Woods.

### 10.1.1 Population cycles and Chitty's hypothesis

In 1937, at a time when the scientific research of nature and ecology was still (largely) in its descriptive stage, Dennis Chitty had been inspired by Charles Elton's publications to carry out a study of the 3–4 year population cycles of field voles (*Microtus agrestis*) in the grassland around Lake Vyrnwy in Wales. Crucial to this investigation was the ringing technique for identification of animals in wild populations (Chitty 1937) and also the live trap which had been devised by Chitty and Kempson (1949). Chitty (1952) considered that:

'...some form of intra-specific strife seems to be the process controlling population density. We may reasonably suppose that the action of the factors capable of producing this control is "governed by the density of the population", or is "density dependent".'

He postulated that the following five steps contribute to the regulation of the size of populations:

1 Strife during the breeding season, results in
2 the early death of the young and physiological derangements among adults.
3 The progeny of these adults survive, but
4 are abnormal from birth, and thus more susceptible to various mortality factors.
5 These constitutional defects, in a more severe form, can be transmitted to the next generation.

This comprehensive, innovative hypothesis was to anticipate Chitty's later proposal that population cycles result in some genotypes being better adapted to survival at different points of the cycle than others (Chitty 1970). Chitty's hypothesis was to provide a guiding framework for the investigation of Wytham's small mammal populations. Gillian Godfrey (1955) studied two field vole populations and concluded that her results did not '...provide sufficient grounds to reject Chitty's hypothesis'. Further studies suggested that sudden large population losses in the spring are the consequence of intra-specific competition affecting the mortality rate and/or immigration of younger individuals (Chitty and Phipps 1966). Chris Richards (1985) analysed the numbers of field

# Charles Sutherland Elton (1900–1991)

Charles Elton.

Charles Elton is a founding figure in ecology and conservation and was among the most important and influential ecologists of the twentieth century. In the early 1920s, while still an undergraduate at Oxford, he took part in three expeditions to the Arctic island of Spitzbergen. This experience helped to shape many of his later ideas. As an undergraduate, he and others rebelled against the continuing emphasis in zoology on comparative anatomy and descriptive embryology. His interests were in the behavior and life histories of living animals. His early zest for natural history was actively encouraged and directed by his eldest brother, Geoffrey, to whom he dedicated his first and best known book, *Animal Ecology*.

*Animal Ecology*, was published in 1927 and soon became a classic. It turned natural history into a science concerned with the quantitative and experimental study of living organisms in relation to their environments. It has remained in print in later editions ever since. Another research interest, fuelled by an Oxford University expedition to Lapland in 1930, was into the regular fluctuations in numbers of certain animals and the importance of movements of their populations. Elton and his colleagues had been carrying out intensive investigations into the population dynamics of wood mice and voles in Bagley Wood, near Oxford, from 1923 to 1931. This work led to another classic book in 1942, *Voles, Mice and Lemmings: Problems in Population Dynamics*. In 1958 his years of study of another important subject, biological invasions, resulted in a third influential title: *The Ecology of Invasions by Animals and Plants*, generally acknowledged as the cornerstone work in that field.

As well as these three books, Elton is remembered for founding, in 1932, and editing for nearly 20 years the *Journal of Animal Ecology*; for his studies on animal community patterns (including the community-oriented concepts of the 'Eltonian niche', and the pyramid of numbers), and for a number of contributions to economic biology and conservation studies.

At Wytham, he started the Wytham Ecological Survey in 1950. He envisaged a continuing ecological survey which would study all groups of animals and habitats, involving plant as well as animal ecologists. The survey was recorded on a large card-index of sites matched by a collection of 'voucher' specimens to confirm the identification of species. The Survey, and the subsequent analysis of communities and changes, supported the claim that Wytham is amongst the most studied bits of land in the world. In his book, *The Pattern of Animal Communities* (1966), Elton set out to assess all that had been revealed using these methods over the previous 20 years (Macfadyen 1992) (Southwood and Clarke 1999).

Besides bird and mammal populations, communities in many kinds of minor habitats had thrown light on the larger canvas of interactions that he referred to as a 'girder system' of partly interlocking community units which might show 'ecological resistance'. Elton developed the then revolutionary insight that the landscape is divided into

> ecological units based on the structure of the habitat, irrespective of the species of green plants that are present, because most animals are not specialist herbivores. His insights led to a revision of the methods of selection and management of British nature reserves, which had initially been dominated by botanists and vertebrate specialists. He stressed to his colleagues on the Scientific Policy Committee of the Nature Conservancy the importance of invertebrate surveys to the management of protected areas.

voles (*Microtus agrestis*) in the Woods and showed that there are annual changes in population size with peaks at various times in the year. Intriguingly, however, in Wytham, field vole populations have never been observed to show the sorts of longer-term cycles observed in extensive grasslands elsewhere in Britain.

Chitty's hypothesis also emphasizes the crucial role aggressive behaviour can play in population cycles. Working in open-air enclosures in Wytham, John Clarke was able to test empirically for the consequences of antagonistic interactions. Weekly censuses, over an 18 month period, revealed frequent aggressive encounters (Clarke 1956) resulting in demonstrable physiological consequences characteristic of stress (Clarke 1953); voles housed at peak density had a lower life expectancy and fertility than those kept at low densities (Clarke 1955).

Extending this line of investigation to Wytham's wood mice showed that aggressive interactions hinder immigration and juvenile survival, leading to population decline in spring and summer, and the prevention of the autumnal rise in population size (Flowerdew 1974). This 'Adult Aggression Hypothesis' necessitates the active aggression of adult group-members resulting in juvenile mortality. By contrast, Susanne Plesner-Jensen (1996) concluded that the small number of juvenile wood mice found in the breeding season could be explained better by the Habitat Saturation Hypothesis, which postulates that the population is regulated by available resources in a density dependent way. Aggression, both within and between the sexes, is high in wood mice, and this can present males in reproductive condition with a behavioural dilemma (Stopka and Macdonald 1998): the only way a male can ascertain female reproductive condition is to sniff her anogenital region. However, females tend to avoid this contact either by shifting their position or through aggressive displays. In order to sniff her rear, the male has to groom her. This helps to reduce her parasites, but increases his risk of acquiring parasitic infection through licking nematode eggs off her fur (Stopka and Macdonald 1999).

Some parasitic nematodes, especially *Heligmosomoides polygyrus*, which occur in the intestine, can have significant effects on wild wood mice populations. Whilst these nematodes have no effect on juvenile mouse survival until the age of one month, the survival rates of older individuals infected with *H. polygyrus* are lower than those of uninfected mice (Gregory 1991). In general, males and heavier individuals appear to have a higher susceptibility for *Heligmosomoides* (Brown *et al.* 1994). Liz Brown and her colleagues

# Dennis Hubert Chitty (1912–2010)

Dennis Chitty. Part of a photo taken by Denys Kempson at Lake Vyrnwy in 1948.

Dennis Chitty, having graduated in Biology from the University of Toronto, came to Oxford in September 1937, joining Charles Elton's recently-established Bureau of Animal Population, alongside the Department of Zoology and Comparative Anatomy in Oxford University. He has carried out a wide range of studies of great originality bearing on mammalian population ecology. This included examination, with Charles Elton, of records of fluctuations in numbers of snowshoe hares. During the Second World War he, H.N. Southern, and others carried out a thorough study of rat and mouse infestations seriously affecting food-supplies in war-time Britain. Before and after that he made an extensive and intensive study of populations of field voles in grassland around Lake Vyrnwy, Montgomeryshire. He put forward a radical theory to explain field vole population cycles, postulating that these regular changes are brought about primarily by intraspecific factors and not by so-called classic external, environmental ones, which neglect behaviour. He devised a ringing technique for small mammals and designed, with D.A. Kempson, a new trap.

These technical improvements have had widespread application not only in Wytham Woods but worldwide. With collaborators he tested the implications of his hypothesis using field vole populations in the Woods as well as in Wales. He promoted and supervised students' fieldwork in Wytham Woods, including the experimental manipulation of populations of small rodents as well as a long-term investigation of the demographic parameters of two experimental populations of field voles maintained by John Clarke in two open-air cages. But his ideas were the stimulus not only for a considerable body of research on animal populations in the Woods, but also for the study of mammalian population dynamics in Europe, North America, and elsewhere.

He emphasized always the crucial importance of putting forward *testable* hypotheses in the search for explanations of the relationship of organisms to each other and to the environment. He returned to Canada in 1961, having been appointed Professor of Zoology at the University of British Columbia, where he mostly gave up his own research in favour of teaching undergraduates and supervising PhD. students.

were able to show that infection is correlated with significant behavioural changes in wood mice: infected individuals not only moved significantly further and faster than uninfected mice, but also spent more time on the move, resulting in significantly larger home ranges.

Small mammals have both seasonal breeding periods and multiple litters, the implications of which have reviewed extensively by Clarke (1981), who studied the accompanying changes in their gonads and reproductive hormones. Clarke's research involved field voles from a laboratory breeding colony developed from wild-type animals. He demonstrated that fertility and the limitation of breeding to spring and summer are regulated by changes in day length (Baker and Ranson 1933), which was also found to influence sexual development in bank voles (*Clethrionomys glareolus*) and wood mice (*Apodemus sylvaticus*) (Clarke *et al.* 1981). The occasional occurrence of winter breeding in natural populations of field and bank voles as well as wood mice, however, showed that other physical and social environmental factors also contribute to the timing and duration of these breeding seasons. For example, further work discovered that the sexual maturation of Wytham's field voles was mediated by the proximity of sexually mature males (Spears and Clarke 1986), that ovulation in voles is not spontaneous, but is induced by mating (Clarke *et al.* 1981), and that it can also be provoked by other stimuli, such as pheromones (Clarke and Hellwing 1977). A number of pheromones, such as the secretions from sebaceous glands on the tail of wood mice and the hindquarters of field voles, are now thought likely to play a role in reproduction and territorial marking (Flowerdew 1971; Clarke and Frearson 1972).

With advances in genetic science, Chitty's hypothesis was later modified (Chitty 1996) to include the role of quite rapid genotypic changes in a population, supported by research showing that sexual development and maturity are also determined genetically (Spears and Clarke 1988).

## 10.1.2 The effects of habitat management on small mammals

Long-term monitoring of species distribution and abundance as well as understanding their uses of, and preferences for, particular habitats can inform conservation and management strategies. The wider ecological importance of small mammals is twofold. Firstly, they support a number of predators, for example in Wytham Woods, stoats (*Mustela erminea*), weasels (*Mustela nivalis*), red foxes (*Vulpes vulpes*), barn owls (*Tyto alba*), tawny owls (*Strix aluco*), and buzzards (*Buteo buteo*) that are vulnerable to changes in the abundance and distribution of their rodent prey. Secondly, small mammals are good environmental indicators for agro-pollutants and pesticide residues, as well as general habitat degradation.

Over the past 65 years, small mammals have been live-trapped twice yearly (at the end of May/beginning of June, and in December) at the same two sites located in secondary deciduous woodland in Wytham Great Wood (Flowerdew and Ellwood 2001). Marking, releasing, and recapturing the same individuals within one trapping session allows their numbers to be estimated, and reveals the impacts of predation and environmental factors on population density. H.N. Southern's long-term research on the predation of wood mice and bank voles by tawny owls in Wytham Woods (Southern 1970, discussed in Chapter 9), which created this dataset, has previously been described as 'the best long-term quantitative study of the population dynamics of an avian predator and its vertebrate prey' (Whitaker 1987).

180 Wytham Woods

Over the past 25 years, these long-term data show a marked overall decline in bank voles, dwindling from an abundance where their numbers always exceeded those of wood mice (1948–80) to a situation where their numbers were considerably lower than those of wood mice, while wood mouse numbers, though subject to inter-annual abundance cycles, maintained the same range over the years (Fig. 10.1).

The start of this bank vole decline coincided with the erection of a deer fence around the perimeter of the Woods in 1989 and is probably linked to the resulting increase in deer-grazing pressure within the Woods (Flowerdew and Ellwood 2001). This has been illustrated neatly by comparing bank vole densities in the open woodland with those found in four deer exclosures in the Woods, which were established in 1998. A two-metre high deer fence excludes all deer, but wood mice and bank voles can move freely through the fence. Thus, grazing pressure is minimized, resulting in a much denser understorey, especially in increased bramble cover (Morecroft *et al.* 2001). Whereas wood mice have keen senses that can alert them to the threat of potential predators, and the agility to evade capture, bank voles rely heavily on thick understorey for protection against predators. In 2002, research showed approximately 5 times higher bank vole population densities inside the exclosures than in the open woodland, whereas wood mouse numbers were evenly distributed. However, in recent years stringent deer control has led to a recovery of the forest understorey, and thus bank voles have begun to spread into the broader woodland.

Christina Buesching *et al.*'s (2008) work in Wytham has also shown that dense understorey is correlated with increased arboreality in small mammals. Whilst wood mice, particularly males, have been found to use the three-dimensional habitat throughout the year, climbing to a height of 12 m, bank voles appear to climb trees and bushes

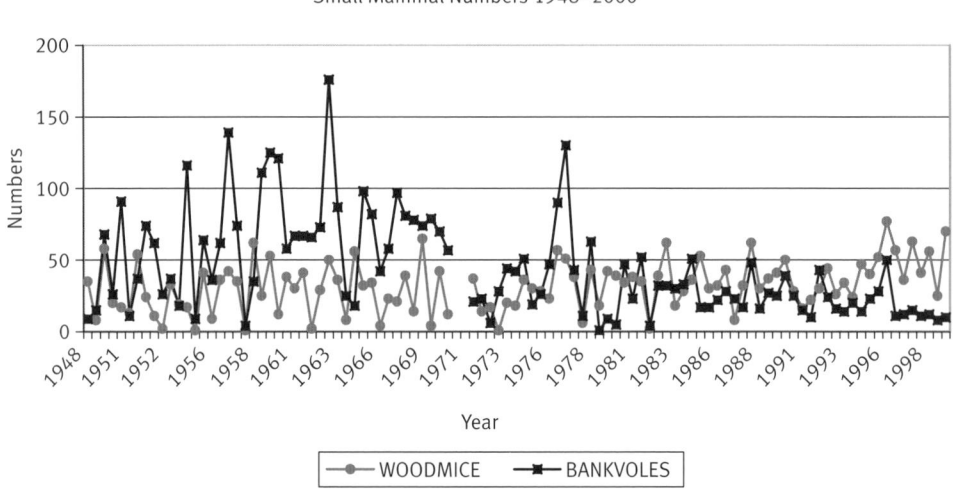

**Figure 10.1** Fluctuations in numbers of wood mice and bank voles 1949–98.

predominantly during times of food shortage and high population densities, and generally to a lesser height (ca. 60 cm). The mating system of wood mice involves one male territory overlapping that of several females (female-defence polygyny: Tew and Macdonald 1994). Hence it is likely that male territory holders use trees not only during foraging trips, but also as direct routes and over-passes between different female territories, which need not be contiguous.

Live-trapping small mammals, however, does not always give us a full picture of habitat use and activity patterns and it can be affected by the provision of the baits and the effects of odours left on traps. Using behavioural observations and radio-tracking, Tew and Macdonald's research at Wytham established that wood mice are not only capable of detecting and remembering rich food patches (Macdonald *et al.* 2006), but that the size of their territories is also heavily affected by food resources. Before the invention of radio-tracking technology, Gillian Godfrey (1954) tracked field voles in Wytham Woods by attaching a capsule containing a radio-active isotope to their legs. Using a Geiger-Müller counter she estimated that, on average, field voles ranges span 235 square yards (194 m$^2$). In comparison, the average home range of wood mice comprises 0.72 acres (300 m$^2$) in males and one third of that in females, although during the breeding season males roam over much greater distances. Behavioural observations of Wytham's wood mice in captivity carried out by Pavel Stopka and David Macdonald (2003) shed light on how these animals learn to navigate their environment: if put in an unfamiliar enclosure, wood mice use a land mark as centre-point for their explorations. Once they have explored, mapped, and scent-marked the immediate vicinity of this land mark, they move on to start mapping out the next sector of their enclosure.

In order to maximize both their energy intake and minimize the risk of predation, wood mice select specific food sources at different times of day. Miller was one of the first to note that most feeding occurs around dawn and dusk. Seeds, a ready source of sugars, are consumed early in the night providing energy for subsequent nocturnal activity (Miller 1954; Plesner-Jensen 1993). In a fascinating contrast, Evans (1973) noted that field voles increased their nutritional intake simply by eating more, rather than by selecting plants with a higher nutritional value.

Field vole abundance has also been monitored in Wytham since 1948 (Southern 1970). While they need a thick layer of ground vegetation or grass tussocks to conceal them from predators, they also benefit from good supplies of young grass growth, and thus extensively grazed sheep pastures appear to be their preferred habitat in Wytham. Flowerdew (1973) found that a good crop of acorns resulted in healthier wood mice and bank voles, and higher migration from the neighbourhood. Laboratory studies have shown that nutrition has an important role in field vole reproductive development (Spears and Clarke 1987).

### 10.1.2.1 *Shrews*

Although of similar size to mice and voles, shrews have a different biology to other small mammals, and they belong to a different Order, the Eulipotyphla (formerly part of Insectivora). Field research on shrews was once considered difficult—if not virtually impossible—due to the perceived fragile constitution of these animals. However, in the

1950s, Peter Crowcroft published several papers on his work on shrews in Wytham Woods, kick-starting research in this field (Crowcroft 1956, 1957). Wytham Woods is home to the common shrew, *Sorex araneus*, and the pygmy shrew, *Sorex minutus*. Although under lab conditions both species feed on ground invertebrates, without much preference for different species, in the wild their diets may differ considerably (Pernetta 1977), with pygmy shrews taking prey preferentially from the soil surface and leaf litter, whilst common shrews specialize in burrowing invertebrates. Irrespective of food abundance, however, shrew density in Wytham (as elsewhere) is influenced predominantly by nematode infection with larval *Porrocaecum* sp. (Buckner 1969). Both males and females establish home ranges in autumn and winter. In spring, males adopt one of two different mate-searching strategies: they either occupy relatively small territories during March, and make repeated long-distance excursions to visit females around the time they become sexually mature in April, or males establish large home ranges in areas of high female density in March, which they then also maintain in April (Stockley *et al.* 1994).

Shrews have an unusual and particularly interesting mating system amongst mammals in that they produce (often at great cost to the female) more offspring per litter than the number that can be reared to the point of weaning (Stockley and Macdonald 1998). The most likely explanation for this seemingly wasteful reproductive strategy is that females promote sibling competition for maternal investment, and hence ensure selection for the most genetically fit young (Stockley *et al.* 1996).

### 10.1.3 Small mammals and their role in the ecosystem: interactions with other species

No species exists in isolation. Hence, many studies at Wytham started with one species and ended with related work on another. For example, work on small mammals led to studies of their predators—such as tawny owls (e.g. Appleby *et al.* 1997, 1999) and weasels (see Section 10.3), or led to investigations of the more complex ecological effects of, for example the behavioural actions of one species on the survival chances of another (for example, the effects of deer grazing on small mammal numbers, see below).

## 10.2 Deer

Wytham has three species of deer: roe (*Capreolus capreolus*), fallow (*Dama dama*), and Reeves muntjac (*Muntiacus reevesi*). Roe are native to Britain, but were largely extirpated by the mid-18th century, through over-hunting. They didn't return to Oxfordshire until the 1970s (Ward 2005) and remained at relatively low densities in Wytham until 2001 (Ellwood 2007). Fallow were probably introduced to Britain by the Normans and are now naturalised, existing in low numbers in Wytham until the 1980s (Perrins and Overall 2001). By contrast, muntjac were introduced recently. Native to south-east China and Taiwan, muntjac escaped from Woburn Abbey Park c. 1900, and spread as far as

Wytham by the mid-1960s. By the late 20th century, all three species were probably having ecological impacts on Wytham's birds (Perrins and Overall 2001), plants (Morecroft *et al.* 2001) and small mammals (Flowerdew and Ellwood 2001).

Warm winters, afforestation, and winter crops, probably caused the increase in Wytham's populations of fallow and muntjac in the 1980/90s – a phenomenon observed across all deer species in the northern hemisphere (Fuller and Gill 2001). Herds of around 200 fallow were observed raiding crops on neighbouring farmland, prompting the construction of a 7km deer fence (1987–1989) encircling the woodland. This effectively created closed populations of fallow and roe (muntjac could still get through). Within the woodland, fallow and muntjac numbers rose to the point where a series of winter culls were required to reduce both the ecological impact of deer, and the risk of them starving. Between 1998 and 2003 the culls provided an opportunity for research focusing on the effect of deer population density on the animals themselves, and the methods used to count deer and thus detect changes in population size (Ellwood 2007).

Empirical deer counts were first conducted in Wytham by Graham Taylor (Taylor and Morecroft 1997). Between 1998–2003, Stephen Ellwood (2007) compared the statistical precision (reliability) and accuracy (proximity to known population size) of two deer count methods, finding that: 1) direct observation of deer, using 'Distance sampling', was far more precise and accurate than dung counting (quantity of dung found in surveys equates to the number of animals defecating); 2) at low densities, using dung counting, muntjac might never be counted as reliably as fallow; 3) if standard dung count techniques are used then more 'man-hours' (sampling effort) may be required to generate a given level of reliability than is generally accepted. These findings have implications for management, conservation and ecological research wherever the measurement of ungulate population density is important, and suggests that poor reliability may render standard dung-counting techniques inadequate in woodlands similar to Wytham.

Winter culling (1998–2003) reduced the fallow population from c. 440 to 40 animals, and muntjac from c. 200 to 15. Roe were not culled until 2001 when their numbers more than doubled, to c. 40 animals, in a single year. Cull data suggested that: 1) fallow and muntjac pregnancy rates were unaffected by population density – they were maximal; 2) in general, animals lost body condition over winter, although fallow fawns and yearlings continued to grow; 3) fallow fawns were smallest at high densities; 4) warm and wet winter weather benefited fallow body condition; 5) there was no evidence for any effects on muntjac, although data for this species were few (Ellwood 2007).

The fact that roe remained at relatively low densities until fallow and muntjac numbers were considerably reduced could be the result of inter-specific competition for food, particularly during the winter months when the diets of all three species are likely to overlap. Ellwood's work (2007) suggests that the warm and wet winters predicted by global warming could allow temperate woods, like Wytham, to support far higher densities of healthy deer than was previously thought. This would increase the need for deer management if the ecological and financial impacts of deer are to be minimized.

## 10.3 Weasels

The advantages of working at a well-researched site, such as Wytham, and sharing data with colleagues, are exemplified by the simultaneous studies of small mammals and their predators. The two key predators of small mammals in Wytham's woodland community are tawny owls (*Strix aluco*) and common weasels (*Mustela nivalis*). Investigating their interactions with wood mice and voles requires data on both predators and prey, which is normally too much work for one person. Carolyn (Kim) King had the good fortune to be studying weasels in the same years (and under the same supervisor, Mick Southern) that John Flowerdew was studying wood mice, *Apodemus sylvaticus*. They were able to swap information on how many mice were available for weasels to eat, and roughly what proportion of those mice were in fact eaten, every month over a year. The resulting calculation provided a preliminary estimate of the predation impact of weasels on voles and mice: on average, 8–10 per cent of the bank vole and wood mouse population per month (King 1980)—a much smaller effect than that of the tawny owls which Southern had studied many years previously (Southern and Lowe 1968). However, whilst weasels switched to raiding the nests of great and blue tits when small mammals were scarce (Dunn 1977; King 1980), the owls had no other source of food to fall back on and in years of poor food supply during the nesting season laid smaller clutches or, in really bad years, did not breed at all (Southern 1970).

Live traps have to be inspected twice a day, and since weasels have large home ranges, trap rounds can be time-consuming and hard work. During the years (1968–70) when King observed weasels in Marley Wood, small mammal numbers were in the region of 21–39 individuals per hectare, and male weasels used home ranges of 7–15 ha with females about 1–4 ha. Several of these home ranges also extended into parts of the adjoining Pasticks plantation (then still young, with open canopy) and farmland. In both, the grass was thick and field voles were more common than the bank voles while wood mice were more typical of the woodland. Droppings collected from resident weasels reflected this in their prey content, and revealed differences between those that did or did not range outside the wood (King 1980).

The distribution and abundance of weasels is influenced strongly by the availability of their mammalian prey, and by the extent of cover from other predators. When Hayward (1983) planned to repeat King's study 15 years later, using exactly the same trap sites and methods, he could find only 3 weasels in her original study area (compared to the 36 individuals, King caught there during her entire study period) and concluded that Marley Wood's weasel population had collapsed because small mammals were not abundant enough to sustain it. By 2000–2005 at least some weasels were back, because Buesching and Newman caught several incidentally in their extensive small mammal monitoring programme (Newman *et al.* 2003). On the northern side of Wytham Hill, Macdonald *et al.* (2004) used radio-tracking rather than trapping to record male weasels moving over some 113 ha, and females over 28 ha, of the farmland and linear hedgerow corridors adjacent to the Great Wood.

All of these studies indicated that weasels are well aware of the activities of their neighbours, and can, on the one hand, avoid direct contact with them, and yet on the other, quickly recolonise vacant home ranges after a previous owner has died or disappeared. King observed this happening most often during spring, when food was short and resident weasels were vulnerable to drastic weight loss. Weasels are always small and thin even when well fed, because small size gives them entry into the runways and nests of their prey, but keeping a small body warm on a cold day in early spring, while also finding the extra energy needed for finding mates and producing young, is often simply too much. The weasel way of life is a fine balance of competing risks, and nowhere has that been better demonstrated than in Wytham.

## 10.4 Foxes

The red fox *Vulpes vulpes* is the most widely distributed carnivore on earth, but although it figures in many ancient fairy tales and folklore, surprisingly little was known about its general biology and natural history before David Macdonald started his research in Wytham Woods in 1972 (Macdonald 1987). His early work in the Woods came to little, because his attempts to catch foxes for radio-tracking were thwarted by the large numbers of badgers that filled his traps. Thus, he moved to the adjoining farmland, where badgers were fewer. From his van, with an aerial mounted on the roof, Macdonald radio-tracked the fox population in and around Wytham, across the adjoining farmland and, particularly, through the gardens that emerged as their richest foraging habitat. This was only the second mammal radio-tracking study conducted in Britain, and was enhanced by other innovations, such as the use of infra-red night vision equipment to reveal the fox's unexpected penchant for hunting earthworms (Macdonald 1980a). Macdonald made TV history by producing the first-ever broadcast quality infra-red night-vision images in the BBC's *Night of the Fox* in 1976.

Macdonald's research revealed the local foxes to be living in small (50–200 ha) group territories, occupied by a dog fox and up to five related vixens, arranged in a social hierarchy in which the subordinates acted as helpers to the dominant's cubs. Whilst sons left their parents' territory before they reached sexual maturity, daughters generally appeared to stay and became helpers with their mother's next litter (Macdonald 1979, 1983). Playback experiments and vocal analyses confirmed that fox calls differ from individual to individual and that members of the same social group can recognize each other by their calls. In 1974 Macdonald hand-reared his first fox cub, and Niff, together with her many successors, walked on long leashes in the wild, providing insight into their feeding and scent marking behaviour (Macdonald 1987). It transpired that foxes store surplus food in caches, which the owner appears to find again later by remembering their exact location rather than following scent-trails or recognizing visual cues of food stashes. Thus, only the original owner can find stashed food and benefit from these savings for leaner times, whereas other foxes can walk straight past caches without detecting them.

In the late 1970s the invention of a new type of trap, the Canadian Novak leg snare, together with a special permission from the then Minister of Agriculture to test this trap in Wytham (as a potential tool for rabies control), enabled Macdonald to start catching

**Figure 10.2** David Macdonald with fox cubs.

foxes in the Woods themselves (these traps largely avoiding the interference of badgers). He caught over 100 foxes in the Woods and the surrounding countryside, tracking them together with his doctoral students Heribert Hofer (1988) and Mike Fenn. Simultaneously, they collected over 9300 fox scats (faeces) from their study area for dietary analyses. Their results showed clearly that foxes are extremely opportunistic foragers, with the shape and size of their territories being influenced predominantly by the distribution of their food resources. The woodland foxes occupied home ranges averaging, 60 ha (ranging from 20–100 ha), lived in groups of one male and two or three females, and fed heavily on rabbits (45 per cent), which they were observed (especially in Wytham Park) to hunt either by a running charge or, occasionally, using sit-and-wait tactics.

Thus, two more general results emerged from this fox work. The first was the idea that fox society could be interpreted as an emergent property of the spatially and temporally patchy dispersion of their food. This idea developed from jottings by Hans Kruuk in 1974. This Resource Dispersion Hypothesis (Macdonald 1983) was first applied to badgers by Kruuk (1978) and foxes by Macdonald and still forms the basis of the current work on Wytham's badgers. The second output of this fox research was more applied. At the time, foxes had been widely considered solitary and wide-ranging. This over-simplified understanding of fox behaviour had led to thinking about rabies control in ways that ignored the emerging ecological reality. Macdonald (1980b) introduced a biologist's perspective, based on his Oxford fox work, which culminated in the successful replacement of killing campaigns with oral vaccination (e.g. Bacon and Macdonald 1980). Part of this thinking was that the killing campaigns so disrupted the activity of survivors that they behaved in ways that increased the spread of the disease. This simple idea, which Macdonald named the Perturbation Hypothesis, has important implications for controlling bovine tuberculosis in badgers (see below).

Foxes are often blamed for losses of gamebirds, chickens, lambs, and wild ground-nesting birds. Generalized aversion is based on the idea that animals can learn to avoid certain foods, if these are repeatedly treated with an aversive substance, which is undetectable without tasting. Field experiments in Wytham (Macdonald and Baker 2004) indicate that foxes can develop a generalized aversion against certain food sources if they are treated with denatonium benzoate (sold as 'Bitrex', the most bitter-tasting substance ever developed).

## 10.5 Wytham's Badgers

Today, one of the most comprehensive long-term studies of medium-sized carnivores anywhere in the world is that of the badgers of Wytham Woods. In 1972, when Hans Kruuk returned from his pioneering studies of the Serengeti's hyaenas, he joined Niko Tinbergen's Animal Behaviour Research Group at Oxford, and took on David Macdonald as his doctoral student, and the two of them began work together in Wytham, respectively on the badgers and foxes. The catalyst was to develop Kruuk's ideas about the evolution and functional role of sociality in carnivores (Kruuk 1975). Kruuk had already realized the importance of the observation that several species of carnivores lived in social groups but did not seem to benefit from the advantages hitherto normally associated with carnivore sociality (group hunting, collective vigilance—in those days awareness of the role of cooperative breeding was only just emerging). Badgers conspicuously illustrated this paradox, being abundant and living in groups, but seemingly doing nothing cooperatively. For example, badgers in lowland England were known to feed predominantly on earthworms. Kruuk (1978) confirmed this using infra-red binoculars, that were purchased using Niko Tinbergen's Nobel Prize money. However there was no evidence that they hunted these worms cooperatively. Similarly, although it would subsequently emerge that Wytham's badgers make 17 distinct vocalizations (Wong *et al.* 1999), there was no evidence of cooperative warning calls or division of labour over vigilance.

By the time Hans Kruuk left Oxford in 1977, he had revolutionized our knowledge of badgers (Kruuk, 1978, 1989), mapping the boundaries of the 12 group territories that then existed in Wytham, and creating a foundation on which David Macdonald was to build when he took on his first badger student, Heribert Hofer, in 1983. A milestone came in 1987, when, at the suggestion of his then student, Jack da Silva, Macdonald initiated the badger population-monitoring programme in Wytham Woods, which has run continually since.

Although badgers live in groups and share a communal sett, the existence of alloparental behaviour (that is, cubs being cared for by additional individuals other than their parents) remains tentative (Woodroffe and Macdonald 2000) and the social integration of cubs into their natal group appears to be driven by the cubs themselves rather than by adult group members (Fell *et al.* 2006). From genetic evidence (Dugdale *et al.* 2008) it appears that badger social groups are highly related. Genetic analyses proved to be difficult in Eurasian badgers, as many usually variable DNA regions show a monomorphic pattern in this species, indicating that populations suffered a severe bottleneck in their distribution in the recent past. Several females can breed within one social group, and multiple paternity within one litter is often observed (Dugdale *et al.* 2008). Approximately 40 per cent of offspring are fathered by extra-territorial matings, leading to close relatedness also between neighbouring social groups. Reproductive success declines with the age of the parent in both males and females (Fig. 10.2)

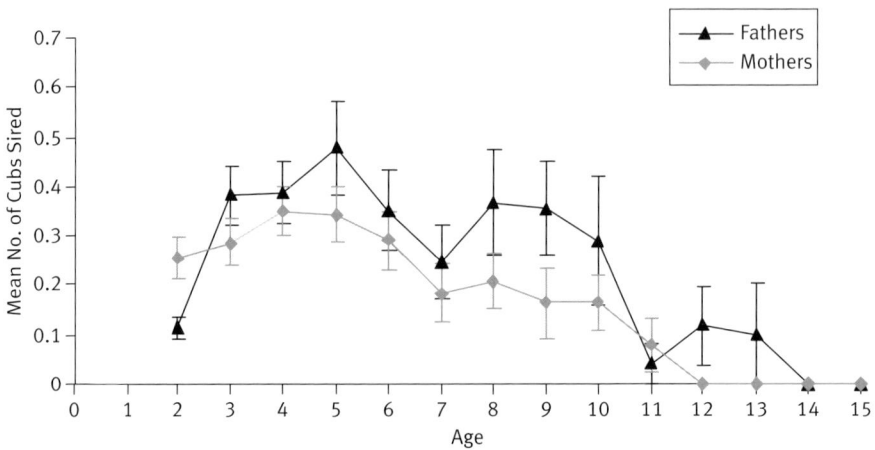

**Figure 10.3** Mean number of badger cubs sired by age of parents.

Several litters of cubs can be raised in one sett at the same time and some females occasionally appear to suckle cubs that are not their own offspring; however, the extent to which communal breeding really benefits badgers remains unclear. The availability of more potential helpers in larger groups or in groups with more than one breeding female does not increase cub survival rates, and group-living appears to incur costs (Macdonald and Newman 2002).

These divergent badger social dynamics present a fascinating evolutionary paradox. As Kruuk first realized, traditional explanations of sociality cannot be applied to badgers, which forage alone and have no major predators. Although it is well known that badgers in some habitats, such as Wytham, live in large, mixed sex social groups, assiduous study has revealed little compelling evidence of cooperation, or of any clear social hierarchy amongst them (Macdonald *et al.* 2002).

### 10.5.1 Territoriality and the Resource Dispersion Hypothesis

As mentioned in the context of foxes, the Resource Dispersion Hypothesis (RDH) states that groups may develop where resources are dispersed, such that the smallest economically defensible territory for a pair can also sustain additional animals (Blackwell and Macdonald 2000). In the case of badgers, Kruuk saw this smallest economically defensible territory being defined by the dispersion and richness of patches of available earthworms. Earthworms, for which badgers forage alone, become available only when they surface in response to optimal microclimatic conditions. Patches of earthworms therefore become available in different habitats at different times, so the principle is that badgers require several blocks of worm-rich habitat to provide a reliable food source. When patches are irregularly shaped and widely dispersed, the minimum territory required by an individual may be highly contorted and defense will therefore be costly. However, coalescence of several individual territories to form a single group territory reduces individuals' defense costs while still providing access to several habitat blocks.

In order to test this hypothesis and to investigate the spatial distribution of badger group territories in Wytham, Kruuk performed bait-marking surveys, during which he delivered a sticky mixture of peanuts, syrup, and inert plastic beads daily to each badger sett for a two-week period. Each sett was supplied with a unique colour of beads. This readily consumed confection is digested in-part by the badgers, but the coloured marker beads are passed in the badgers' faeces, which are neatly deposited in pits at well-used latrine sites, marking the periphery of each group's territory. These bait-marking surveys are still being carried out biannually and continue to provide a wealth of information on badger spatial behaviour. The 478 latrine sites currently on record for the Wytham badger population, when compared to surveys in previous years, reveal that the population has consistently utilized an area of around 6 km$^2$ during the past two decades. Variation among territories in the availability of food-rich habitats is reflected in the reproductive rates and body weights of the badger groups that inhabit them (da Silva *et al.* 1994).

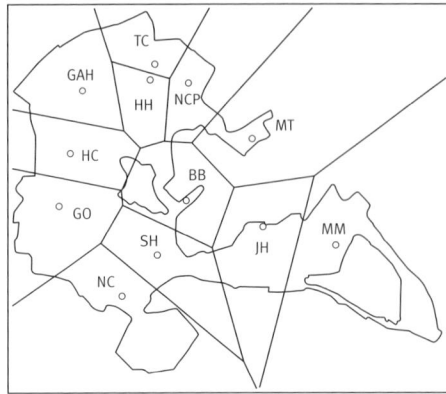 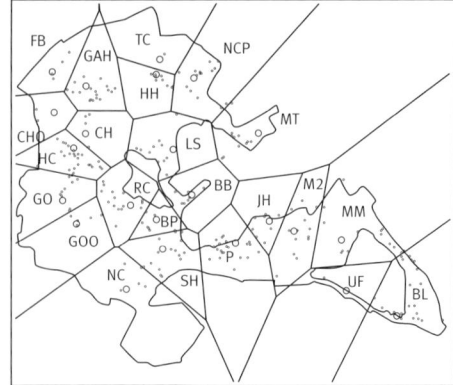

**Figure 10.4** Territories of social groups of badgers in Wytham Woods in 1974 (left) and in 2008 (right). The number has nearly doubled, from 12 to 23 during this period.

The 'Ideal Flea Distribution Hypothesis' (Johnson *et al.* 2004) provides another, but not exclusive, explanation for the evolution of sociality in badgers, and is based on the occurrence of extensive grooming amongst individuals: ectoparasites impose fitness costs on their hosts (e.g. in badgers, transmission of diseases and blood parasites: see below). As mobile ectoparasites follow an ideal free distribution amongst their host individuals (that is, all hosts, who come in contact with each other, end up with similar parasitic loads), allo-grooming can lower these parasite burdens amongst group-members.

Whilst the area used by badgers in Wytham has remained constant since the beginning of the study, the number of territories and the number of badgers inhabiting Wytham has increased significantly over the past 30 years. Since Hans Kruuk first surveyed Wytham in 1974, the number of social groups increased from an initial 12 to 20 by 1993 (Macdonald and Newman 2002). Recent bait-marking surveys reveal that the present territory count stands at 23 (see Fig. 10.4), though ultimately the subdivision of the 6 km² range steadfastly utilized by the population limits the viability of the smaller territories long-term.

While there is a limit to the number of social groups that can fit into an area, the number of places where badgers have dug holes has proliferated significantly, in proportion to growing population size. Macdonald *et al.* (2004) recorded 279 active sett sites within Wytham, comprising 1130 entrances, 80 per cent of which were active. In terms of habitat and geography, setts are non-randomly distributed within Wytham. Badgers preferentially select sites with sandy, well-drained soils, situated on north-west-facing, convex, and moderately inclined slopes at moderate altitude over other habitats.

Importantly, these new setts did not appear measurably inferior to older established ones. Doncaster and Woodroffe (1993) proposed that suitable sett sites might be limiting and influence the social structure of badgers. However, more recent evidence from Wytham indicates that sett sites are readily available and not a driving factor in social configurations (see Blackwell and Macdonald 2000).

Behaviourally, the increased socio-spatial dispersion of the population may also be significant. Bite-wounding, resulting from aggressive encounters between individuals, increases in frequency as well as in severity with increasing population density and social instability within groups (Delahey *et al.* 2006). Cresswell *et al.* (1992) proposed that the presence of annex setts correlated with increased productivity in younger sows; the proposed mechanism being that additional sett sites enable younger sows and their cubs to avoid the aggression of other individuals. In Wytham, increased social spacing appears to circumvent sociological feedback mechanisms that otherwise limit group sizes and cub productivity, allowing the badger population to sustain a higher threshold density than when fewer sett-sites were present (Macdonald and Newman 2002).

This research into Wytham's badger territories has stimulated thinking on carnivore territorial defense and the function of territory borders. Badgers from different social groups are frequently observed to meet at boundary latrines and pass by each other amicably into their neighbour's territory (Stewart *et al.* 2002). However, latrine marked borders clearly serve an important function in badger society by defining group ranges, and are frequently reinforced by all adjacent groups with fresh faeces, urine, and sub-caudal scent-marks. The Passive Range Exclusion hypothesis proposes that badgers setting out to forage from their setts will deplete their available food supply as they go, creating a gradient as they move away from the sett, leaving a depleted zone behind them. Ultimately, badgers foraging in this way will reach the advancing front of badgers from their neighbouring group. Should these badgers pass each other and forage beyond this contact zone they will enter into an area already depleted, and thus the point at which the two groups interface represents an isopleth (or 'contour') of peak food abundance, which is defined by latrine marking. Therefore, foraging beyond the group boundary is not necessarily an adversarial transgression. However, to approach a neighbouring sett directly generally sparks an aggressive response.

## 10.5.2 Badger numbers and the role of climate change

Since its inception in 1987, the four annual badger trap-ups in Wytham have resulted in over 8600 captures of more than 300 individuals. Trapping protocols are continuously monitored and refined to ensure the highest welfare standards (Thornton *et al.* 2005). As all animals get marked with an individual tattoo at first capture (usually as cubs), this trapping regime allows each individual to be followed through its lifetime, almost from 'womb to tomb'; thus, facilitating measures of population processes such as survival patterns, dispersal rates, and a variety of morphometric parameters. Population dynamics, as revealed by the trapping regime, indicate that at the inception of the capture–mark–release–recapture study in 1987 the Wytham population comprised 60 adults and 23 cubs; already a significant population size. However, by 1996, this population had risen to a peak of 235 adults, with 45 cubs born that year, maintaining this high-density through the late 1990s and the early 2000s, giving a maximum total badger density of 46.6 badgers per km$^2$ (Macdonald and Newman 2002, Macdonald *et al.* 2002). Currently, the population comprises some 220 adults and around 45 cubs per annum, creating a badger density of 37 adults per km$^{-2}$, which remains the highest published density in the world. Groups can be centred on one large single sett or between a number of dispersed outliers within a territory. These groups are generally of mixed sex composition, though all-male groups have been observed on rare occasion.

The survival dynamics of the population vary annually, and each cohort is affected differently. In this context, parasitic infections are of paramount importance. The blood parasites *Babesia missirolii* and *Trypanosoma pestanai* have both been recorded (Macdonald *et al.* 1999), with a tendency for young badgers to be most susceptible to infection, although with no pronounced pathology. However, a similar trend in juvenile infection was also seen with intestinal coccidian parasites. Both *Eimeria melis* and *Isospora melis* are present in the population (Anwar *et al.* 2000), but it is *E. melis* that is most significant: cubs are especially prone to parasitic infections and *Eimeria* prevalence is 100 per cent in cubs sampled immediately after first emergence. From this age onwards weather conditions over the summer are critical to the cubs' chances of survival (Woodroffe and Macdonald 2000). If food and water is limited, cubs will die of this coccidiosis, but with wetter, more earthworm-productive, summers cub survival is much improved (Newman *et al.* 2001) resulting in near-complete adult immunity. Moderated by weather conditions, juvenile coccidiosis is potentially one of the most important factors controlling cub survival in the Wytham population, and thus ultimately governs population recruitment. On average 50 per cent cohort mortality occurs by the age of 2.5 years. Badgers surviving juvenile years typically live to 8 or 9. Around 2 per cent of badgers make it into their teenage years, while Wytham's oldest known badger survived to 17 years old.

So detailed are these population data that it has been possible to model their dynamics using very responsive ecological models. The mean generation time is 5.8 years and the mean individual lifetime reproductive output was 1.4 offspring (Macdonald and Newman 2002). Analyses of the relative order of importance of demographic parameters reveal that fertility has the greatest influence on population growth, followed closely by adult survival, juvenile survival, and age at first reproduction, while age at last reproduction is of minor importance.

Climate is one of the primary controls on species diversity and distribution globally, and anticipated changes in global and regional climates have significant implications for species and habitat conservation (Root *et al.* 2007). During mild winters badgers lose less weight than during colder ones, and as heavier females are more likely to rear cubs, the increasingly mild weather throughout the 1990s was correlated with an increase in Wytham's badger population (Macdonald and Newman 2002). Rainfall also influences survival dynamics, with more cubs dying in dry than in damp summers; and wetter weather in spring, and specifically in May, is beneficial for juvenile survival (Woodroffe and Macdonald 2000).

The Wytham study makes another important contribution to our understanding of badger populations. Badger numbers have increased nationally and some have claimed that this was due to increased levels of protection. However the Wytham numbers have

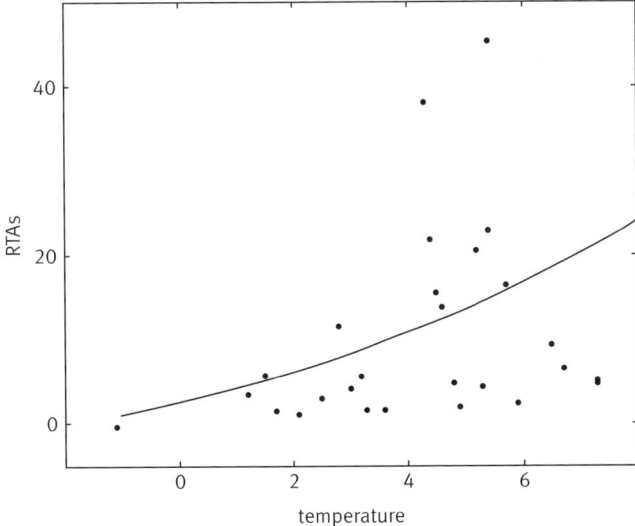

**Figure 10.5** The relationship between badger mortality in numbers of road traffic accidents and temperature.

increased at a similar rate without a change in protection, showing that the increases can be explained by climate not changes in protection.

Badgers are highly susceptible to road traffic accidents (approximately 50 per cent of Wytham's annual badger mortality can be attributed to road traffic accidents). The Wytham studies also show that the number of road traffic accidents increases as temperature (especially in February) increases, because badgers are more likely to leave their sett for foraging trips when temperatures are milder (see Fig 10.5).

This observation only underlines the intricate ways in which changing climate might interact with anthropogenic agents to influence the fortunes of a species.

### 10.5.3 Badgers as a model for the evolution of sociality

Eurasian badgers are timid, nocturnal, and individuals look rather similar. The opportunity to study their social behaviour and ecology was limited until WildCRU's work in Wytham pioneered remote controlled infra-red video surveillance equipment combined with a fur-clipping method (Stewart *et al.* 1997), to identify individuals on film. This allowed privileged insights into the behaviour of individually known animals.

Most badger social behaviour follows a tit-for-tat pattern, indicative of a species that has evolved only basic forms of sociality. Perplexingly, there is no clear social hierarchy in either a feeding or a mating context. In general, badgers show very few cooperative behaviours. Whilst they spend a considerable proportion of their time grooming each other (Stewart and Macdonald 2003), this behaviour is usually directed to areas of their bodies which are hard to reach by themselves (Johnson *et al.* 2004), and grooming partners do not give one another any credit, they stop grooming within one second of each other. Sett digging and maintenance is predominantly the responsibility of males, potentially as a means of attracting females (Stewart *et al.* 1999).

As badgers are nocturnal, visual communication signals are essentially non-existent and badgers communicate predominantly through scent (Buesching and Macdonald 2001). They possess a unique subcaudal gland, which they use to convey information about their individuality, fitness, and group-membership (Buesching *et al.* 2002). Whilst males scent-mark objects, such as the sett, paths, and feeding sites, predominantly during the mating season, object-marking by females peaks during the period in spring when the cubs come above ground for the first time and learn the geography of the group territory (Buesching and Macdonald 2004). Whereas object-marking appears to serve predominantly as an individual advertisement signal, sequential allo-marking on the body of other badgers on the other hand appears to serve also as an appeasement behaviour between group-members (Buesching *et al.* 2003). The necessity to include individual-specific information in a scent-profile that also encodes group-membership leads to an olfactory dilemma in badgers, as the chemical composition of the subcaudal gland secretion is largely dependent on the bacteria present in the subcaudal pouch, metabolizing the primary gland products. To generate a shared group-odour, members of the same social group will perform a behaviour termed 'mutual allo-marking'. During this, two badgers will back up pouch-to-pouch and exchange small amounts of subcaudal secretion, and thus bacteria, which will lead in turn to an assimilation of the bacterial pouch flora amongst group-members (Buesching *et al.* 2003), facilitating group recognition and thus cohesion.

### 10.5.4 Intra-guild competiton between badgers and other species

Since the demise of the wolf, badgers in the UK have had no natural predators, but due to their strong bite and locking jaw are formidable and can inflict severe injuries. Their characteristic white face with black eye stripes is likely to serve as a warning signal to other predators so as not to mistake them as a potential prey animal (Newman *et al.* 2004). The relationship, and indeed the existence of competition, between badgers and other members of their community raises interesting questions. The most obvious competitors in Wytham Woods are foxes (Hofer 1988) and the two species often time-share the use of dens. In food-related encounters, the much flightier foxes usually avoid direct contact with badgers.

## 10.6 Hedgehogs

Hedgehogs (*Erinaceus europaeus*), also share many food items with badgers, such as earthworms, snails, and the eggs of ground-nesting birds. While hedgehogs are protected from most potential predators by their spines, badgers readily roll them over to attack their unprotected underside (Doncaster 1994). Thus, hedgehogs show a discontinuous distribution around Wytham Woods, which is negatively correlated to badger numbers (Young *et al.* 2006), although separate pockets are usually within dispersal distance of each other. Hedgehog home ranges are approximately 40 ha (Doncaster *et al.* 2001), although they react to changes in food resource distribution with shifts in their activity patterns. If faced with a choice, hedgehogs will avoid areas marked with badger faeces, but despite their fear of badgers will return to high-yield feeding sites within 24 hours if no fresh faeces are deposited in the meantime (Ward *et al.* 1997).

## 10.7 Squirrels

The native red squirrel (*Sciurus vulgaris*) has long been replaced in Wytham by the American grey squirrel *Sciurus carolinensis*. Introduced by the Victorians, the grey squirrel spread rapidly throughout most parts of Britain, wiping out the native reds by transmitting a fatal disease, the parapox virus. There is some niche separation (Bryce *et al.* 2002; Macdonald *et al.* 2001) which can lead to co-existence of both species in some habitats, especially mature pine forests. Whilst red squirrels eat predominantly the nuts and the seeds of coniferous trees, the greys also strip the bark of deciduous trees to lick the sweet sap, killing the tree if it becomes ring-barked. In Wytham, grey squirrels predominantly attack sycamore (Burgess 1957) and younger beech trees, especially those in plantations. Their numbers have been estimated by drey (spherical squirrel nests) counts (Don 1985), and their numbers are commonly controlled in Wytham by the provisioning of specially designed poison hoppers with warfarin-treated grain.

## 10.8 Amateur volunteers as biodiversity monitors

To understand the complex network of inter-specific relationships between the different mammal species as well as their place in the wider ecosystem of plants and other animals, (and how this might be affected by a changing environment due to human intervention and climate change), detailed monitoring is required. Biological monitoring is time- and labour-intensive, but in many cases requires only limited scientific training (Macdonald *et al.* 1998). To evaluate appropriate training methods, and to test the validity of data collected by novice volunteers, WildCRU implemented a Mammal Monitoring Project in collaboration with the Earthwatch Institute (Europe) in April 2000. Although different volunteers show different strengths and weaknesses, in general most monitoring tasks, such as standing faecal crop counts for deer and field sign transect surveys, can be taught effectively with half a day of practical field training if the task is concise and is explained sufficiently (Newman *et al.* 2003).

Public participation in research, or 'citizen science', is also encouraged at Wytham, for example volunteers are invited to help with the annual badger census, held during the first week of May each year since 1976, when all badger setts within Wytham Woods are

**Figure 10.6** Amateur volunteer biodiversity monitors at work.

watched simultaneously to count the number of cubs and adults present at each sett over the course of three evenings. Interestingly, people see consistently 65 per cent of the badgers known to live at any one sett from the trapping records, a result that provides a valuable calibration of this method often used by badger groups and wildlife trusts to estimate badger numbers in an area. To further implement WildCRU's mission to engage, enthuse, and involve the local community in conservation research, residents from the Ley Community Drugs and Alcohol Rehabilitation Centre are chosen on merit to assist with manual tasks on the badger project, an initiative for which WildCRU and Earthwatch were awarded the Charity Award in 2001.

# 11

# Conservation Management of Wytham Woods

N. Fisher, N.D. Brown, and P.S. Savill

## 11.1 Conservation management before 1900

The idea of setting an area aside exclusively for the protection, enjoyment, and study of nature is a peculiarly modern idea. Protected areas have existed in Britain for over a thousand years but these have been to protect nature from the many, for harvesting and hunting by the few. Wytham was no exception, providing game, timber, coppice, and other resources, first for Abingdon Abbey and then for the earls of Abingdon.

It was not until the mid-eighteenth century in Britain that nature, alive and untamed, began to be valued aesthetically (Evans 1997). As the population became increasingly urban and the countryside was transformed into a patchwork of fields and hedges, so the concept of a benign and beautiful nature came to the fore.

> To her fair works did Nature link
> The human soul that through me ran;
> And much it grieved my heart to think
> What man has made of man.
> 'Lines Written In Early Spring', William Wordsworth, 1798.

In Wytham, it was not until the Enclosure Act of 1814 allowed a new structure of land ownership that traditional forms of land-use were swept away and modern patterns of estate management were laid down. The fifth Earl of Abingdon, following the romantic fashion of his time, sought to create sweeping curves, irregular outlines, and broken masses (Grayson and Jones 1955). Wytham Park was created by cutting away parts of Marley Wood and Mount Wood to leave a stunning open vista running uphill, with a scattering of standard oak trees. Ornamental belts were planted, as were clumps of trees for aesthetic purposes. A mix of native and non-native tree species such as horse chestnut, sycamore, and turkey oak were included. Interestingly, these nineteenth century landscapes, which blurred the boundary between nature and the garden, had a conservation motive at their heart; for it was believed that the moral and spiritual decline that caused so much grief to Wordsworth could be reversed by reconnecting the human soul with nature. Nineteenth century landscape design was about conservation of cultural values through the medium of nature.

The late nineteenth century was a period of widespread agricultural decline, and saw the demise of many great estates as a succession of bad harvests, competition from foreign imports, taxes, and falling rents ate into meagre profits. As the prosperity of the Wytham Estate declined, the Woods fell into a state of relative neglect and shooting became an increasingly important driver of management. Up until 1831 the right to hunt game was restricted, by law, to those of certain social status. The Game Act abolished this restriction, and made the right to kill game conditional only on the possession of a game licence. This, combined with the development of reliable and safe shotguns, led to a spectacular increase in the popularity of 'field sports'. Conservation management was seen as important to avoid a situation where hunting endangered the existence of, or seriously diminished the stock of, game. Lords of the manor were authorized to appoint professional gamekeepers to manage game and protect their estates against poachers. Conservation was very specifically directed at promoting shooting, often at the expense of non-game species. Game-keepers had a major impact on birds of prey populations at this time. All of Wytham's ponds were created, for shooting purposes, during the second half of the nineteenth century, as were its broad sweeping rides.

## 11.2 Twentieth century conservation

Protective management of species and habitats—our modern conception of nature conservation—did not begin explicitly until the end of the nineteenth century in Britain. The accelerating pace of damage to natural habitats by urban and industrial development inevitably raised concerns and helped to thrust the fledgling science of ecology into the mainstream. This was aided by developments in biology which recognized the relationship between species, populations, and the environment (Bowler 1992). Management of the natural world, rather than simple exploitation, required detailed understanding of the processes governing the abundance and distribution of species and this was the explicit focus of ecology. The Victorian passion for collecting and cataloguing nature as a means to achieve order, understanding, and control, was replaced by a conviction that the study of ecology would provide the insights necessary to manage nature for human benefit. This was no longer the much disparaged 'stamp-collecting' type of natural history, but a true science which examined communities of organisms in context in order to identify what factors drove change in numbers. In 1913 the British Ecological Society was formed to provide professional organization for the newly developing science of ecology, and there was rapid improvement in the methods of mensuration and enumeration.

These attitudes are strongly reflected in one of the first ecological studies of Oxford and its surrounds, Church's (1922) *Introduction to the Plant Life of the Oxford District*. Church castigates as 'unscientific' the time-honoured method pursued by past generations of amateurs, who tramped the country collecting all the plant species to complete a 'herbarium'. His book, which was designed as a text for Oxford University biology students, sought instead to explore the:

'...intimate relation of the plants to their conditions and each other, as determined by continued observation and experiment, constituting what has become known as their Ecology'

(Church 1922).

A similar conceptual revolution was occurring in zoology. In 1932 the Bureau of Animal Population was established at Oxford University by Charles Elton (see Chapter 8, Section 8.2) to examine the controls on animal species numbers. The research work of the Bureau was seen to be of great practical relevance:

If adequate records are available, scarcity due to persistent over-destruction can be readily distinguished from the purely temporary scarcity due to a 'crash' in a normal cycle of fluctuation.

(Anon. 1937)

Sporting interests were, however, still strong. One of the first studies carried out by the Bureau was a survey of partridge numbers on the major estates across the whole of the country.

This new cadre of professional ecologists took the lead in campaigning for and then in designing the policies necessary to create a new national system of reserves. They argued that there was a need for areas to be set aside for nature preservation, its scientific study, and for education. These reserves would provide the laboratories and classrooms in which the research and education necessary for a science-based conservation management could be carried out. In 1947 the Report of the Committee on the Conservation of Nature in England and Wales (Huxley 1947) concluded:

In the long run, the success of wild life conservation must depend predominantly on the completeness with which wild life problems and their implications are known and on the extent to which public opinion accepts them and therefore gives support to the necessary measures of conservation.

It was seen as futile to try to preserve nature when there was only a very partial understanding of how natural ecosystems worked and when conservation initiatives were not supported by the general public (Sheail 1995). But although it was important that the public understood the rationale for conservation management, there was a widespread view amongst the scientific community that public access was inimical to proper protection. Nature reserves were to be for scientific study and education, not for recreation.

Nature-sensitives admiring the beauties of the local flora are not happy unless they can take away as much as they can carry with them; leaving behind in exchange ginger-beer bottles and orange-peel.

(Church 1922)

At the same time, momentum was developing for the creation of a series of national parks. These had a very different motive—that of improving access to the countryside for a steadily growing urban population and to protect traditional rural landscapes. The post-war aspiration to create a new world 'fit for heroes' saw the countryside as a resource that everyone should be able to enjoy. Battered bodies and damaged minds, it was thought, would be healed by peaceful and health-giving out-door recreation. National parks could also allow new planning controls to prevent insidious urban sprawl and conserve traditional farming use.

Whilst many would see these different motives for conservation as complementary, there was clearly tension between those who wanted to restrict access to reserves so that they could be set aside for nature, free from human influence, and those who saw that the countryside had been shaped by centuries of human use and who felt that it should be made widely accessible for public enjoyment.

## 11.3 Management by Oxford University

These would have been very live issues at the time that Raymond ffennell made his gift of Wytham Woods to the University in 1942 (Chapter 1, Section 1.6), particularly in Oxford where so many of the protagonists lived and worked. ffennell specified that all reasonable steps should be taken to protect the beauty of Wytham and that it should be used for education and to provide facilities for research. The mining magnate was keen to provide a natural laboratory for the university but also to promote education both for students and underprivileged youngsters from urban areas. His vision of Wytham was clearly one of a wilderness area, free from the impacts of pollution and production management.

*The Times*, reporting the gift, commented that:

...the University has acquired a colony, so fortified that it is inviolable by the pressure of traffic and industry, in which the life of learning and contemplation may be finding its best opportunities of enrichment for a thousand years to come.

(*The Times*, 11 February 1943)

But ffennell, whilst aiming to provide a facility for the University, was also determined that the Woods should be 'thrown open to public resort' thus providing access to University and city alike to the enjoyment of the serene beauty of rural England. He was renowned as a philanthropist and generous benefactor of public education. It is clear that he would have held little truck with the idea of excluding casual ramblers and bird-watchers in order to privilege research. His vision of accommodating research, nature conservation, and public access was prescient of twenty-first century approaches to conservation management.

There were more conflicts to come over the management of the Woods. The job of running Wytham Woods was handed to the University Forestry Department, and for the next twenty years the focus of their actions was to improve the quality and quantity of timber production. There was much to do. Almost all valuable timber had been 'creamed' between 1900 and 1918 (Osmaston 1959). Jones (1945) made a preliminary description of timber crops and found poor stocking rates, poor quality standards, a neglect of coppice management, large areas of sycamore regeneration, and severe rabbit damage. The wood pasture had reverted to secondary woodland through natural successional processes. The Oxford foresters saw Wytham as a derelict man-made woodland (Biological Advisory Committee minutes 1948), and Grayson and Jones (1955) had 'never seen a more neglected, devastated and derelict wood'. There was a yawning ideological chasm between their views and those of the Oxford ecologists.

When donating the Woods to the University, ffennell had made it a condition that the natural flora and fauna were to be preserved, with 'bird and flower sanctuaries to be set aside if necessary'. A Biological Advisory Committee was formed to give guidance on these aspects of estate management. It soon became clear that the ecologists felt that it was imperative that there were areas set aside for no management intervention. Rather than neglect and dereliction, they saw Wytham becoming ever wilder and offering opportunities to study the interactions between species unaffected by human influence. The importance of these wild areas as 'controls' in research activity was deemed to be crucial. The head of the Bureau of Animal Population commented that by the time the University became owners, the Woods were 'pleasantly unkempt' (Elton 1966).

By 1947 the Biological Advisory Committee had negotiated a compromise between the two sides that allowed the demarcation of certain biological reserves. These were excluded from forestry operations and evaluated as protected areas. Originally 40 ha were set aside, but ultimately this was increased to 75 ha. The decision to create wilderness reserves in Wytham mirrored the approach to conservation that was being taken in the wider countryside at this time. SSSI status was bestowed on Wytham in 1950 and was linked to maintaining areas as 'reserves' rather than encouraging active management.

However, minutes from a 1948 meeting of the Committee note the 'irreconcilable conflict of interest between running an estate to the general accepted principles of estate management and keeping part of it as a nature reserve for scientific and ecological studies'. The ideological differences between foresters and ecologists were never far from the surface for at least 40 years of Wytham's history.

## 11.4 Wytham's role in national conservation policy

Church (1922) described the local Oxfordshire flora as 'characteristically commonplace... with no special developments in any direction, and with little to attract the visitor from other more favoured districts'. Nevertheless, whilst nature conservation was still in its infancy Wytham was of seminal importance because it was the laboratory in which important concepts were developed. The fact that it was not exceptional meant that it faced many of the typical problems experienced by conservation managers in the wider countryside.

One example of the importance of Wytham based research was the Wytham Ecological Survey initiated in 1943 by the Bureau of Animal Population (see Chapter 8, Section 8.2). Elton and his team had concluded that in order to understand population processes it was important to have a full measure of community relationships and the survey in Wytham was designed to improve the methods of recording these. The survey was partly supported by the Nature Conservancy as a pilot study for developing methods that would be applicable to any nature reserve (Elton and Miller 1954). This survey was to continue for over 20 years and became the basis for Elton's (1966) book *The Pattern of Animal Communities*. It was no vain claim that Wytham was more intensively studied by animal ecologists than any other area in the country, or indeed the world. By 1964 the Ecological Survey had identified 3800 species and Elton (1966) estimated that as much as a fifth of the British fauna was to be found within Wytham.

Wytham has continued to play a leading role in conservation biology research. Long-term studies of tit and badger populations by the Edward Grey Institute and the Wildlife Conservation Research Unit have permitted comparisons with environmental data sets providing key insight into human impacts including climate change. However, some would argue that despite half a century of scientific research effort and a phenomenal increase in our knowledge and understanding of biological systems, practical conservation has made only halting steps forward. The idea, espoused in 1947 that successful conservation would spring from detailed biological information has been found wanting. There is a growing realization that success will often depend on our ability to make educated guesses when information is incomplete. It will also depend on incorporating insights from social sciences and humanities and recognizing that conservation must take place in a human-dominated landscape rather than a protected wilderness (Robinson 2006). For nearly a century, Wytham has been a hotspot for the robust, long-term, and often innovative science on which high quality conservation depends. The frustration is that science often plays a minor role in setting national conservation policy.

## 11.5 Conservation management: The last 60 years

Of the 390 ha woodland area in Wytham 137 ha (35 per cent) is ancient semi-natural woodland, 90 ha (23 per cent) secondary woodland, and 163 ha (42 per cent) recent plantations of which 25 ha are on ancient woodland sites. The plantations were mostly established in the 1950s. Those not on the ancient woodland sites were established on non-calcareous grasslands and arable fields. Almost 25 ha of limestone grasslands also exist with a range of histories, ages, and conservation values. Wytham thus has an assortment of habitats that range from ancient semi-natural woodland and abandoned wood pasture to conifer-dominated plantations, and from arable weed plots to semi-natural calcareous grassland. There is also a valley-side mire and several ponds. The 1977 SSSI citation states that the site has an exceptionally rich flora and fauna, with over 500 species of vascular plants and 800 species of butterflies and moths having been recorded.

The transfer of responsibility for managing the Woods from the Forestry Department to the University Land Agent in 1962 (see Chapter 5) led to research becoming the main driver of woodland management practices. The order and tidiness of the previous traditional forestry operations gave way to promoting natural regeneration and an increase in the quantities of senescent trees and lying dead wood. Active management steadily diminished and non-intervention became normal. In the 1980s and 1990s regular thinning regimes became increasingly rare even amongst the plantations.

Gibson (pers. comm.) summarized the key historical events that have had major influences on the Woods since 1943. These included the creation of plantations until 1963 and the last coppicing also occurring in 1963, as well as the occurrence of myxomatosis in 1954. Up to 1954 rabbit grazing had kept most grassland and scrub regrowth in check. Deliberate intervention on these areas was very seldom necessary. However, by the mid-1970s many areas of grassland had been invaded by scrub and some limited restoration was started. The main driving force for this was Dr Gibson who recognized the ecological and research values of these sites. Before 1980 SSSI designation actually did very little to protect habitats. It was only after the Wildlife and Countryside Act of 1981 that the Nature Conservancy Council's powers to influence detrimental management increased. Agreements for positive conservation management though were still on a voluntary basis.

It was only towards the end of the century that a comprehensive Site Management Statement, written in conjunction with English Nature, linked together forestry, conservation, and research into a single, coherent plan. By this time conservation in Britain had seen a change of emphasis from the promotion of protection and preservation of specific sites to more flexible agreements that set clear management objectives. Conservation grants changed from being based on an assessment of profit forgone to subsidizing some of the costs of carrying out conservation operations in the field (Kirby 2003). Similarly, woodland grants changed in emphasis from regarding timber production as overwhelmingly important to a giving at least equal, if not greater, weight to public access and management for conservation objectives, including observing all the provisions of the UK Woodland Assurance Scheme (Forestry Commission 2004).

## 11.6 Woodland management

The great advantage that Wytham has over many sites is that the size of the Woods allows a variety of techniques to be tested. Thus management varies from minimum intervention, through intensive thinning, to a limited amount of coppicing.
All current management aims to:

1 minimize major disturbances to fauna,
2 allow regeneration of a range of native plants,
3 improve the range of age classes of trees, and
4 minimize adverse visual impacts both from within the woodland and from external viewpoints.

A policy of non-intervention is now practised in the great majority of the ancient, semi-natural and secondary woodlands in Wytham. A small amount of coppicing might be carried out in places in order to meet conservation objectives. Bean Wood has been the favoured location.

It is hoped that a start can soon be made at restoring all the plantations to something that resembles more locally native woodland, including the 25 ha of post-Second World War plantations on ancient woodland sites (known as PAWS). This will be done by following the practical guidance written for the purpose by the Forestry Commission. Essentially, the approach is to thin the trees, following normal forestry practices. In thinning, locally native species are favoured for retention, including ash, oak, and beech. The non-native species will eventually be removed completely. These include the coniferous components and non-native broadleaves. In many of the plantations, 60–70 per cent of the stems will be removed. Attempts to remove sycamore might prove more difficult (see below). Natural regeneration would be encouraged and, if it fails, the compartment will be planted. Although deer numbers were probably as low in 2009 as they have been since 1980, temporary deer fencing might be required to assist regeneration in becoming established. The effects of climate change on natural regeneration and woodland composition are currently unknown, but in the relatively near future decisions will have to be made about whether to try to maintain the *status quo* or whether areas are allowed to move into a new 'adapted' state.

This type of management will eventually result in a predominantly irregular broadleaved high forest and should confer the sort of scientific and amenity benefits advocated by Elton in 1961 (13 January 1961. Notes for the Wytham Committee, unpublished). Ultimately, old growth forest structures will develop, including substantial standing and fallen deadwood communities.

In a report to English Nature, Mountford (2000) included Wytham within a provisional list of sites for minimum intervention baseline recording and long-term monitoring. Monitoring has been practiced periodically in the Dawkins vegetation plots since 1974

**Figure 11.1** A recently thinned beech plantation in Wytham Woods.

and continues under the auspices of the Environmental Change Network (ECN) and Natural England.

Sycamore is generally disliked by ecologists, being a non-native tree that casts a very dense shade. In Wytham, it has been a cause for concern for many years because it appeared to be invading much of the ash woodland rather aggressively. It is thoroughly naturalized and regenerates plentifully and successfully in most places where it occurs. Its status in Wytham is monitored as part of the recording of the Dawkins plots. It has been the subject of at least two scientific investigations. Waters and Savill (1991) suggested that alternation of regeneration was likely to occur between ash and sycamore generations and that an equilibrium between the two would eventually become established. Morecroft *et al.* (2008), like Walters and Savill, thought that sycamore is unlikely to become a problem within undisturbed ancient semi-natural woodland and may well decline in undisturbed areas where it is found in competition with ash. The predicted summer droughts may also limit its spread.

A management issue that is being considered at the time of writing is whether to plant oak trees in the Woods. It was noted from the time of Osmaston's 1950s management plan that while there were a reasonable number of mature oaks, there was very little regeneration. This situation has persisted and been noted in subsequent inventories ever since.

## 11.7 Veteran trees

The significance of deadwood habitats within Wytham has been recognized since at least the 1960s, for example Elton (1966) (who suggested that perhaps a fifth of Wytham's fauna is in fallen timber and slightly decayed trees: see Chapter 3), Paviour-Smith and Elbourn (1993), and Kirby *et al.* (1998). English Nature's Parkland and Wood Pasture Veteran Tree Survey (of 16 July 2004), placed Wytham in the top 30 sites in southeast England for deadwood habitats. A selection of veteran trees in the Woods was identified by Cleveland (1997). A programme of 'haloing', or the gradual removal of competing trees and other woody vegetation from round the large old veteran wood pasture trees, is progressing slowly and the veteran tree management techniques proposed by Read (2000) are playing an increasing role in preserving the old beeches (Plate 14) in particular.

The veteran trees are mostly found in the quite extensive areas of former wood pasture. They all now have closed woodland canopies resulting from infilling with natural regeneration, but they offer the prospect of restoration. Consideration is being given to converting about 7 ha in Little Ash Hill back to true wood pasture, which could be done without adversely affecting any research activities. Similarly Wytham Park, whose trees are known to be important for invertebrates (Elton 1966) has lost many through agricultural intensification, the use of fertilizers, and simple ageing. A replanting programme started in 2005, coinciding with plans to convert the Park to the Soil Association's organic status.

**Figure 11.2** Broad oak, one of the largest and oldest oak trees in Wytham Woods. It is a remnant from an old hedgerow. Photograph: Fred Topliffe.

## 11.8 Deer

Deer browsing within the Woods became a serious problem in the 1980s (see Chapter 10, Section 10.2 and Fig. 5.5) to the extent that the ground flora and tree regeneration both suffered very badly. The need for controlling their numbers was stressed in a 1990 Report. The enclosure of the Woods by a deer fence in 1989 led to a significant increase in grazing pressures within them, because the deer could not get out to the neighbouring farmland. The ecological aspects of the rise and subsequent control of deer populations are described in Chapter 5 Section 5.6 and further information is given in papers in *Forestry* (Volume 74, 2001), including the effects of deer on invertebrates, butterflies, and birds.

**Figure 11.3** A large veteran beech tree in Wytham Woods. Crown reduction is in progress.

Because of the extent of browsing caused by high deer populations and the dense shade levels resulting from the suspension of thinning in many parts of the Woods, signs of regeneration of trees have been very scarce since the mid 1980s. Marley Wood, which was separately fenced in 2001, has been almost deer free since 2002. Future surveys of the Dawkins vegetation plots should provide an indication as to the extent to which and how quickly the flora returns. Similarly, work by the Edward Grey Institute will analyse changes in bird populations. One of the key research questions is whether the vegetation will revert to its former state or whether selective grazing, nutrient deposition via dung, and the unknown state of remnant seed banks will lead to a different vegetation community once deer populations can be maintained at acceptably reduced levels.

In 2009 the deer density in the Woods was probably at its lowest level since 1980, at around 20 fallow, 20 roe, and scarcely any muntjac deer: these are the population targets set for the end of March each year. At their highest, in 1998 there were estimated to be about 420 fallow deer in the Woods, as well as roe and muntjac. Numbers of the latter are uncertain. From a deer management and conservation perspective, three management options are available for keeping numbers down:

1 concentrate on reducing overall deer numbers,
2 protect key conservation sites by internal fencing, and
3 manage habitats to alleviate pressures on the more sensitive sites.

The latter is currently favoured with the creation of deer glades and managed shooting lanes as key components. Areas of rotationally mown (forage harvested) grasslands provide deer grazing areas. It is hoped that restructuring the Woods by widening rides and opening plantations by felling conifers will reduce grazing pressures elsewhere. This technique not only permits more effective deer control and enhances habitats but also, importantly, is less costly than internal fencing.

Internal fencing to create small exclosures has been employed on a very limited scale to allow coppice re-growth along managed rides and also to protect blackthorn regeneration. The latter is important for a black hairstreak butterfly (*Satyrium pruni*) management programme which was formulated by Gibson (1988).

Deer numbers have been significantly reduced since 1998 and target numbers (see above) have also been steadily reduced. Target numbers are set every three years on

**Figure 11.4** Widening Mint Ride, to meet nature conservation and deer management objectives.

**Figure 11.5** Coppice re-growth in coppice-with-standards, Bean Wood.

monitoring of the deer population and the recovery of the vegetation. The eventual aim is to find and maintain a level at which deer do not adversely affect the woodland ecosystem. It is unclear whether the current population targets will achieve these conservation objectives.

## 11.9 Grey squirrels

The American grey squirrel was introduced to Britain several times between 1879 and 1920, and is now widespread over southern England and Wales. In the 1950s and before, the squirrel was regarded as a pest of secondary importance to the rabbit, but the significance of the problem it poses has grown throughout the last 50 years, as attempts to control its spread have failed. Squirrels have become common in mixed deciduous woodland, and are well known to foresters on account of their bark-stripping activities (see Chapter 10.7). Squirrels can cause serious damage to a wide range of broadleaved tree species, and are now regarded as one of the biggest problems facing broadleaved forestry in Britain.

Different trees are damaged in different ways. Beech is particularly susceptible as a tree of 8–15 years old. All above ground parts are at risk, as beech does not lay down thick bark. Loss of the leading shoot is a common consequence of squirrel damage, and promising young trees can be turned into 'witches' brooms' by repeated bark stripping at the growth point. Where bark is gnawed from the base of trees, growth can be seriously affected. In extreme cases, the whole trunk is ring-barked and the tree killed. More usually, patches of bark are removed, and the wounds become focal points for decay or weak points in the stem. Partial bark damage to beech often leads to fluting as the wounds heal. Bark removal from canopy branches in mature trees can lead to dieback of

**Figure 11.6** Damage caused by grey squirrels to a beech tree.

the tree tops. Oak is principally susceptible as a young tree, 10–20 years old, when its leading shoot is often attacked at a height of 3–4 m. Ring barking of the leader results in forking, severely compromising the potential of the tree to produce good quality timber. Older oaks are less susceptible on account of their thick bark, and damage is limited to the very young branches in the crown. Pole stage sycamore is also targeted by squirrels, which typically peel long strips down the stem. However the severity of squirrel damage to mature sycamore trees is unclear. Only ash and cherry appear to be relatively immune to bark stripping by squirrels.

Numbers of squirrels can increase very quickly. They are controlled in Wytham by laying out specially designed hoppers that contain warfarin-treated grain from April to July each year when cambial activity and sap flow are high. This is the period they do most damage. Because of fears of secondary poisoning the use of warfarin is limited to the plantations within Pasticks and small areas within Radbrook Common. Shooting is carried out throughout the rest of the Woods. Trapping is not done because of time constraints. Since 2008, when Forest Stewardship Council certification was gained, derogation for warfarin use has to be obtained from the EC. Earthwatch volunteers have assisted with the monitoring of tree damage.

## 11.10 Grasslands

The calcareous grasslands are one of the most intensively studied habitats at Wytham. Lloyd (1974) noted that there had probably been no active management over the majority of them since the Second World War. Scrub invasion became prolific as rabbit numbers declined due to myxomatosis. This, together with the spread of bramble and coarse grasses such as tor grass (*Brachypodium pinnatum*) and false oat-grass (*Arrhenatherum elatius*), began to dominate communities. A range of management options were suggested and in 1988 Gibson devised a management and monitoring regime for them (Gibson *et al.* 1987b, c). Initially this involved a limited amount of grazing by sheep on a small number of sites, whilst on other sites swiping occurred yearly. Some areas had scrub cut on a 7–10 year rotational basis. Further restoration work started on other areas of neglected grassland from 2001, with grazing being introduced to the majority of the sites. Rotational coppicing of scrub regrowth adjacent to grasslands will be introduced on a 7–10 year cycle.

The majority of restored grasslands are now (2009) maintained by spring and autumn grazing by sheep. Spring grazing is used to reduce the regrowth of scrub and brambles, though whether this and the resulting nutrient flush favours coarser grasses has not yet

**Figure 11.7** Restoring grassland on Lower Seeds.

been established. Grazing by cattle would be beneficial in that it would assist in breaking up the *Brachypodium* tussocks, but it has proved impossible to locate a suitable cattle herd. Gibson (pers. comm.) suggested flailing in March and using sheep to graze the subsequent fresh *Brachypodium* regrowth.

Since the decline of rabbit populations, grazing by large numbers of deer has influenced grassland succession. It appears that it was only the high density of deer, and in particular fallow deer, that was keeping successional processes in check. Since the phased reduction in deer numbers from 1998 the ungrazed areas of Wytham have shown marked increases in scrub regeneration. Deer tend not to graze areas used by sheep and probably have to find other suitable sites in the Woods. Attempts have been made to find out where these sites are through GPS tracking, but the technology has proved not to be sufficiently advanced to make this possible in terms of costs that can be afforded.

In 1982, Upper Seeds was taken out of arable production and an experimental process of reversion back to calcareous grassland began (see Chapter 7, Section 7.3). This has greatly improved knowledge about the successional processes among grassland plants, though much still has to be learnt about invertebrates and soil organisms. It is becoming increasingly apparent that traditional management practices are excellent for maintaining established habitats but are not necessarily ideal as restoration practices.

A small area of Upper Seeds was left for arable weed conservation. It is ploughed every year and then left alone. Details relating to it can be found in the annual reports of the Oxford Rare Plants Group.

## 11.11 Management of rides and other woodland edges

Over 30 km of woodland rides are linked into the grasslands. They span all the major soil types found in the Woods. A research project was started in 2005 to monitor vegetation changes on rides that had been subjected to a variety of mowing regimes and widening programmes. This work supplements research being undertaken by the Forestry Commission in various British woodlands (Ferris and Carter 2000). As the majority of Wytham is composed of closed canopy high forest the importance of ride communities cannot be overstated.

## 11.12 Ponds and Fen

Wytham has a nationally scarce valley-side fen, and the three ponds were created in the nineteenth century for shooting. Wormstall, which is in Wytham Park, was restored in 2002 and the Upper Radbrook pond in 2007, though some scrub cutting had been done in the 1970s. Scrub regrowth around them is cut on about a seven year cycle.

The fen within Marley Wood has been unmanaged since at least the 1950s. It is unclear when drainage outlets were added. Extensive coring work for pollen analysis has been undertaken in the past.

## 11.13 Conflicts between Research and Conservation

Conservation managers at Wytham have to be aware of potential conflicts with research interests. The creation of the (now abandoned) woodland and grassland Biological Reserves during the 1940s actually physically demarcated key research sites. During the 1960s and 1970s research priorities dominated over forestry and conservation objectives. Though 'neglect', or minimum intervention, has huge benefits for long-term ecological research, it was only in the late 1970s and 1980s that a few individuals, including C.W.D. Gibson, saw the potential for linking positive conservation management and research.

In 1998 a more flexible management regime was adopted (SMS 1998). This was based on the consensus of all those with research and conservation interests in the Woods. This still involves maintaining minimum intervention zones for long term research with management being concentrated on the remaining 150 ha. At present, one of the Conservator's objectives is to link conservation measures with research. Thus current research studies are investigating grassland restoration, ride management, veteran trees, and coppicing. A Geographic Information System (GIS) enables conservation and woodland management activities to avoid sensitive research sites. In an attempt not to disturb protected species, such as badgers and great crested newts, licences have to be obtained. No tree felling occurs during the birds' nesting season between the end of March and the beginning of July.

## 11.14 Promoting Public Access

When the Woods were bequeathed to the University in 1942 one of the conditions of the bequest was that they should be available for enjoyment by the residents of Oxford. Though there are no rights of way, anyone applying for a permit to walk in the Wood is given one. Permits are considered an important means of ensuring that visitors keep only to paths in order to minimize disturbance to research and vulnerable habitats. Although many people see Wytham as an area of unspoilt 'natural' countryside, fortunately the visitors' enjoyment has been enhanced by the majority of conservation projects, including, for example, grazing on the calcareous grasslands and coppicing. Efforts are made to explain the potentially more controversial management operations such as deer culling, felling dangerous trees, and woodland thinning. Unfortunately many of the outstanding veteran trees are adjacent to major access routes and could pose a danger of falling branches if they did not receive sympathetic attention from the woodland staff.

## 11.15 Monitoring the effectiveness of conservation management

Although Wytham Woods have been very intensively studied, and there has been an enormous increase in knowledge and understanding of biological systems in them, advances in practical conservation have been much slower and rather faltering.

Conservation management requires clear and measurable objectives to be set that are based on the results of long-term monitoring regimes. Regular monitoring enables trends to be established and allowances to be made for natural population fluctuations. One of the failings of the Wytham data sets is that many of the research priorities have not coincided with management requirements. Obvious exceptions exist, such as Gibson's grassland studies and Dawkins vegetation monitoring plots, but even these have limitations, such as the fact that, for example, there may not be sufficient plots within Marley Wood to pick up changes in vegetation soon enough, following the exclusion of deer (see Section 11.8).

Kirby and Thomas (1999, 2000) noted that conservation managers must decide whether they are concerned with flora (individual species present) or vegetation (abundance and distribution), the latter requiring more intensive sampling. They recommended management systems that reduce the rate of change but allow species losses in one area to be balanced by gains elsewhere in the woodland. They also added that in order to implement conservation objectives effectively, knowledge based on monitoring is required to indicate changes in the ground flora as well as in the tree and shrub layers. Kirby and Thomas (2000) stated that monitoring both rare and common species is necessary, and that stand and whole site surveys will often pick out different types of change.

In terms of conservation management Corney *et al.* (2008) reinforce these comments. They suggest that the flora of ancient woodlands has been moving along an 'undesirable' trajectory away from the site's nature conservation objectives. The drivers for it are deer grazing, increasing canopy density, and nutrient dynamics. They point out that 'the likely difficulty of reversing such changes...has important consequences for practical conservation management'.

In spring 2009 the new Forestry Commission Woodland Grant Scheme (WGS) not only requires exclosures to be created and monitored to indicate grazing pressures, (this has been done since 1992 in Wytham), but also for deadwood and photographic monitoring to be undertaken. In terms of species monitoring, many groups, such as butterflies and bats, are already dealt with by organizations such as Natural England (formerly English Nature—the Dawkins plots), the Environmental Change Network (ECN, and the University's Wildlife Conservation Research Unit (WildCRU). The grassland monitoring plots for newly restored grasslands set up by Gibson in 2003 will also be continued. A base line survey of all of Wytham's veteran trees will also be completed as part of the WGS. Nevertheless, many species are still relatively neglected. Data on rare or scarce plants are sporadically available whilst invertebrate records are often decades out of date.

## 11.16 Future challenges

In relation to the wider landscape, Wytham became an island of biodiversity during the latter part of the twentieth century. The surrounding farmers followed the general fashion of maximizing profits by applying the intensive farming methods of modern agriculture. This process, combined with the abandonment of thinning, coppicing, and pollarding within the Woods, placed extreme pressures on many species groups (Gibson 1985 unpublished). It was only at the turn of the century that a Countryside Stewardship Scheme for the land on the Wytham side of the Woods started to reverse this trend. This, in association with the increasing structural and habitat diversity that is slowly developing within the Woods means that the ability of species to survive and migrate through the estate should improve.

## 11.17 Climate change

Although the ECN has monitored many biological variables closely, little is understood about how any particular species on a specific site will be respond to climate change. A species' ability to survive is strongly dependent on micro-climates within particular sites. Wytham's many aspects, altitudes, and soil types might ameliorate some short- to medium-term changes. Management systems will also be crucial, though many questions remain to be answered. Planned thinning regimes in plantations will minimize climatic variations within these stands, but it is not known how climate change will alter stand structure and composition. Future experiments may include investigating how thinning regimes and coupe sizes affect biological change, and how extreme climatic events such as gales or drought might drive the ecological change. Lessons also need to be learnt on how veteran trees will survive. Climate change will not only affect a species but probably different age classes within the species.

## 11.18 Conclusion

The long and stable tenure of Wytham Woods has allowed continued long-term ecological research in many fields of biology. The data accumulated is increasingly being applied to conservation management regimes. The strength of research carried out in Wytham Woods lies in the number of long-term data sets that exist not only in terms of particular species, but increasingly for a range of habitats and management techniques.

# 12

# Wytham in a changing world

M.D. Morecroft and M.E. Taylor

Changing land use and management have been the main environmental changes in historical times at Wytham, as elsewhere in Britain, and have been documented in Chapter 3. We have, however, entered a new phase. In the last 40 years, it has become clear that the impact of people on the natural world has extended far beyond their deliberate intentions. In the 1960s the full impacts of agro-chemicals, such as DDT, began to emerge. In the 1970s and 1980s the dangers of 'acid rain' became a source of public concern, eventually leading to the regulation of power station and vehicle emissions. As Chapter 2 has already shown, these regulations have been successful in reducing the acidity of rainfall at Wytham, as elsewhere. Nitrogen deposition, particularly in the form of ammonia, which is mostly released from intensive livestock rearing, remains a cause of concern, however. Although a nutrient, nitrogen inputs to a plant community can lead to the loss of species adapted to low fertility conditions as the growth rate of other more competitive species increases.

From the late 1980s onwards one environmental issue has grown to be predominant in research, policy, and public debate: climate change. Global temperatures have risen by approximately 0.7 °C over the last century and there is scientific consensus that this is largely the result of rising concentrations of carbon dioxide and other 'greenhouse' gases in the atmosphere (Solomon *et al.* 2007). This rate of warming is expected to increase over the course of the next century under all realistic scenarios.

Climate change has become a major theme for research at Wytham in recent decades. The global rise in temperature is mirrored in the local Oxford temperature records (see Chapter 2) and a natural first step towards understanding the ecological impacts of this warming has been to look for changes in long-term ecological data sets. The Wytham great tit study (Chapter 9) has provided a classic example of climate change impacts detected in a long-term record, with a clear relationship between temperature and date of egg laying and a trend towards earlier laying.

## 12.1 The Environmental Change Network

The value of long-term ecological studies to understanding the impacts of environmental changes, particularly climate change and air pollution, was clear by the early 1990s and plans were developed to establish a new network of monitoring sites, throughout the UK. The result was the UK Environmental Change Network (ECN), which was established in 1992 to provide detailed environmental monitoring at a series of research sites, using common protocols. The Network monitors physical changes in the environment, such as climate and air pollution, together with their impacts on soils, hydrology, vegetation, and selected animal populations (Table 12.1). The measurements themselves range from continuous automatic measurements of climate and stream flow, to the skilled recording of plant species in plots, and counts of animal species, such as butterflies and bats, along transects. Land management is also recorded, so the effects of local changes can be separated from larger scale changes in climate and air pollution. ECN monitoring therefore captures a uniquely broad picture of the state of the ecosystem at any given time and allows the investigation of relationships between the physical and biological environment, as well as simply detecting change over time.

Wytham Estate was one of the 8 founding sites when the ECN was established. There are currently 12 terrestrial sites like Wytham in the network, together with 44 freshwater ones (rivers and lakes—which are less intensively monitored). The terrestrial sites include many of the UK's major ecological research sites, such as Rothamsted research

Table 12.1 Environmental Change Network measurements.

**Meteorology**. Hourly data for: air and soil temperature, humidity, precipitation, total solar radiation, net solar radiation, albedo, wind direction, wind speed, soil water, surface wetness.

**Atmospheric Chemistry**. Nitrogen dioxide, ammonia*, heavy metal concentration (lead, copper, nickel, cadmium, zinc, mercury and platinum) in air and rain water *, mercury in air and rain water*

**Water chemistry** for precipitation, stream water and soil water at 165 mm and 272 mm at Ten Acre Copse: pH, Conductivity, Alkalinity, Concentrations of Sodium, Potassium, Calcium, Magnesium, Iron, Aluminium, $PO_4$-Phosphate, $NH_4$-ammonium, $NO_3$-Nitrate, Chloride, $SO_4$-Sulphate, Dissolved Organic Carbon.

**Stream flow**. 15 minute interval volume flow.

**Soils**. Baseline survey, permanent plots sampled at 5 and 20 year intervals for Carbon concentration and all main nutrients. Sampled by depth and horizon.

**Vegetation**. Baseline survey, permanent quadrats recorded on 1*, 3, or 9 year time intervals, tree growth, pasture productivity.

**Animal populations**. Rabbits, deer, bats, butterflies, moths, ground beetles, spiders, spittle bugs.

**Weekly fixed point photography***

**Site Management**.

**Phenology***. Following the methodology of the UK Phenology Network (www.phenology.org.uk).

*indicates measurements which are not strictly part of the ECN programme but are managed with it. Full details of ECN protocols were published by Sykes and Lane (1996) and are available at www.ecn.ac.uk.

station, Moor House National Nature Reserve, and Alice Holt Forest. The Wytham ECN site includes both the Woods and the farmland of Northfield and Home Farms. ECN is a collaboration between a wide range of environmental research and funding organizations, but the programme at Wytham is operated by the Centre for Ecology and Hydrology (CEH). ECN researchers work closely with Oxford University and the programme has given rise to a succession of jointly supervized research student projects and research grants.

ECN data have provided valuable insights into many areas of research at Wytham and references to them can be found throughout this book. However, detecting the impacts of environmental change, particularly climate change, was its primary purpose. Most actual monitoring started in 1993, giving 15 year runs of data at the time of writing. In this time, the ECN automatic weather station has recorded a significant increase in mean annual temperature of almost 1 °C (Chapter 2, Fig. 2.4). The warmest temperature on record, 34.4 °C, was on 19 July 2006 at 15:00 GMT. The rise in mean temperature is substantially higher than the global trend over the same time period (Solomon *et al.* 2007), but this probably reflects the particular circumstances of the years in which the station has operated. Interannual variability in climate is large relative to the trend. The variability is even larger when seasonal data are analysed, and although there are upward trends in summer (June–August) and winter (December–January) temperatures, neither is significant. This is important as mean annual temperature is not usually the most relevant aspect of temperature for biological processes. Other temperature variables such as mean temperature during the growing season or the number of days above a threshold temperature in the spring are more important for specific contexts.

Precipitation has varied greatly in the period 1993–2007 (see Chapter 2). Annual precipitation fluctuated between 499 mm (1996) and 923 mm (2000), with even more dramatic seasonal changes. Particularly important for ecological impacts are summer droughts and flooding episodes. Summer rainfall varied from 35 mm in 1995 to 248 mm in 2007. 1995 was a serious drought episode, whereas 2007 brought summer flooding. It is not uncommon for the floodplain of the Thames at Wytham to flood during winter months, when the soil is saturated, but in 2007 it flooded in July, which was unprecedented.

Many biological processes are sensitive to climate and phenology—the timing of seasonal phenomena—is one of the clearest examples. Phenology records for the last ten years at Wytham show large year-to-year variability and there is evidence of plants flowering earlier than would historically be expected—primroses in January as opposed to early March and bluebells flowering regularly in March rather than April. Despite this, ECN data on populations and communities show little evidence of major ecological change associated with the rising temperatures of the last two decades at Wytham. However, most species in the Wytham area are not close to their climatic limits and might not be expected to show early signs of change, in the way that has been shown, for example, in studies of altitudinal range shifts (Rosenzweig *et al.* 2007). A wider analysis of ECN data across the whole series of sites, showed that changes consistent with an impact of rising temperatures were more common in upland sites, where there was

220 Wytham Woods

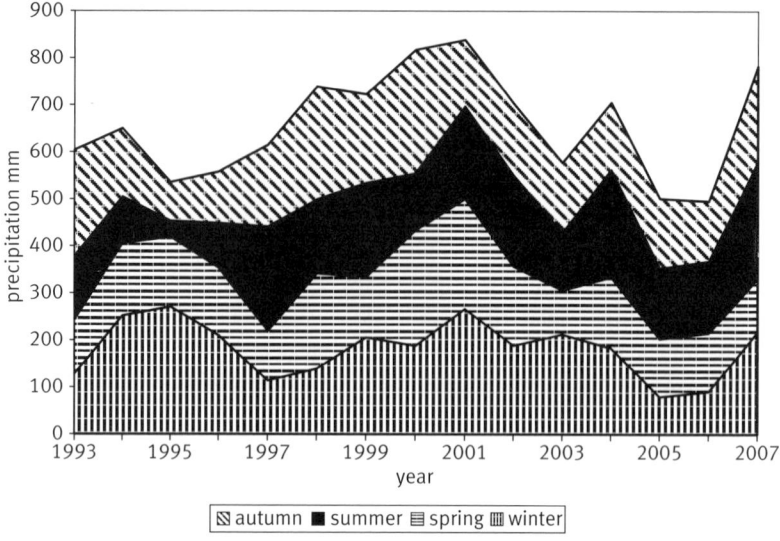

**Figure 12.1** Trend in precipitation at Wytham by seasons (Spring: March–May; Summer: June–August; Autumn: September–November; Winter: December in previous year to February in the marked year).

evidence of butterfly species colonizing new areas, whilst some upland Carabid beetle species declined (Morecroft *et al.* 2009).

In contrast to temperature, the effects of interannual variability in precipitation can be clearly seen in the ECN results from Wytham. Two prolonged dry periods can be identified in the climate record: summer 1995–spring 1997 and winter 2004/5–autumn 2006. The summer of 2003 was also relatively dry. Although the 2004–6 drought was drier overall, the ecological impacts were most noticeable in the mid-1990s. This was because the 2004–6 drought was largely a result of dry conditions in winter, when most species are not water-limited, whereas the summer of 1995 was unusually hot and dry. Projections of climate change for the UK (Jenkins *et al.* 2009) suggest that drier summers are likely to become more frequent in future, so observations during this period may offer clues as to what the future may hold. However these projections are not infallible and they do not imply that all years will be dry. Future climates may be more variable and more extreme droughts and wet periods may well occur, such as the two unusually wet summers that occurred in 2007 and 2008. Nevertheless, an examination of the effects of drought illustrates the sensitivity of ecosystems to climatic events.

Changes in both physical and biological aspects of the environment were seen in the summer of 1995 and the following 18 months. Morecroft *et al.* (2000) reported increased nitrate concentrations in soil water and streams as a result of both a concentrating effect of reduced stream flow and increased release of inorganic forms of nitrogen (mineralization) within the soil following drought. Many tree species lost their leaves early in 1995

(Morecroft 2000), with trees on clay soils being particularly badly affected. Elder and sycamore were the species that showed most early leaf loss.

Some of the woodland ground flora, notably dog's mercury (*Mercurialis perennis*), wilted in the summers of 1995 and 2003. However, the vegetation recovered the following year and no change in composition was detected in monitoring plots. This contrasted with some of the grasslands at Wytham, where there was an increase in ruderal (weedy) species the following year. These types of plants characteristically produce many seeds, disperse widely, and grow quickly; the smooth goat's beard (*Crepis capillaris*) is a good example. They are not, however, good competitors and 'outbreaks' of ruderals in grasslands reflect the adverse impact of drought on many grass species, which are shallow-rooted. The grasses die back, creating gaps in the sward in which ruderals can thrive briefly, before wetter conditions return and competition from the grasses increases again. This phenomenon was particularly observed in agriculturally improved grasslands and ones which had been allowed to revert to a semi-natural state after agricultural use, such as Upper Seeds (Morecroft and Paterson 2006). It was not observed in old, species rich grasslands protected for conservation.

The 1995–97 drought affected various invertebrates differently. Most butterfly and moth species increased through the dry period, both at Wytham and at other ECN sites (Morecroft *et al.* 2002). There were, however, exceptions and the speckled wood butterfly (*Pararge aegeria*) in particular decreased during the drought. Ground beetles (Carabidae; Coleoptera) showed a wider range of responses than the butterflies and moths, with approximately equal numbers of species increasing and decreasing. Overall, Morecroft *et al.* (2002) concluded that species typical of relatively warm or dry conditions tended to increase during drought, whereas those of damp or cold conditions decreased, as might be expected. There was also evidence that more mobile species, able to travel further, tended to increase, whereas less mobile species decreased, suggesting that the capacity to disperse to new locations favoured survival during drought.

## 12.2 Upper Seeds climate change experiment

At the same time that ECN was established, the Natural Environment Research Council also started a new research programme, the Terrestrial Initiative in Global Environmental Research (TIGER). TIGER funded a number of large collaborative projects on the ecological impacts of climate change; two of these consortium projects were based at Wytham, which was designated a 'flagship' site for TIGER. One of these built on the long-term research into great tits, winter moth caterpillars, and oak trees, particularly looking at the trophic interactions between them and the potential impact of rising temperatures on synchrony between the different organisms. The results are discussed in other chapters in context of other work on these systems. The other established an experiment manipulating climate in grassland at Upper Seeds.

The Upper Seeds climate change experiment simulated the effects of increasing winter temperature using soil warming cables, increasing or decreasing summer rainfall (the former by applying water using a hosepipe and the latter by mobile polycarbonate 'rainshelters' which covered experimental plots during rainfall in July and August), and the interactions between them. The impacts on soils, plants, and invertebrates were monitored. The full combination of treatments was carried on for five years. Subsequent to this, the experiment continued with just the water manipulations. A number of organizations were involved in the experiment over the period, including Reading University, Imperial College, London, CABI Bioscience, and CEH.

Some of the earliest results from the experiment concerned soil processes, especially mineralization of nitrogen—the release of inorganic nitrogen components during the decomposition of organic matter. The main finding was a surprising one: that drought stimulated the mineralization of nitrogen in the autumn following summer drought (Jamieson et al. 1998). Dry conditions themselves are conventionally understood to inhibit decomposition processes. However, the release following drought may reflect a release of nutrients following the death of soil micro-organisms. This experimental finding helps to explain the pulse of nitrate in soil and stream water detected in ECN monitoring data following the 1995–7 drought.

Total vegetation cover increased with increasing rainfall in the preceding summer (Morecroft et al. 2004). The plant community showed a range of responses to the climate change treatments and interactions with weather conditions and successional processes. The phenology of some species was advanced by warming (Fox et al. 1999). Drought had a major effect on the species composition of the plant community (Sternberg et al. 1999; Morecroft et al. 2004). Deep-rooted species, such as wild parsnip (*Pastinacca sativa*) tended to increase with drought, whereas shallower rooted species, particularly many of the grasses, tended to decrease. Ruderal species also increased following drought, as was found in ECN monitoring plots (see above). There was, however, evidence for an interaction between summer drought and winter rainfall, in that the effects of the drought treatment were offset in years when autumn and winter were particularly wet.

A similar experiment was set up near Buxton in the Derbyshire Peak District and the two sites presented an interesting contrast. Both were on calcareous grassland, but the Buxton experiment was on an old, relatively undisturbed grassland, whereas the Wytham site was formerly an arable field which had been allowed to revert to a semi-natural state from 1982 onwards. Comparison with the Buxton experiment showed that the Wytham community was much more responsive to the experimental treatments (Grime et al. 2000). This probably reflects the different types of species associated with the different successional states of the two communities. The grassland at Buxton is dominated by 'stress tolerant' plants, which grow and reproduce slowly but survive adverse conditions for plant growth, such as drought and low nutrient status. These species only slowly colonize communities, and the Upper Seeds grassland is still composed mainly of competitive species which are faster growing but less stress tolerant, together with ruderals. The sensitivity of newly restored grasslands to climate change will need to be taken increasingly into account in future schemes to restore the biodiversity of agricultural land.

The interactions between the plant community and their insect herbivores have been a large part of the research associated with the climate change experiment. The picture that emerged was complex, with different species, sometimes even closely related ones with similar life histories and other characteristics, responding differently to the same treatments. There were also complex interactions between species and treatments. Staley *et al.* (2006) investigated the effects of summer rainfall on leaf mining insects and found a diversity of responses in what is ostensibly a rather similar 'guild' of species. Complex interspecific interactions may account for some these differences. The larvae of the moth, *Stephensia brunnichella*, feed as 'leaf miners' within the leaves of wild basil (*Clinopodium vulgare*), they are however parasitized by minute wasp larvae, which can lead to the death of the moth larvae. A higher incidence of parasitism by these wasps was found in the experimental drought treatment. Staley *et al.* (2007a) went on to demonstrate an interaction between the same moth larva (*S. brunnichella*) and another herbivore: wireworms, which are beetle (*Agriotes* sp.) larvae and feed on the roots of the wild basil. The abundance and performance of leaf miner, and extent to which it was parasitized, were both reduced when wireworm were present on the roots. This interaction did not, however, occur on droughted plants, which were smaller. Staley *et al.* (2007b) also demonstrated differences in the response of the two most common root herbivores in the plots to the summer rainfall manipulations. Wireworm larvae were more numerous under enhanced rainfall in both the spring and autumn, but the scale insect *Lecanopsis formicarum* (Coccoidea) was unaffected by the rainfall manipulations.

Earlier work by Masters *et al.* (1998) found that the abundance of the leaf, plant, and frog hoppers (Auchenorrhyncha) increased with added summer rainfall, reflecting plant productivity. Abundance did not, however, decrease with summer drought, possibly because an increase in nutritional quality under drought conditions may have compensated for the effects of lower water supply. Temperature affected phenology, with egg hatch and the termination of nymphal hibernation advanced in warmed plots. Fox *et al.* (1999) also found effects of phenological change caused by warmer winter temperatures, in that they stimulated earlier growth of the perforate St. John's wort (*Hypericum perforatum*) and reduced damage by gall-forming and sap sucking insects in spring. In this study, summer drought reduced reproductive success of the St John's wort, as a result of modified interactions with herbivorous insects and interactions between summer drought and winter warming.

The climate change experiment demonstrated that it is important to understand the effects of changing water supply, not simply the effects of rising temperatures, a message which also came out clearly from ECN monitoring. It also showed that ecological responses to climate change may be influenced by a complex series of interactions between organisms.

## 12.3 Tree growth and interactions with the atmosphere

Tree survival, growth, and reproduction are sensitive to climate change and this has been a subject of study at Wytham. Trees may also have an impact on climate change, through the uptake of carbon dioxide from the atmosphere and its storage in wood. The role of forests in the global carbon cycle has been a topic of much international scientific interest and concern. On a global scale, British woodlands have a very minor influence on the atmospheric carbon balance. However, the history of research and scientific infrastructure at Wytham has made it an ideal place to study the processes controlling the carbon cycle.

ECN monitoring of tree growth has revealed interesting trends in growth rates (Fig. 12.2)—particularly that sycamore is growing more slowly than its main competitor, ash, and is more sensitive to dry conditions (Morecroft et al. 2008). In view of projections for more frequent summer droughts, this suggests that sycamore is likely to decline in future at places like Wytham. This is consistent with evidence from a modelling-based study by Broadmeadow et al. (2005) of tree growth responses to climate change in the UK and the observed European distribution of sycamore, which is within principally damp, montane conditions.

The construction of a canopy walkway, in 1993, with funding from the TIGER programme opened up new opportunities for research in the forest canopy itself. The canopy is where carbon is taken up from the atmosphere in photosynthesis and where water is lost in transpiration. It is also where much of a woodland's biodiversity is to be found. It is, however, rarely accessible for study and much remains unknown. A programme of measurements of photosynthesis and transpiration in oak and sycamore leaves in 1996 revealed some fascinating results (Morecroft and Roberts 1999). The photosynthetic rate of sycamore was substantially lower than that of oak, which was surprising in what is usually viewed as a fast

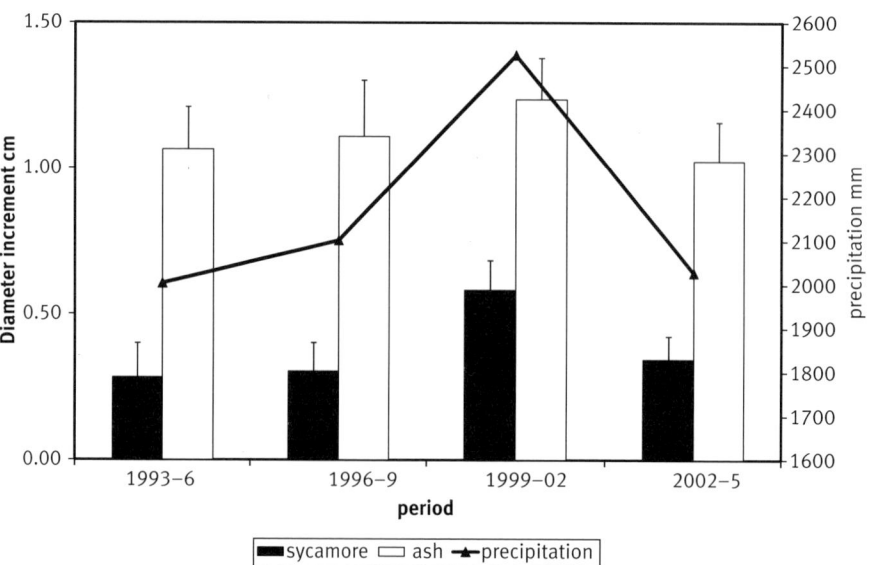

**Figure 12.2** Diameter growth in sycamore and ash trees in successive three-year intervals, with contrasting total precipitation. Diameter at Breast Height (DBH—height at 1.3 m) is measured in ECN monitoring plots throughout Wytham Woods.

growing, invasive species. Although photosynthesis provides the raw materials for the trees' growth, other factors including the number of leaves and balance between leaves, wood, and roots also have an effect on growth rate. Another interesting finding was that the leaves of oak trees developed their maximum photosynthetic rates in the middle of June, a quarter of the way through the time the leaves were on the trees. Typically, oaks are one of the last trees to come into leaf, but physiologically they are even slower, and development of maximum photosynthetic rates lags some weeks behind the structural development of the leaves. This slow development of photosynthetic capacity was surprising as it means that leaves are not fully photosynthetically functional during May and early June when the day length is long and much potential for growth is therefore missed.

These findings were followed up with further studies from 1999 to 2001 in collaboration with Essex University (Stokes 2002). The low rate of photosynthesis in sycamore, which had been observed in 1996, was not found when similar measurements were made in 1999 and 2000 (Morecroft *et al.* 2008). 1996 was a dry year preceded by the dry summer of 1995 and the soil was dry. In contrast, 1999 and 2000 were much wetter; it therefore appears that the low rate of photosynthesis in 2006 was a result of drought. Oak tree photosynthesis was less sensitive to drought conditions, which may be a result of its deeper rooting. The explanation for the slow developmental rate of oak remains something of a puzzle. One possible explanation is that oak is a ring porous, rather than diffuse porous, species, in which most wood is laid down in the spring, rather than throughout the growing season, so full function may not be possible until vascular tissue in the trunk has been fully formed (Morecroft *et al.* 2003). The lag between budburst and attainment of full photosynthetic capacity varied between years, indicating climate sensitivity in these processes, but the lag was still substantial. Stokes *et al.* (2006) also used the walkway to investigate the factors that affect the boundary layer conductance (ease of transfer of gases and heat from the leaf surface to the wider atmosphere) of oak and sycamore leaves, and properties of the leaves themselves, such as nitrogen content and specific leaf area. These data help to provide mechanistic understanding of the processes controlling photosynthesis and transpiration and allow the development of simulation models.

Understanding water loss from forests through transpiration is important for water resource management and catchment hydrology, but it is also important to understanding the impact of climate change on carbon balance and biodiversity. Where water is limiting, carbon dioxide uptake is typically reduced as stomata (pores in the leaf) normally shut to prevent excess water loss. This is particularly important in light of the projections of more frequent summer droughts in future (Jenkins *et al.* 2009). Wytham has been the site of some important studies of water loss from forest edges. Herbst *et al.* (2007) used measurements of tree sap flow to measure canopy transpiration. The flow of water up the stem is the direct result of evaporation at the leaf surface and can be measured using probes installed in the stem, which record flow rate by measuring how fast a heat pulse to the sap from a probe is dissipated. At both the northern and southern edges of the wood, water loss was substantially higher than in the middle. This reflects the greater leaf area of trees at the edge and indicates that the edges of woodlands are likely to be more prone to soil drying during drought. It also follows from this that smaller woods with proportionately more edge will be more vulnerable to drought, with implications for conservation and carbon sequestration into the future. Herbst *et al.* (2008) also used sap flow to demonstrate that water loss by closed canopy forest at Wytham was similar to that of another deciduous woodland, Grimsbury Wood, Berkshire, with different species composition. This suggests that at the canopy scale species composition is relatively less important for water relations than fragmentation.

The construction of a 'flux tower' in Great Wood (Fig. 12.3), by CEH, has allowed researchers to start investigating the exchange of carbon dioxide and water over large areas of woodland. The tower was established in 2006 and wind speed and net radiation

**Figure 12.3** A Flux Tower, established in 2006, allows researchers to investigate the exchange of carbon dioxide and water, wind speeds, and net radiation.

data from above the canopy started to be collected soon after then (Herbst *et al.* 2008). Net radiation is the balance between incoming solar radiation and that reflected or re-emitted by canopy; including the infrared wavelengths as well as visible 'light'. Reliable monitoring of carbon and water fluxes was established over the course of 2007, with 2008 the first complete growing season to be monitored. Results are yet to be fully analysed, but preliminary results clearly show the expected seasonal rise in carbon uptake as the canopy develops in spring. At the same location as the flux tower, a detailed study of the component processes controlling forest carbon fluxes is taking place (K. Fenn, unpublished DPhil research). Stem growth, root growth, litter production, and respiration loss from soils are all being monitored in detail to allow us to understand more about the carbon cycle and the seasonal and climatic factors that influence it.

Climate change is the focus of three important new projects on carbon and water fluxes and their interrelationships with biodiversity. One is the establishment of a large (18 ha) plot in Great Wood in which every tree's diameter has been measured; this will form part of a new international network of plots coordinated by the Smithsonian Institution. The other two both look further at the effects of fragmentation and the differences between forest edges and the centre of the woods, extending the work across the local landscape. All of these are collaborations between Oxford University and CEH; one is particularly unusual in that it is funded by HSBC Bank, and researchers are working with the Earthwatch Institute to use volunteers from HSBC in carrying out the work. Another is extending our ability to study forest canopies by using 'cherry pickers' (mobile elevated platforms; Fig. 12.4) to gain access to the canopy in a range of

**Figure 12.4** Recording data from the platform of a 'Cherry Picker' from an ash canopy, using an Infra Red Gas Analyser.

new places, especially at the forest edge, linking ground based measurements with remote sensing.

## 12.4 Conclusions

Does all this research help us to understand how Wytham and other similar woodlands may change as the climate changes? Does it suggest ways in which we may adapt management plans to minimize adverse changes? It is worth stepping back from the detail of the specific studies and considering the wider messages from the research that has taken place at Wytham over the years. One striking feature which is rarely reported (as journals are rarely interested in 'no change' stories) is that the ancient woodland plant communities show little evidence of change in response to fluctuating weather conditions. They are very resistant to change, compared to, for example, younger grassland communities like that on Upper Seeds. This is not to say that they will continue unchanged indefinitely, but it does mean that it is realistic to think in terms of maximizing their resistance and resilience to change. The fact that Wytham is a large woodland area means that edge effects are a less serious issue than for many British woodlands. The varied topography with north and south facing slopes and spring lines also makes for a resilient system, better able to withstand climatic variability than many places. The point at which canopy trees die is the point at which change in the wider ecosystem is most likely. This may well be when the woods experience an extreme event, such as a drought or a gale, although it may happen more gradually over a period of time. The modelling work of Broadmeadow *et al.* (2005) suggests that of the range of species at Wytham, sycamore, birch, and beech are likely to be the most sensitive to climate change because of their sensitivity to drought. Findings at Wytham bear this out, particularly for sycamore as we have seen. The other common canopy species at Wytham, particularly ash and oak, are less vulnerable to drought. As sycamore is non-native, the beech is all planted, and birch only occasional at Wytham, a gradual replacement with ash and the occasional oak could be viewed as a change for the better by many conservationists. However, sycamore is such an abundant species that any sudden decline following an extreme drought would leave large gaps in the canopy, with the potential for colonization by new species in both ground and canopy layers.

One important issue, which has been widely recognized by others (e.g. Hopkins *et al.* 2007; Mitchell *et al.* 2007) is that adapting to climate change begins with protecting biodiversity from other adverse changes. In the case of Wytham, the impact of deer has been the single largest influence on ground flora and the animal life it supports. Controlling their numbers may prove to be the most successful climate change adaptation measure! Whether to start to take more proactive measures to adapt to climate change, for example, planting provenances of trees from warmer, drier climates, has been debated. It would be interesting to do this on an experimental scale, and indeed Boshier and Stewart (2005) have done this with ash at Wytham, but a larger scale planting exercise would be premature. We simply do not know what the future holds in any detail and we do not know the limits of the resilience of the Woods. There is, however, a deeper reason, perhaps rather unique to Wytham, for allowing nature to take its course. Wytham Woods are a rare place where we can watch processes as they unfold, learning through observation, and building on the knowledge of our predecessors. The insight and understanding that this offers is something of value not merely for deciding how Wytham may best be managed, but for guiding management and conservation across the whole temperate region and beyond. Maybe it can also give us something of an understanding of the place of people in the ecosystem—what it has been, how it is now, and how it might be in future.

# References

Alexander, K.N.A. (1988). The development of an index of ecological continuity for deadwood associated beetles. In: R.C. Welch, compiler: Insect indicators of ancient woodland. *Antenna*, **12**: 69–70.

Anon. (1937). The Oxford University Bureau of Animal Population. *Science*, **85**: 575.

Anon. (1950). *Working Plan for 1949/50 – 1959/60 for the Woods of Hazel, Wytham, Berkshire*. Oxford Forestry Institute Library, unpublished MS.

Anwar, M.A., Newman, C., Macdonald, D.W., Woolhouse, M.E.J., and Kelly, D.W. (2000). Coccidiosis in the European badger (*Meles meles*) from England, an epidemiological study. *Parasitology*, **120**: 255–60.

Appleby, B. (1996). *The behaviour and ecology of the tawny owl*, Strix aluco. D.Phil thesis, University of Oxford.

Appleby B.M., Yamaguchi N., Johnson P.J., and Macdonald, D.W. (1999). Sex specific territorial responses in Tawny Owls *Strix aluco*. *Ibis*, **141**: 91–99.

Appleby, B.M., Petty, S.J., Blakey, J.K., Rainey, P., and Macdonald, D.W. (1997). Does variation of sex ratio enhance reproductive success of offspring in tawny owls (*Strix aluco*)? *Proceedings of the Royal Society of London Series B*, **264**: 1111–16.

Arkell, W.J. (1945). The geology of Wytham Hills. In F.C. Osmaston, ed. (1959). *Wytham Woods Forest Working Plan 1960–69*. Mimeogr. pp. 124–34.

Arkell, W.J. (1947). *The geology of Oxford*. Clarendon Press, Oxford.

Askew, R.R. (1962). The distribution of galls of *Neuroterus* (Hym., Cynipidae) on oak. *Journal of Animal Ecology*, **31**: 439–55.

Atkinson, P.M., Foody, G.M., Gething, P.W., Mathur, A., and Kelly, C.K. (2007). Investigating spatial structure in specific tree species in ancient semi-natural woodland using remote sensing and marked point pattern analysis. *Ecography*, **30**: 88–104.

Avery, B.W. (1980). Soil classification for England and Wales. Soil Survey, Harpenden, *Technical Monograph* No.14.

Bacon, P.J. and Macdonald, D.W. (1980). To control rabies: vaccinate foxes. *New Scientist*, **87**: 640–45.

Baines, M., Hambler, C. Johnson, P.J., Macdonald, D.W., and Smith, H.E. (1998). The effects of arable field margin management on the abundance and species richness of Araneae (spiders). *Ecography*, **21**: 74–86.

Baker, J.R. and Ranson, R.M. (1933). Factors affecting the breeding of the field mouse (*Microtus agrestis*). Part III. Locality. *Proceedings of the Royal Society of London Series B*, **113**: 486–95.

Bateson, M. (2003). Interval timing and optimal foraging. In: W.H. Meck, (ed). *Functional and Neural Mechanisms of Interval Timing*, pp. 113–41. CRC Press, Boca Raton, Florida.

Bazeley, D.R., Myers, J.H., and Burke, Da Silva K. (1991). The response of numbers of bramble prickles to herbivory and depressed resource availability. *Oikos*, **61**: 327–36.

Beaver, R.A. (1966). The development and expression of population tables for the bark beetle *Scolytus scolytus* (F.). *Journal of Animal Ecology*, **35**: 27–41.

Beaver, R.A. (1967). The regulation of population density in the bark beetle *Scolytus scolytus* (F.). *Journal of Animal Ecology*, **36**: 435–51.

Bell, J.R., Johnson, P.J., Hambler, C. Haughton, A.J., Smith, H., Feber, R.E., Tattersall, F.H., Hart, B.H., Manley, W., and Macdonald, D.W. (2002). Manipulating the abundance of *Lepthyphantes tenuis* (Araneae: Linyphiidae) by field margin management. *Agriculture, Ecosystems and Environment*, **93**: 295–304.

Blackwell, P.G. and Macdonald, D.W. (2000). Shapes and sizes of badger territories. *Oikos*, **89**: 392–98.
Boshier, D. and Stewart, J.L. (2005) How local is local? *Forestry*, **78**: 135–43.
Bowler, P.J. (1992). *The Fontana History of the Environmental Sciences*. Fontana Press, London.
Braithwaite, M.E., Ellis, R.W., and Preston, C.D. (2006). *Change in the British flora 1987–2004*. Botanical Society of the British Isles, London.
Brereton, J. Le G. (1957). The distribution of woodland Isopods. *Oikos*, **8**: 85–106.
Broadmeadow, M.S.J., Ray, D., and Samuel, C.J.A. (2005). Climate change and the future for broadleaved tree species in Britain. *Forestry*, **78**: 145–61.
Brooks, M. (1999). Effects of protective fencing on birds, lizards, and black-tailed hares in the Western Mojave Desert. *Environmental Management*, **23**: 387–400.
Brown, A.J.F. and Oosterhuis, L. (1981). The role of buried seed in coppicewoods. *Biological Conservation*, **21**: 19–38.
Brown, E.D., Macdonald, D.W., Tew, T.E., and Todd, I.A. (1994). *Apodemus sylvaticus* infected with *Heligmosomoides polygyrus* (Nematoda) in an arable ecosystem: Epidemiology and effects of infection on the movements of male mice. *Journal of Zoology*, **234**: 623–40.
Brown, V.K. and Gange, A.C. (1989). Differential effects of above- and below-ground insect herbivory during early plant succession. *Oikos*, **54**: 67–76.
Brown, V.K., Gibson, C.W.D., and Kathirithamby, J. (1992). Community organisation in leaf hoppers. *Oikos*, **65**: 97–106.
Brown, V.K., Jepsen, M., and Gibson, C.W.D. (1988) Insect herbivory: effects on early old field succession demonstrated by chemical exclusion methods. *Oikos*, **52**: 293–302.
Bryce, J.M., Johnson, P.J., and Macdonald, D.W. (2002). Can niche use in red and grey squirrels offer clues for their apparent coexistence? *Journal of Applied Ecology*, **39**: 875–87.
Buckley, G.P. (1992). *Ecology and Management of Coppice Woodlands*. London, Chapman and Hall.
Buckner, C.H. (1969). Some aspects of the population ecology of the common shrew, *Sorex araneus*, near Oxford, England. *Journal of Mammalogy*, **50**: 326–32.
Buesching, C.D. and Macdonald, D.W. (2001). Scent-marking behaviour of the European badger (*Meles meles*): Resource defence or individual advertisement? In A. Marchlewska-Koj, J.J. Lepri, and D. Müller-Schwarze, (eds). *Chemical Signals in Vertebrates*, pp. 321–28. Plenum Press, Oxford.
Buesching, C.D. and Macdonald, D.W. (2004). Variations in object-marking activity and over-marking behaviour of European badgers (*Meles meles*) in the vicinity of their setts. *Acta Theriologica*, **49**: 235–46.
Buesching, C.D., Newman, C., and Macdonald, D.W. (2002). Variations in colour and volume of the subcaudal gland secretion of badgers (*Meles meles*) in relation to sex, season and individual-specific parameters. *Zeitschrift für Säugetierkunde*, **67**: 1–10.
Buesching, C.D., Newman, C., Twell, R., and Macdonald, D.W. (2008). Reasons for arboreality in wood mice *Apodemus sylvaticus* and Bank voles *Myodes glareolus*. *Mammalian Biology*, **73**: 318–24.
Buesching, C.D., Stopka, P., and Macdonald, D.W. (2003). The social function of allo-marking behaviour in the European badger (*Meles meles*). *Behaviour*, **140**: 965–80.
Bulmer, M.G. (1973). Inbreeding in the Great Tit. *Heredity*, **30**: 313–925.
Burgess, J.M. (1957). Damage to bark of sycamore by grey squirrels. *Journal of the Oxford University Forestry Society*, **45**: 15–16.
Burn, J.L. (1996). *Polygyny and the wren*. D. Phil thesis, University of Oxford.
Buse, A., Dury, S.J., Woodburn, R.J.W., Perrins, C.M., and Good, J.E.G. (1999). Effects of elevated temperature on multi-species interactions: the case of Pedunculate Oak, Winter Moth and Tits. *Functional Ecology*, **13**: 74–82.

Cain, A.J. and Sheppard, P.M. (1954). Natural selection in *Cepaea. Genetics*, **39**: 89–116.
CEH - Centre for Ecology and Hydrology (2007). *UK National Ammonia Monitoring Network Sites: Wytham Woods*. Website. http://www.cara.ceh.ac.uk/cgi-bin/sitesearch.pl?site=Wytham+Woodsandlistsubmit=Go. 7 December 2007.
Chamberlain, D.E., Hatchwell, B.J., and Perrins, C.M. (1995). Spaced-out nests and predators: an experiment to test the effects of habitat structure. *J. Avian Biol.*, **26**: 346–349.
Charmantier, A., McCleery, R.H., Cole, L.R., Perrins, C.M., Kruuk, L.E.B., and Sheldon, B.C. (2008). Adaptive phenotypic plasticity in response to climate change in a wild bird population. *Science*, **320**: 800–803.
Chitty, D. (1937). A ringing technique for small mammals. *Journal of Animal Ecology*, **6**: 36–53.
Chitty, D. (1952). Mortality amongst voles (*Microtus agrestis*) at Lake Vyrnwy, Montgomeryshire in 1936–39. *Philosophical Transactions of the Royal Society B*, **236**: 505–552.
Chitty, D. (1970). Variation and population density. *Mammal Review*, **1**: 31.
Chitty, D. (1996). *Do Lemmings Commit Suicide? Beautiful Hypotheses and Ugly Facts*. Oxford University Press, New York.
Chitty, D. and Kempson, D.A. (1949). Prebaiting small mammals and a new design of live trap. *Ecology* **30**, 536–42.
Chitty, D. and Phipps, E. (1966). Seasonal changes in survival in mixed populations of two species of vole. *Journal of Animal Ecology*, **35**: 313–31.
Chitty, D., Pimentel, D., and Krebs, C.J. (1968). Food supply of overwintered voles. *J. anim. Ecol.* **37**, 113–20.
Church, A.S. (1922). Introduction to the plant life of the Oxford District. 1. General review. *Botanical Memoirs*, **13**: 1–103.
Churchill, T.B. and Ludwig, J.A. (2004). Changes in spider assemblages along grassland and savanna grazing gradients in northern Australia. *Rangeland Journal*, **26**: 3–16.
Clarke, J.R. (1953). The effect of fighting on the adrenals, thymus and spleen of the vole (*Microtus agrestis*). *J. Endocrinol*, **9**, 114–126.
Clarke, J.R. (1955). Influence of numbers on reproduction and survival in two experimental vole populations. *Proceedings of the Royal Society of London B*, **144**: 68–85.
Clarke, J.R. (1956). The aggressive behaviour of the vole. *Behaviour*, **9**: 1–23.
Clarke, J.R. (1981). Physiological problems of seasonal breeding in eutherian mammals. In C. A. Finn, (ed). *Oxford Reviews of Reproductive Biology*, **3**: 244–312. Clarendon Press, Oxford.
Clarke, J.R. and Frearson, S. (1972). Sebaceous glands on the hindquarters of the vole, *Microtus agrestis. Journal of Reproduction and Fertility*, **31**: 477–81.
Clarke, J.R. and Hellwing, S. (1977). Remote control by males of ovulation in bank voles (*Clethrionomys glareolus*). *Journal of Reproduction and Fertility*, **50**: 155–58.
Clarke, J.R., Baker, J., Craven, R.P., Feito, R., Mansard, F.-X., Petts, V., Stewart, J., and Wong, S. (1981). Seasonal breeding in voles and wood mice: coarse and fine adjustments. In *Photopériodisme et Reproduction Chez les Vertebrés. Les Colloques de L'INRA* **6**: 291–317. INRA Publishing.
Clarke, J.R., Clulow, F.V. and Greig, F. (1970). Ovulation in the bank vole, *Clethrionomys glareolus. Journal of Reproduction and Fertility*, **23**: 531.
Cleveland, C. (1997). A survey of the veteran trees in Wytham Woods with a view to future management. Unpublished MSc dissertation. Oxford Forestry Institute.
Cooke, A.S. (2007). *Monitoring muntjac deer Muniacus reevesi and their impact in Monks Wood National Nature Reserve*. English Nature, Research Report 681, Peterborough.
Cooper, N.S. (2000). How natural is a nature reserve? An ideological study of British nature conservation landscapes. *Biodiversity and Conservation*, **9**: 1131–52.
Copeland, T. (2002). *Iron Age and Roman Wychwood*. Wychwood Press, Charlbury.

Corney, P.M., Kirby, K.J., Le Duc, M.G., Smart, S.M., McAllister, H.A., and Marrs, R.H. (2008). Changes in the field-layer of Wytham Woods - assessment of the impacts of a range of environmental factors controlling change. *Journal of Vegetation Science*, **19**: 287–98.

Corry, J.S. (1984). Aspects of the ecology of predatory ground and rove beetles as related to their pest control potential. Unpublished D.Phil. thesis, University of Oxford.

Cowie, R.H. and Jones, A.S. (1987). Ecological interactions betwen *Cepaea nemoralis* and *Cepea hortensis*: competition, invasion but no displacement. *Functional Ecology*, **1**: 91–97.

Crampton, A.B., Stutter, O., Kirby, K.J., and Welch, R.C. (1998). Changes in the composition of Monks Wood National Nature Reserve (Cambridgeshire, UK) 1964–1996. *Arboricultural Journal*, **22**: 229–45.

Cresswell, W.J., Harris, S., Cheeseman, C.L., and Mallinson, P.J. (1992). To breed or not to breed: an analysis of the social and density-dependent constraints on the fecundity of female badgers (*Meles meles*). *Philosophical Transactions of the Royal Society of London*, **338**: 393–407.

Crowcroft, P. (1956). On the life span of the common shrew (*Sorex araneus* L.). *Proceedings of the Zoological Society of London*, **127**: 285–92.

Crowcroft, P (1957). *The Life of the Shrew*. Reinhardt, London.

Crowcroft, P. (1991). *Elton's Ecologists. A History of the Bureau of Animal Population*. The University of Chicago Press, Chicago.

Da Silva, J., Macdonald, D.W., and Evans, P.G.H. (1994). Net costs of group living in a solitary forager, the Eurasian badger. *Behavioural Ecology*, **5**: 151–58.

Da Silva, J., Woodroffe, R., and Macdonald, D.W. (1993). Habitat, food availability and group territoriality in the European badger, *Meles meles. Oecologia*, **95**: 558–64.

Davies, N.B. (1978). Territorial defence in the speckled wood butterfly (*Pararge aegeria)*: the resident always wins. *Animal Behaviour*, **26**: 138–47.

Dawkins, H.C. (1971). Techniques for long-term diagnosis and prediction in forest communities. In: E. Duffey and A.S. Watt, (eds): *The Scientific Management of Plant and Animal Communities*, pp. 33–44, Blackwell Scientific Publications, Oxford.

Dawkins, H.C. and Field, D.R.B. (1978). *A Long-Term Surveillance System for British Woodland Vegetation*. Commonwealth Forestry Institute, Oxford, Occasional Paper No.1.

Dawkins, R. (1982). *The Extended Phenotype. The Gene as the Unit Of Selection*. W.H. Freeman and Co., Oxford.

Day, S.P. (1991). Post-glacial vegetation history of the Oxford region. *New Phytologist*, **119**: 445–70.

Defra/FC (2005) *Keepers Of Time—A Statement Of Policy For England's Ancient And Native Woodland*. Defra/Forestry Commission, London.

Defra/FC (2007). *A Strategy For England's Trees, Woods And Forests*. Defra/Forestry Commission, London.

Delahay, R.J., Walker, N.J., Forrester, G.J., Harmsen, B., Riordan, P., Macdonald, D.W., Newman, C., and Cheeseman, C.L. (2006). Demographic correlates of bite wounding in Eurasian badgers, *Meles meles* L., in stable and perturbed populations. *Animal Behaviour*, **71**: 1047–55.

Dempster, J.P. (1977). The scientific basis of practical conservation: factors limiting the persistence of populations and communities of animals and plants. *Proceedings of the Royal Society of London Series B*, **197**: 69–76.

Den Boer, P.J. and Reddingius, J. (1996). *Regulation and Stabilization Paradigms in Population Ecology*. Chapman and Hall, London.

Dennis, P., Skartveit, J., McCracken, D.I., Pakeman, R.J. Beaton, K., Kunaver, A., and Evans, D.M. (2008). The effects of livestock grazing on foliar arthropods associated with bird diet in upland grasslands of Scotland. *Journal of Applied Ecology*, **45**: 279–287.

Don, B.A.C. (1985). The use of drey counts to estimate grey squirrel populations. *Journal of Zoology*, **206**: 282–86.

Doncaster, C.P. (1994). Factors regulating local variations in abundance: field tests on hedgehogs, *Erinaceus europaeus. Oikos*, **69**: 182–92.
Doncaster, C.P. and Woodroffe, R. (1993). Den site can determine shape and size of badger territories - implications for group-living. *Oikos*, **66**: 88–93.
Doncaster, C.P., Rondinini, C., and Johnson, P.C. (2001). Field tests for environmental correlates of dispersal in hedgehogs *Erinaceus europaeus. Journal of Animal Ecology*, **70**: 33–46.
Duffey, E. (1956). Aerial dispersal in a known spider population. *Journal of Animal Ecology*, **25**: 85–111.
Duffey, E. (1962a). A population study of spiders in limestone grassland. Description of study area, sampling methods and population characteristics. *Journal of Animal Ecology*, **31**: 571–99.
Duffey, E. (1962b). A population study of spiders in limestone grassland. The field-layer fauna. *Oikos*, **13**: 15–34.
Duffey, E. (1966). Spider ecology and habitat structure (Arach., Araneae). *Senckenbergiana Biologica*, **47**: 45–49.
Duffey, E. (1974). *Nature Reserves and Wildlife*. Heinemann, London.
Duffey, E. (1975). The efficiency of the Dietrick vacuum sampler (D-vac) for invertebrate population studies in different types of grassland. *Bulletin d'Ecologie*, **11**: 421–31.
Duffey, E. and Watt, A.S. (1971). *The Scientific Management of Animal and Plant Communities for Conservation*. Blackwell Scientific Publications, Oxford.
Dugdale, H.L., Macdonald, D.W., Pope, L.C., Johnson, P.J. and Burke, T. (2008). Reproductive skew and relatedness in social groups of European badgers *Meles meles. Molecular Ecology*, **17**: 1815–27.
Dunn, E.K. (1977). Predation by Weasels (*Mustela nivalis*) on breeding tits (*Parus* spp.) in relation to the density of tits and rodents. *Journal of Animal Ecology*, **46**: 633–52.
Dupouey, J.L., Dambrine E., Laffite, J.D., and Moares, C. (2002). Irreversible impact of past land use on forest soils and biodiversity. *Ecology*, **83**: 2978–84.
Edlin, H.L. (1949). *Woodland Crafts In Britain*. Batsford, London.
Edlin, H.L. (1966). *Trees, Woods And Man*. (2nd Edn.) London: Collins New Naturalist.
Efford, I.E. (1965). Ecology of the watermite *Feltria romijni* Besseling. *Journal of Animal Ecology*, **34**: 233–51.
Elbourn C.A. (1966). Wytham ecological survey. An inventory: indexing and specimens, original data, literature, aerial photographs, maps. Unpublished manuscript prepared by the then Curator of the Survey and privately circulated.
Elbourn, C.A. (1970). Influence of substrate and structure on the colonisation of an artefact simulating decaying oak wood on oak trunks. *Oikos*, **21**: 32–41.
Ellwood, S.A. (2007). *Evaluating deer monitoring methods and the density dependence and independence of skeletal size and body condition of fallow and muntjac deer in a UK lowland wood*. D.Phil. thesis, Department of Zoology, University of Oxford.
Elton, C.S. (1924). Periodic fluctuation in the number of animals: their cause and effects. *Brit. Journal of Experimental Biology*, **2**: 119–63.
Elton, C.S. (1927). *Animal Ecology*. Sidgwick and Jackson, London.
Elton, C.S. (1939). On the nature of cover. *Journal of Wildlife Management*, **3**: 332–38.
Elton, C.S. (1942). *Voles, Mice and Lemmings: Problems in Population Dynamics*. Clarendon Press, Oxford.
Elton, C.S. (1958). *The Ecology of Invasions by Animals and Plants*. Methuen, London.
Elton, C.S. (1966). *The Pattern of Animal Communities*. Chapman and Hall, London.
Elton, C.S. and Miller, R.S. (1954). The ecological survey of animal communities: with a practical system of classifying habitats by structural characters. *The Journal of Ecology*, **42**: 460–96.

Elton, C.S., Ford, E.B., Baker, J.R., and Gardner, A.D. (1931). The health and parasites of a wild mouse population. *Proceedings of the Zoological Society of London, Part 3*, 657–721.

English Nature (1998). *UK Biodiversity Group: Tranche 2 Action Plans* (volumes 1 and 2). English Nature, Peterborough.

Evans, D. (1997) *A History of Nature Conservation in Britain* (2nd edition). Routledge, London.

Evans, D.M. (1973). Seasonal variation in the body composition and nutrition of the vole *Microtus agrestis*. *Journal of Animal Ecology*, **42**: 1–18.

Fager, E.W. (1967). The ecology of Wytham Hill. *The Quarterly Review of Biology*, **42**: 514–16.

Fager, E.W. (1968). The community of invertebrates in decaying oak wood. *Journal of Animal Ecology*, **37**: 121–42.

Fairbairn, W.A. (1980). Obituary of F.C. Osmaston. *Forestry* **53**: 99–100.

Falinski, J.B. (1986). *Vegetation Dynamics in Temperate Lowland Primeval Forest*. Geobotany **8**, Junk, Dordrecht, Netherlands.

Farmer, A. (1995). *Changes in Soil Chemistry at Wytham Wood (Oxfordshire) 1974–1991*. English Nature, Peterborough.

Farmer, A.M. (1995). Soil chemistry change in a lowland English deciduous woodland 1974–1991. *Water, Air and Soil Pollution*, **85**: 677–82.

Farnham, T.J. (2007). *Saving nature's legacy: origins of the idea of biological diversity*. Yale University Press, New Haven.

Feber, R.E., Asteraki, E.J., and Firbank, L.G. (2006). Can farming and wildlife coexist? In D.W. Macdonald and K.M. Service, (eds) *Key Topics in Conservation Biology*, pp. 239–52. Blackwell Publishing, Oxford.

Feber, R.E., Smith, H.E., and Macdonald, D.W. (1996). The effects on butterfly abundance of the management of uncropped edges of arable fields. *Journal of Applied Ecology*, **33**: 1191–1205.

Feber, R.E., Smith, H.E., and Macdonald, D.W. (1999). The importance of spatially variable field margin management for two butterfly species. *Aspects of Applied Biology*, **54**: 155–62.

Feeny, P. (1970). Seasonal changes in oak leaf tannins and nutrients as a cause of spring feeding by Winter Moth caterpillars. *Ecology*, **51**: 565–81.

Feeny, P. (1976). Plant apparency and chemical defense. In: J.W. Wallace and R.L. Nansel (eds). *Biological Interactions Between Plants and Insects. Recent Advances in Phytochemistry* **10**, pp. 1–40. Plenum Press, New York.

Fell, R.J., Buesching, C.D., and Macdonald, D.W. (2006). The social integration of European badger (*Meles meles*) cubs into their natal group. *Behaviour*, **143**: 683–700.

Ferris, R. and Carter, C. (2000). Managing rides, roadsides and edge habitats in Lowland Forests. *Bulletin* **123**. Forestry Commission, Edinburgh.

Flowerdew, J.R. (1971). The subcaudal glandular area of *Apodemus sylvaticus*. *Journal of Zoology*, **165**: 525–27.

Flowerdew, J.R. (1973). The effect of natural and artificial change in the food supply on breeding in woodland mice and voles. *Journal of Reproduction and Fertility, Supplement*, **19**: 259–69.

Flowerdew, J.R. (1974). Field and laboratory experiments on the social behaviour and population dynamics of the wood mouse (*Apodemus sylvaticus*). *Journal of Animal Ecology*, **43**: 499–511.

Flowerdew, J.R. and Ellwood, S.A. (2001). Impacts of woodland deer on small mammal ecology. *Forestry*, **74**: 277–87.

Ford, E.B. (1971). *Ecological Genetics* (3rd edn). Chapman and Hall, London.

Forestry Commission (2001). *National Inventory of Woodland and Trees: England*. Forestry Commission, Edinburgh.

Forestry Commission (2002). *National Inventory of Woodland and Trees – Oxfordshire*. Forestry Commission, Edinburgh.

Forestry Commission (2004). *UK Forestry Standard*. Forestry Commission, Edinburgh.

Foster, D.R. and Aber, J.D. (2004). *Forests in Time*. Yale University Press, Yale.
Foster, J. and Godfrey, C. (1950). A study of the British Willow Tit. *British Birds* **43**: 351–61.
Fowler, A.D. (1994) *Plant recruitment in relation to land management.* Unpublished PhD thesis, University of London.
Fox, L.R., Ribeiro, S.P., Brown, V.K., Masters, G.J., Clarke, I.P. (1999). Direct and indirect effects of climate change on St. John's wort, *Hypericum perforatum* L. (Hypericaceae). *Oecologia* **120**: 113–22.
Fuller, R.J. (1995). *Birdlife of Woodland and Scrub*. Cambridge University Press, Cambridge.
Fuller, R.A. and Gill, R.M.A. (2001). Ecological impacts of deer in woodland. Forestry (special issue), **74**, 189–309.
Fuller, R.J. and Warren, M.S. (1993). *Coppiced Woodlands: their Management for Wildlife* (2nd edn.). JNCC, Peterborough.
Garcia, E. (1981). *An experimental and observational study of interspecific territoriality between the Blackcap Sylvia atricapilla (Linnaeus) and the Garden Warbler Sylvia borin (Boddaert)*. D. Phil thesis, University of Oxford.
Gates, S., Feber, R.E., Macdonald, D.W., Hart, B.J., Tattersall, F.H., and Manley, W.J. (1997). Invertebrate populations of field boundaries and set-aside land. *Aspects of Applied Biology*, **50**: 313–22.
Gaston, A.J. (1973). The ecology and behaviour of Long-tailed Tits. *Ibis*, **115**: 330–351.
Geer, T.A. (1979). *Sparrowhawk* (Accipiter nisus) *predation on tits* (Parus spp.). D. Phil thesis, University of Oxford.
Geiger, R. (1965). *The Climate Near The Ground*. Harvard University Press, Cambridge, MA.
Gentle, L.K. and Gosler, A.G. (2001). Fat reserves and perceived predation risk in the great tit, *Parus major*. *Proc. Roy. Soc. B*: **268**: 487–91.
Gibb, J. (1954). Feeding ecology of tits with notes on treecreeper and goldcrest. *Ibis*, **96**: 513–43.
Gibb, J.A. 1950. The breeding biology of the Great and Blue Titmice. *Ibis* **92**: 507–39.
Gibson, C.W.D. (1980). Niche use patterns among some Stenodemini (Heteroptera: Miridae) of limestone grasslands, and an investigation of the possibility of interspecific competition between *Notostira elongata* Geoffroy and *Megaloceraea recticornis* Geoffroy. *Oecologia*, **47**: 352–64.
Gibson, C.W.D. (1986). Management history in relation to changes in the flora of different habitats on an Oxfordshire Estate, England. *Biological Conservation* **38**: 217–32.
Gibson, C.W.D. (1988). The distribution of ancient woodland plant species amongst areas of different history in Wytham Woods, Oxfordshire. In: K.J. Kirby and F.J. Wright, (eds). *Woodland Conservation and Research in the Clay Vale of Oxfordshire and Buckinghamshire*, pp. 32–40. Nature Conservancy Council, Peterborough (Research and survey in nature conservation 15).
Gibson, C.W.D. (1995). *Chalk grasslands on former arable land: a review*. Unpublished report, Bioscan (UK) Ltd, St. Clements, Oxford.
Gibson, C.W.D. and Brown, V.K. (1985). Plant succession: theory and applications. *Progress in Physical Geography*, **9**: 473–93.
Gibson, C.W.D. and Brown, V.K. (1991a) The effects of grazing on local colonisation and extinction during early succession. *Journal of Vegetation Science*, **2**: 291–300.
Gibson, C.W.D. and Brown, V.K. (1991b) The nature and rate of development of calcareous grassland in southern Britain. *Biological Conservation*, **58**: 297–316.
Gibson, C.W.D. and Brown, V.K. (1992). Grazing and vegetation change: deflected or modified succession? *Journal of Applied Ecology*, **29**: 121–31.
Gibson, C.W.D., Bazely, D.R., and Shore, J.S. (1993). Response of brambles, *Rubus vestitus*, to herbivory. *Oecologia*, **95**: 454–57.
Gibson, C.W.D., Brown, V.K., and Jepsen, M. (1987a). Relationships between the effects of insect herbivory and sheep grazing on seasonal changes in an early successional plant community. *Oecologia*, **71**: 245–53.

Gibson, C.W.D., Dawkins, H.C., Brown, V.K., and Jepsen M. (1987b). Spring grazing by sheep: effects on seasonal changes during early old field succession. *Vegetatio*, **70**, 33–43.

Gibson, C.W.D., Watt, T.A., and Brown, V.K. (1987c). The use of sheep grazing to recreate species-rich grassland from abandoned arable land. *Biological Conservation*, **42**: 165–83.

Gibson, C.W.D., Brown, V.K., Losito, L., and McGavin, G.C. (1992a). The response of invertebrate assemblies to grazing. *Ecography*, **15**: 166–176.

Gibson, C.W.D., Hambler, C., and Brown, V.K. (1992b). Changes in spider (Araneae) assemblages in response to succession and grazing management. *Journal of Applied Ecology*, **29**: 132–42.

Gill, R.M.A. and Fuller, R.J. (2007). The effects of deer browsing on woodland structure and songbirds in lowland Britain. *IBIS* (supplement) 119–127.

Glen, N.W. and Perrins, C.M. (1988). Co-operative breeding by Long-tailed Tits. *Brit. Birds*, **81**: 630–641.

Godfrey, G.K. (1953a). A technique for finding *Microtus* nests. *J. Mammalogy*, **34**, 503–505.

Godfrey, G.K. (1953b). The food of *Microtus agrestis hirtus* (Bellamy, 1839) in Wytham, Berkshire. *Säugetierkundliche Mitteilungen* **1**, 148–151.

Godfrey, G.K. (1954). Tracing field voles (*Microtus agrestis*) with a Geiger-Müller counter. *Ecology*, **35**: 5–10.

Godfrey, G.K., (1955). Observations on the nature of the decline in numbers of two *Microtus* populations. *Journal of Mammalogy*, **36**: 209–14.

Godwin, H. (1975). *The History of the British Flora*. Cambridge University Press.

Goodfellow, S. and Peterken, G.F. (1981). A method for survey and assessment of woodlands for nature conservation using maps and species lists: the example of Norfolk woodlands. *Biological Conservation*, **21**: 177–95.

Gosler, A.G. (1990). The birds of Wytham - an historical survey. *Fritillary* 1: 29–74.

Gosler, A.G. (1996). Environmental and social determinants of winter fat storage in the great tit Parus major. *Journal of Animal Ecology*, **65**: 1–17.

Gosler, A. G., Higham, J. P., and Reynolds S. J. (2005). Why are birds' eggs speckled? *Ecology Letters*, **8**: 1105–13.

Gosler, A.G., Greenwood, J.J.D., and Perrins, C. (1995). Predation risk and the cost of being fat. *Nature*, **377**: 621–623.

Gove, B., Power, S.A., Buckley, G.P., and Ghazoul, J. (2007). Effects of herbicide spray drift and fertilizer overspread on selected species of woodland ground flora: comparison between short-term and long-term impact assessments and field surveys. *Journal of Applied Ecology*, **44**, 374–84.

Gray, I.L. (1987). *The feeding ecology of the Sparrowhawk* (Accipiter nisus) *outside the breeding season*. D. Phil thesis, University of Oxford.

Grayson, A.J. and Jones, E.W. (1955). *Notes on the History of the Wytham Estate with Special Reference to the Woodlands*. Pp 28, plus map. Imperial Forestry Institute, Oxford.

Green, T. (2005). Is there a case for the Celtic maple or the Scots plane? *British Wildlife*, **16**: 184–88.

Gregory, R.D. (1991). Parasite epidemiology and host population growth: *Heligmosomoides polygyrus* (Nematoda) in enclosed wood mouse populations. *Journal of Animal Ecology*, **60**: 805–21.

Gregory, R.D.,Wilkinson, N.I., Noble, D.G., Robinson, J.A., Brown, F., Hughes, J., Proctor, D., Gibbons, D.W., and Galbraith, C. (2002). 'The population status of birds in the United Kingdom, Channel Islands and Isle of Man: an analysis of conservation concern 2002–2007. *British Birds*, **95**: 410–448.

Grieg, J. (1982). Past and present lime woods of Europe. In: M. Bell and S. Limbrey, (eds), *Archaeological Aspects of Woodland Ecology*, pp. 23–25. BAR International Series, Oxford.

Grime, J.P., Brown, V.K., Thompson, K., Masters, G.J., Hillier, S.H., Clarke, I.P., Askew, A.P., Corker, D., and Kielty, J.P. (2000). The response of two contrasting limestone grasslands to simulated climate change. *Science*, **289**: 762–65.

Grime, J.P., Hodgson, J.G., and Hunt, R. (2007) *Comparative Plant Ecology*. (2nd edn.) Castlepoint Press, Dalbeattie.

Grove, S.J. (2002). Saproxylic insect ecology and the sustainable management of forests. *Annual Review of Ecology and Systematics*, **33**: 1–23.

Gyan, K.Y. and Woodell, S.R.J. (1987a). Nector production, sugar content, amino acids and potassium in *Prunus spinosa* L., *Crataegus monogyna* Jacq. and *Rubus fruticosus* L. at Wytham, Oxfordshire. *Functional Ecology*, **1**: 251–59.

Gyan, K.Y. and Woodell, S.R.J. (1987b). Flowering phenology, flower colour and mode of reproduction of *Prunus spinosa* L. (blackthorn), *Crataegus monogyna* Jacq. (hawthorn), *Rosa canina* (dog rose) and *Rubus fruticosus* L. (bramble) in Oxfordshire, England. *Functional Ecology*, **1**: 261–68.

Haines-Young, R.H., Barr, C.J., Black, H.I.J., et al. (2000). *Accounting for Nature: Assessing Habitats in the UK Countryside*. London: Department of the Environment, Transport and the Regions.

Hambler, C. (1997). The new Gaia *Bulletin of the British Ecological Society*, **28**: 101–104.

Hambler, C. (2004). *Conservation*. Cambridge University Press, Cambridge.

Hambler, C. and Speight, M.R. (1995). Biodiversity conservation in Britain: science replacing tradition. *British Wildlife*, **6**: 137–47.

Hambler, C. and Speight, M.R. (1996) Extinction rates in British non-marine invertebrates since 1900. *Conservation Biology*, **10**: 892–96.

Hanski, I. (1980). Spatial patterns and movement in coprophagous beetles. *Oikos*, **34**: 293–310.

Harding, P.T. and Alexander, K.N.A. (1993). The saproxylic invertebrates of historic parklands: progress and problems. In K.J. Kirby and C.M. Drake, (eds). Dead wood matters. *English Nature Science*, **7**, pp. 58–73 English Nature, Peterborough.

Harding, P.T. and Rose, F. (1986). *Pasture Woodlands in Lowland Britain*. Institute of Terrestrial Ecology, Huntingdon.

Harris, R., York, A., and Beattie, A.J. (2003). Impacts of grazing and burning on spider assemblages in dry eucalypt forests of north-eastern New South Wales, Australia. *Austral Ecology*, **28**: 526–38.

Haslett, J.R. (1989). Interpreting patterns of resource utilisation: randomness and selectivity in pollen feeding by adult hoverflies. *Oecologia*, **78**: 433–42.

Hassell, M.P. (1968). The behavioural response of a tachinid fly (*Cyzenis albicans* (Fall.)) to its host, the winter moth (*Operophtera brumata* (L.)). *Journal of Animal Ecology*, **37**: 627–39.

Hassell, M.P., Crawley M.J., Godfray H.C.J. and Lawton, J.H. (1998). Top-down versus bottom-up and the Ruritanian bean bug. *Proceedings of the National Academy of Sciences*, **95**, 10661–664.

Hatchwell, B.J., Chamberlain, D.E., and Perrins, C.M. (1996a). The demography of blackbirds *Turdus merula* in rural habitats: is farmland a sub-optimal habitat. *J. Appl. Ecol.*: **33**, 1114–1124.

Hatchwell, B.J., Chamberlain, D.E., and Perrins, C.M. (1996b). The reproductive success of blackbirds *Turdus merula* in relation to habitat structure and choice of nest site. *Ibis*, **138**: 256–262.

Haughton, A.J., Bell, J.R., Gates, S., Johnson, P.J., Macdonald, D.W., Tattersall, F.H., and Hart, B.J. (1999). Methods of increasing invertebrate abundance within field margins. *Aspects of Applied Biology*, **54**: 163–70.

Hayward, G.F. (1983). *The bioenergetics of the weasel*, Mustela nivalis L. D.Phil thesis, Department of Zoology, Oxford University, Oxford.

Henderson, P.A. (in preparation) *Southwood's Ecological Methods* (4th edn.). Wiley-Blackwell, Oxford.

Herbst, M., Roberts, J.M., Rosier, P.T.W., Taylor, M.E., and Gowing, D.J. (2007). Edge effects and forest water use: A field study in a mixed deciduous woodland. *Forest Ecology and Management*, **250**: 176–86.

Herbst, M., Rosier, P.T.W., Morecroft, M.D., and Gowing, D.J. (2008). Comparative measurements of transpiration and canopy conductance in two mixed deciduous woodlands differing in structure and species composition. *Tree Physiology*, **28**: 959–70.

Hill, M.O., Mountford, J.O., Roy, D.B., and Bunce, R.G.H. (1999). *Ellenberg's indicator values for British plants: ECOFACT Volume 2 Technical Annex*. Institute of Terrestrial Ecology, NERC and DETR, London.

Hill, M.O., Preston, C.D., and Roy, D.B. (2004). *PLANTATT: Attributes of British and Irish Plants*. Centre for Ecology and Hydrology, Monks Wood, Huntingdon.

Hill, R.A. and Thomson, A.G. (2005). Mapping woodland species composition and structure using airborne spectral and LiDAR data. *International Journal of Remote Sensing*, **26**, 3763–779.

Hill, R.A., Hinsley, S.A., Bellamy, P.E., and Balzter, H. (2003). Ecological applications of airborne laser scanner data: modelling woodland bird habitats. Proceedings of *Scandlaser: Scientific Workshop on Airborne Laser Scanning of Forests*, 3–4 September 2003, Umeå, Sweden.

Hirons, G.J.M. (1976). *A population study of the Tawny Owl Strix aluco L. and its main prey species in woodland*. D. Phil thesis, University of Oxford.

HMSO (1952). *Census of Woodlands 1947–1949*. HMSO, London.

HMSO (1994). *Biodiversity: the UK Action Plan*. HMSO, London.

Hobson, P.R. and Bultitude, J. (2004). Evaluating biodiversity for conservation: a victim of the Traditional Paradigm. In M. Oksanen, and J. Pietarinen, (eds). *Philosophy and Biodiversity*. Cambridge University Press, Cambridge.

Hodder, K.H., Bullock, J.M., Buckland, P.C., and Kirby, K.J. (2005). Large herbivores in the wildwood and modern naturalistic grazing systems. *English Nature Research Reports*, No. 648. Peterborough.

Hofer H. (1988). Variation in resource presence, utilisation and reproductive success within a population of European badgers (*Meles meles*). *Mammal Review*, **18**: 25–36.

Hone, R., Anderson D.E., Parker, A.G., and Morecroft, M.D. (2001). Holocene vegetation change at Wytham Woods, Oxfordshire. *Quaternary Newsletter*, **94**: 1–15.

Hope-Simpson, J.F. (1940). Studies of the vegetation of the English chalk VI. Late stages in succession leading to chalk grassland. *Journal of Ecology*, **28**: 386–402.

Hopkins, J.J., Allison, H.M., Walmsley, C.A., Gaywood, M., and Thurgate, G. (2007). *Conserving Biodiversity in a Changing Climate: Guidance on Building Capacity to Adapt*. Defra, London.

Horsfall, A.S. and Kirby, K.J. (1985). *The Use of Permanent Quadrats to Record Changes in the Structure and Composition of Wytham Woods, Oxfordshire*. Research and Survey in Nature Conservation 1, Nature Conservancy Council, Peterborough.

Hunter, M.L.O. and Krebs, J.R. (1979). Geographical variation in the song of the Great Tit *Parus major* in relation to ecological factors. *J. Anim. Ecol.* **48**: 759–785

Huxley, J. (1947). *Conservation of Nature in England and Wales*. Report of the Wild Life Conservation Special Committee. Pp. 26. HMSO.

Ingrem, C. and Robinson, M. (2007). in: eds D. Miles, S. Palmer, A. Smith and G. P. Jones *Iron Age and Roman Settlement in the Upper Thames Valley*, pp351–64. Oxford Archaeology Unit, Oxford,

Jamieson, N., Barraclough, D., Unkovich, M., and Monaghan, R. (1998). Soil N dynamics in a natural calcareous grassland under a changing climate. *Biology and Fertility of Soils*, **27**: 267–73.

Jenkins, G.J., Murphy, J.M., Sexton, D.S., Lowe, J.A., Jones, P., and Kilsby, C.G. (2009). UK Climate Projections: *Briefing report*. Met Office Hadley Centre, Exeter, UK.

Jervis, M.A. (ed). (2005). *Insects as Natural Enemies: a Practical Perspective*. Springer, Dordrecht.

Johnson, D.D.P., Stopka, P. and Macdonald, D.W. (2004). Ideal flea constraints on group living: unwanted public goods and the emergence of cooperation. *Behavioral Ecology*, **15**: 181–86.

Jones, E.W. (1945a). The structure and reproduction of the virgin forest of the north temperate zone. *New Phytologist*, **44**: 130–48.

Jones, E.W. (1945b). Biological flora of the British Isles: *Acer pseudoplatanus* L. *Journal of Ecology*, **32**: 220–37.

Jones, E.W. (1945c). *A Preliminary Description Of Crops*. Wytham Woods. Oxford Forestry Institute, unpublished.

Jones, E.W. (1959). Biological flora of the British Isles *Quercus* L. *Journal of Ecology*, **47**: 169–222.

Jones, T., Feber, R., Hemery, G., Cook, P., James, K., Lamberth, C., and Dawkins, M. (2007). Welfare and environmental benefits of integrating commercially viable free-range broiler chickens into newly planted woodland: a UK case study. *Agricultural Systems* **94**: 177–88.

Kadas, G. (2006). Rare invertebrates colonising green roofs in London. *Urban Habitats*. http://www.urbanhabitats.org/v04n01/ Accessed 18/10/08.

Kempson, D.A., Lloyd, M., and Ghelardi, R. (1963). A new extractor for woodland invertebrates. *Pedobiologia*, **3**: 1–21.

Kettlewell, H.B.D. (1973). *The Evolution of Industrial Melanism*. Clarendon Press, Oxford.

Key, R. (1993). What are saproxylic invertebrates? In: K.J. Kirby and C.M. Drake, (eds). Dead wood matters: the ecology and conservation of saproxylic invertebrates in Britain. *English Nature Science*, **7**, p. 5. English Nature, Peterborough.

King, C.M. (1980). The weasel *Mustela nivalis* and its prey in an English woodland. *Journal of Animal Ecology*, **49**: 127–59.

King, K.L. and Hutchinson, K.J. (1983). The effects of sheep grazing on invertebrate numbers and biomass in unfertilized natural pastures of the New England Tablelands (NSW). *Australian Journal of Ecology*, **8**: 245–55.

Kirby, K.J. (1976). *The growth, production and nutrition of bramble Rubus fruticosus in woodland*. Unpublished D.Phil thesis, University of Oxford.

Kirby, K.J. (1988). *A Woodland Survey Hand Book*. Nature Conservancy Council, Peterborough.

Kirby, K.J. (1993). Assessing nature conservation values in British woodland—a review of recent practice. *Arboricultural Journal*, **17**: 253–76.

Kirby, K.J. (2003). Woodland conservation in privately-owned cultural landscapes: the English experience. *Environmental Science and Policy*, **6**: 253–59.

Kirby, K.J. (2004a). A model of a natural wooded landscape in Britain driven by large-herbivore activity. *Forestry*, **77**: 405–20.

Kirby, K.J. (2004b). Changes in the composition of Wytham Woods (southern England), 1974–2002, in stands of different origins and past treatment. In: O. Honnay, K. Verheyen, B. Bossuyt, and M. Hermy, (eds). *Forest Biodiversity: Lesson from History for Conservation*, pp. 193–203. CABI Publishing, Wallingford.

Kirby, K.J. and Drake, C.M. (eds) (1993). Dead wood matters: the ecology and conservation of saproxylic invertebrates in Britain. *English Nature Science*, **7**, English Nature, Peterborough.

Kirby, K.J. and Thomas, R.C. (1999). Changes in ground flora in Wytham Woods, southern England, 1974–1991 and their implications for nature conservation. English Nature Research Report 320.

Kirby, K.J. and Thomas, R.C. (2000). Changes in the ground flora in Wytham Woods, southern England, from 1974 to 1991—implications for nature conservation. *Journal of Vegetation Science*, **11**: 871–80.

Kirby, K.J. and Woodell, S.R.J. (1998). The distribution and growth of bramble (*Rubus fruticosus*) in British semi-natural woodland and the implications for nature conservation. *Journal of Practical Ecology and Conservation*, **2**: 31–41.

Kirby, K.J. and Wright, F.J. (editors) (1988). *Woodland conservation and research in the clay vale of Oxfordshire and Buckinghamshire*. Nature Conservancy Council, Peterborough (Research and survey in nature conservation 15). 132pp.

Kirby, K.J., Bines, T., Burn, A., Mackintosh, J., Pitkin, P., and Smith, I. (1986). Seasonal and observer differences in vascular plant records from British woodlands. *Journal of Ecology*, **74**: 123–31.

Kirby, K.J., Reid, C.M., Thomas, R.C., and Goldsmith, F.B. (1998). Preliminary estimates of fallen dead wood and standing dead trees in managed and unmanaged forests in Britain. *Journal of Applied Ecology*, **35**: 148–55.

Kirby, K.J., Smart, S.M., Black, H.I.J., Bunce, R.G.H., Corney, P.M., and Smithers, R.J. (2005). Long term ecological change in British woodlands (1971–2001). *English Nature Research Reports*, No. 653. English Nature, Peterborough.

Kirby, K.J., Thomas, R.C., and Dawkins, H.C. (1996). Monitoring of changes in tree and shrub layers in Wytham Woods (Oxfordshire), 1974–1991. *Forestry*, **69**: 319–34.

Kitching, R.L. (1971). An ecological study of water-filled tree-holes and their position in the woodland ecosystem. *Journal of Animal Ecology*, **24**: 120–36.

Kitching, R.L. (2000). *Food Webs and Container Habitats: their Natural History and Ecology of Phytotelmata*. Cambridge University Press, Cambridge.

Krebs, J.R. (1970). *A study of territorial behaviour in the Great Tit*, Parus major L. D. Phil. thesis, Oxford University.

Krebs, J.R. and Davies, N.B. (eds). (1978). *Behavioural Ecology: an Evolutionary Approach* (1st edn.). Blackwell, Oxford.

Krebs, J.R., Clayton, N.S., Healy, S.D., Cristol, D.A., Patel, S.N., and Jolliffe, A.R. (1996). The ecology of the avian brain: Food-storing memory and the hippocampus. *Ibis*, **138**: 34–46.

Kruuk, H. (1975). Functional aspects of social hunting by carnivores. In: G. Baerends and A. Manning, (eds). *Function and Evolution in Behaviour*, pp. 119–41. Clarendon Press, Oxford.

Kruuk, H. (1978). Spatial organisation and territorial behaviour of the European badger *Meles meles*. *Journal of Zoology*, **184**: 1–19.

Kruuk, H. (1989). *The Social Badger: Ecology and Behaviour of a Group-Living Carnivore* (Meles meles). Oxford University Press, Oxford.

Lack, D. (1943). *Life of the Robin*. H.F. and G. Witherby.

Lack, D. (1947/48). The Significance of Clutch-size. *Ibis* **89**: 302–52 and **90**: 25–45.

Lack, D.L. (1954). *The Natural Regulation of Animal Numbers*. Oxford University Press, Oxford.

Lack, D. (1968). *Ecological adaptations for breeding in birds*. Methuen, London.

Lack, D. and Lack, E. (1958). The nesting of the Long-tailed Tit. *Bird Study*, **5**: 1–19.

Lamborn, E.A.G. (1943). Wytham. An Historical Account. In *Wytham*. A record issued by the Oxford Preservation trust on the Acquisition of Wytham Abbey and Estate by the University of Oxford 1943, pp 13–23. University Press, Oxford.

Larkin, P.A. and Elbourn, C.A. (1964). Some observations on the fauna of dead wood on live oak trees. *Oikos*, **15**: 79–92.

Lawton, J.H. (1983) Plant architecture and the diversity of phytophagous insects. *Annual Review of Entomology*, **28**: 23–39.

Leps, J. and Stursa, J. (1989). Species-area curve, life-history strategies and succession: a field test of relationships. *Plant Ecology*, **83**: 249–57.

Linhart, Y.B. and Whelan, R.J. (1980). Woodland regeneration in relation to grazing and fencing in Coed Gorswen, North Wales. *Journal of Applied Ecology*, **17**: 827–40.

Lloyd, P.S. (1974). Wytham estate—grassland areas and scrub invasion. Report to the Wytham Management Committee. Unpublished.

Lockie, J.D. (1954). *The feeding ecology of the Jackdaw, Rook, and related Corvidae*. D. Phil thesis, University of Oxford.

MacArthur, R.H. and Wilson, E.O. (1967). *Theory of Island Biogeography*. Princeton University, Princeton.

Mace, R. (1987). The dawn chorus in the great tit *Parus major* is directly related to female fertility. *Nature*, **330**: 745–746.

Macdonald, D.W. (1979). "Helpers" in fox society. *Nature*, **282**: 69–71.

Macdonald, D.W. (1980a). The red fox, *Vulpes vulpes*, as a predator upon earthworms, *Lumbricus terrestris*. *Zeitschrift für Tierpsychologie*, **52**: 171–200.

Macdonald, D.W. (1980b). *Rabies and Wildlife: a Biologist's Perspective*. Oxford University Press, New York.

Macdonald, D.W. (1983). The ecology of carnivore social behavior. *Nature*, **301**: 379–84.

Macdonald, D.W. (1987). *Running with the Fox*. Unwin Hyman, London.

Macdonald, D.W. and Johnson, P.J. (2000). Farmers and the custody of the countryside: trends in loss and conservation of non-productive habitats 1981–1998. *Biological Conservation*, **94**: 221–34.

Macdonald, D.W. and Newman, C. (2002). Badger (*Meles meles*) population dynamics in Oxfordshire, UK. Numbers, density and cohort life histories, and a possible role of climate change in population growth. *Journal of Zoology*, **256**: 121–38.

Macdonald, D.W. and Smith, H.E. (1991). New perspectives on agro-ecology: between theory and practice in the agricultural ecosystem. In: L.G. Firbank, N. Carter, J.F. Darbyshire and G.R. Potts, (eds). *The Ecology of Temperate Cereal Fields (32nd Symposium of the British Ecological Society)* pp. 413–48. Blackwell Scientific Publications, Oxford.

Macdonald, D.W. and Stafford S. (eds) (1997). *The Wytham Communicator*. WildCRU, Oxford.

Macdonald, D.W. and Baker, S.E. (2004). Non-lethal control of fox predation: The potential of generalized aversion. *Animal Welfare*, **13**: 77–85.

Macdonald, D.W., Mace, G., and Rushton, S. (1998). *Proposals for Future Monitoring of British Mammals*. HMSO, London.

Macdonald, D.W., Anwar, M., Newman, C., Woodroffe, R. and Johnson, P.J. (1999). Inter-annual differences in the age-related prevalences of *Babesia* and *Trypanosoma* parasites of European badgers (*Meles meles*). *Journal of Zoology*, **247**: 65–70.

Macdonald, D.W., Bryce, J.M. and Thom, M.D. (2001). Introduced mammals: do carnivores and herbivores usurp native species by different mechanisms? In: H.J. Pelz, D.P. Cowan, and C.J. Feare, (eds). *Advances in Vertebrate Pest Management II*, pp. 11–44. Filander Verlag, Fürth.

Macdonald, D.W., Newman, C., Dean, J., Buesching, C.D., and Johnson, P.J. (2004). The distribution of European badger *Meles meles* setts in a high density area: Field observations contradict the sett dispersion hypothesis. *Oikos*, **106**: 295–307.

Macdonald, D.W., Newman, C., Stewart, P.D., and Domingo-Roura, X. (2002). Density-dependent regulation of body weight and condition in badgers (*Meles meles*) from Wytham Woods, Oxfordshire. *Ecology*, **83**: 2056–61.

Macdonald, D.W., Tew, T.E., Todd, I.A., Garner, J.P. and Johnson, P.J. (2006). Arable habitat use by wood mice (*Apodemus sylvaticus*). 3. A farm-scale experiment on the effects of crop rotation. *Journal of Zoology*, **250**: 313–20.

Macdonald, D.W., Feber, R.E., Tattersall, F.H., and Johnson, P.J. (2000). Ecological experiments in farmland conservation. In: M.J. Hutchings, E,A. John, and A.J.A. Stewart eds. *The Ecological Consequences of Environmental Heterogeneity*, pp. 357–78. Blackwell Scientific Publications, Oxford.

Macfadyen A. (1992). Obituary. Charles Sutherland Elton. *Journal of Animal Ecology*, **61**: 499–502.

MacLeod, A., Wratten, S.D., Sotherton, N.W., Thomas, M.B., (2004). 'Beetle banks' as refuges for beneficial arthropods in farmland: long-term changes in predator communities and habitat. *Agricultural and Forest Entomology*, **6**: 147–54.

Macleod, R., Gosler, A.G., and Cresswell, W. (2005). Diurnal mass gain strategies and perceived predation risk in the great tit *Parus major*. *Journal of Animal Ecology*, **74**: 956–64.

Masters, G.J., Brown, V.K., Clarke, I.P., Whittaker, J.B., and Hollier, J.A. (1998). Direct and indirect effects of climate change on insect herbivores: Auchenorrhyncha (Homoptera). *Ecological Entomology*, **23**: 45–52.

Matthews, J.D. (1963). Factors affecting the production of seed by forest trees. *Forestry Abstracts* (review article), *24 (1)*, i–xii.

May, R.M. (1973). *Stability and Complexity in Model Ecosystems*. Princeton University Press, Princeton.

May, R.M., Levin, S.A., and Sugihara, G. (2008). Complex systems: Ecology for bankers *Nature*, **451**: 893–95.

McCleery, R.H. and Perrins, C.M. (1998). Temperature and egg-laying trends. *Nature*, **391**: 30–31.

McDonald, A.W. (2001). Succession during the re-creation of a flood-meadow 1985–1999. *Applied Vegetation Science*, **4**: 167–76.

McGregor, P.K. and Krebs, J.R., and Perrins, C.M. (1981). Song repertoires and lifetime reproductive success in the Great Tit (*Parus major*). *Amer. Natur.* **118**: 149–159.

Mihok, B. (2007). *Forty-year changes in the canopy and understorey in Wytham Woods*. Unpublished dissertation, Oxford University.

Miles, D., Palmer, S., Smith, A., and Jones, G.P. (2007). *Iron Age and Roman Settlement in the Upper Thames Valley. Excavations at Claydon Pike and other sites within the Cotswold Water Park.* Oxford Archaeology Thames Valley Landscapes Monograph 26, Oxford.

Miller, R.S. (1954). Food habits of the wood mouse, *Apodemus sylvaticus* (Linné 1758) and the bank vole, *Clethrionomys glareolus* (Schreber 1780) in Wytham Woods, Berkshire. *Säugetierkundliche Mitteilungen* Band II, 109–114.

Mitchell, F.J.G. (2005). How open were European primeval forests? Hypothesis testing using paleoecological data. *Journal of Ecology*, **93**: 168–77.

Mitchell, R.J., Morecroft, M.D., Acreman, M, Crick, H.Q.P., Frost, M., Harley, M., Maclean, I.D.M., Mountford, O., Piper, J., Pontier, H., Rehfisch, M.M., Ross, L.C., Smithers, R.J., Stott, A., Walmsley, C.A., Watts, O., Wilson, E. (2007). *England Biodiversity Strategy—towards adapation to climate change. Final report to Defra for contract CRO327.* Defra, 177pp. (Unpublished).

Moore, P.D. (2005). Palaeoecology: Down to the woods yesterday. *Nature*, **433**: 588–89.

Morecroft, M.D. (2000). Integrating climate and biological monitoring - the effects of drought at Wytham Woods. In: K.J. Kirby and M.D. Morecroft, (eds). Long-term monitoring in British woodlands, pp. 120–29. *English Nature Science*, **34**, English Nature, Peterborough.

Morecroft, M.D. and Roberts, J.M. (1999). Photosynthesis and stomatal conductance of mature canopy oak (*Quercus robur*) and sycamore (*Acer pseudoplatanus*) trees throughout the growing season. *Functional Ecology*, **13**: 332–42.

Morecroft, M.D., Bealey, C.E., Howells, O., Rennie, S.C., and Woiwod, I. (2002). Effects of drought on contrasting insect and plant species in the UK in the mid-1990s. *Global Ecology and Biogeography*, **11**: 7–22.

Morecroft, M.D., Burt, T.P., Taylor, M.E., and Rowland, A.P. (2000). Effects of the 1995–1997 drought on nitrate leaching in lowland England. *Soil Use and Management*, **16**, 117–23.

Morecroft, M.D., Masters, G.J., Brown, V.K., Clarke, I.P., Taylor, M.E., and Whitehouse, A.T. (2004). Changing precipitation patterns alter plant community dynamics and succession in an ex-arable grassland. *Functional Ecology*, **18**: 648–55.

Morecroft, M.D., and Paterson, J.S. (2006). Effects of temperature and precipitation changes on plant communities. In: J.I.L. Morison and M.D. Morecroft, (eds). *Plant Growth and Climate Change.* Blackwell Publishing, Oxford, 232. (Biological Sciences Series).

Morecroft, M.D., Stokes, V.J., and Morison, J.I.L. (2003). Seasonal changes in the photosynthetic capacity of canopy oak (*Quercus robur*) leaves: the impact of slow development on annual carbon uptake. *International Journal of Biometeorology*, **47**, 221–26 (Special Issue: Proceedings of International Phenology Conference, Wageningen, Netherlands, December 2001).

Morecroft, M.D., Stokes, V.J., Taylor, M.E., and Morison, J.I.L. (2008). Effects of climate and management history on the distribution and growth of sycamore (*Acer pseudoplatanus*) in a southern British woodland in comparison to native competitors. *Forestry*, **81**: 59–74.

Morecroft, M.D., Taylor, M.E., and Oliver, H.R. (1998). Air and soil microclimates of deciduous woodland compared to an open site. *Agricultural and Forest Meteorology*, **90**: 141–56.

Morecroft, M.D., Taylor, M.E., Ellwood, S.A., and Quinn, S.A. (2001). Impacts of deer herbivory on ground vegetation at Wytham Woods, central England. *Forestry*, **74**: 251–57.

Morecroft, M.D., Bealey, C.E., Beaumont, D.A., Benham, S., Brooks, D.R., Burt, T.P., Critchley, C.N.R., Dick, J., Littlewood, N.A., Monteith, D.T., Scott, W.A., Smith, R.I., Walmsley, C., and Watson, H. (2009). The UK Environmental Change Network: emerging trends in the composition of plant and animal communities and the physical environment. *Biological Conservation*, **142**: 2814–2832.

Morris, M.G. (1971). The management of grassland for the conservation of invertebrate animals. In: E. Duffey and A.S. Watt, (eds). *The Scientific Management of Animal and Plant Communities for Conservation*, pp. 527–52. Blackwell Scientific Publications, Oxford.

Morris, M.G. (2000). The effects of structure and its dynamics on the ecology and conservation of arthropods in British grasslands. *Biological Conservation*, **95**: 129–42.

Mountford, E.P. (2000). A provisional minimum intervention woodland reserve series. English *Nature Research Reports*, No 385.

National Archives (1989). Edmund Brisco Ford FRS; (1901–1988). *National Archives reference NCUACS 14.7.89*. Bodleian Library, Oxford University.

Natural England (2008). *State of Natural Environment*. Natural England, Peterborough.

NCC (1984). *Nature Conservation in Great Britain*. Nature Conservancy Council, Peterborough.

Newman, C., Buesching, C.D., and Macdonald, D.W. (2003). Validating mammal monitoring methods and assessing the performance of volunteers in wildlife conservation - "Sed quis custodiet ipsos custodes?". *Biology and Conservation*, **113**: 189–97.

Newman, C., Buesching, C.D., and Wolff, J.O. (2004). The function of face masks in 'midguild' carnivores. *Oikos*, **108**: 623–33.

Newman, C., Macdonald, D.W. and Anwar, A. (2001). Coccidiosis in the European Badger, *Meles meles* in Wytham Woods: Infection and consequences for growth and survival. *Parasitology*, **123**: 133–42.

Newton, I. (1964). *The ecology and moult of the Bullfinch*. D. Phil thesis, University of Oxford.

Newton, I. (1972). *Finches*. New Naturalist. Collins, London.

Odum, E.P. and Pontin, A.J. (1961). Population density of the underground ant, *Lasius flavus*, as determined by tagging with $P^{32}$. *Ecology*, **42**: 186–88.

Oliver, C.D. and Larson, B.C. (1996). *Forest Stand Dynamics*. Wiley, New York.

Ollerton J (1993). *Ecology of flowering and fruiting in* Lotus corniculatus *L*. Ph.D. thesis, Oxford Brookes University, U.K.

Osmaston, F.C. (1959). *The Revised Working Plan for the Wytham Woods 1959/60 – 1968/69*. Oxford Forestry Institute Library, unpublished MS.

Ozanne, C.M.P. (2005). Techniques and methods for sampling canopy insects. In: S.R. Leather (ed): *Insect Sampling In Forest Ecosystems*, pp. 146–67. Blackwell Science, Oxford.

Ozanne, C.M.P., Hambler, C., Foggo, A. and Speight, M.R. (1997). The significance of edge effects in the management of forests for invertebrate biodiversity. In: N.E. Stork, J. Adis, and R.K. Didham (eds) *Canopy arthropods*, pp. 534–50. Chapman and Hall, London

Paviour-Smith, K. (1960). The fruiting bodies of macrofungi as habitats for beetles of the family Ciidae (Coleoptera). *Oikos*, **11**: 43–71.

Paviour-Smith, K. (1971). Fungi in Wytham Woods, Berkshire. *Proceedings and Reports of the Ashmolean Natural History Society*, pp. 3–13, Oxford.

Paviour-Smith, K. and Elbourn, C.A. (1993). A quantitative study of the fauna of small dead and dying wood in living trees in Wytham Woods, near Oxford. In: K.J. Kirby and C.M. Drake (eds) Dead wood matters. *English Nature Science*, **7**, pp. 33–57, English Nature, Peterborough.

Pernetta, J.C. (1977). Population ecology of British shrews in grassland. *Acta Theriologica*, **22**: 279–96.

Perrins, C.M. (1992). Tits and their caterpillar food supply. *Ibis*, **133** (suppl.1): 49–54.

Perrins, C.M. and Jones, P.J. (1974). The inheritance of clutch size in the Great Tit (*Parus major L.*). *Condor*, **76**: 225–29.

Perrins, C.M. and Overall, R. (2001). Effect of increasing numbers of deer on bird populations in Wytham Woods, central England. *Forestry*, **74**: 299–309.

Peterken, G.F. (1974). A method for assessing woodland flora for conservation using indicator species. *Biological Conservation*, **6**: 239–45.

Peterken, G.F. (1977). Habitat conservation priorities in British and European woodlands. *Biological Conservation*, **11**: 223–36.

Peterken, G.F. (1993). *Woodland Conservation And Management* (2nd edn.). Chapman and Hall, London.
Peterken, G.F. (1996). *Natural Woodland*. Cambridge University Press, Cambridge.
Peterken, G.F. and Gane, M. (1984). Historical factors affecting the number and distribution of vascular plant species in the woodlands of central Lincolnshire. *Journal of Ecology*, **69**: 781–96.
Peterken, G.F. and Jones E.W. (1987). Forty years of change in Lady Park Wood: the old growth stands. *Journal of Ecology*, **75**: 479–512.
Peterken, G.F. and Jones, E.W. (1989). Forty years of change in Lady Park Wood: the young growth stands. *Journal of Ecology*, **77**: 401–29.
Phillipson, J. (1983). Slug numbers, biomass and respiratory metabolism in a beech woodland - Wytham Woods, Oxford. *Oecologia*, **60**: 38–45.
Phillipson, J. and Meyer, E (1984). Diplopod numbers and distribution in a British beechwood. *Pedobiologia*, **26**: 83–94.
Phillipson, J., Abel, R., Steel, J., and Woodell, S.R.J. (1977). Nematode numbers, biomass and respiratory metabolism in a beech woodland - Wytham Woods, Oxford. *Oecologia*, **27**: 141–55.
Phillipson, J., Abel, R., Steel, J., and Woodell, S.R.J. (1978). Earthworm numbers, biomass and respiratory metabolism in a beech woodland - Wytham Woods, Oxford. *Oecologia*, **33**: 291–309.
Phillipson, J., Putman, R.J., Steel, J., and Woodell, S.R.J. (1975). Litter input, litter decomposition and the evolution of carbon dioxide in a beech woodland – Wytham Woods, Oxford. *Oecologia*, **20**: 203–17.
Phillipson, J, Abel, R. Steel, J., and Woodell, S.R.J. (1979). Enchytraeid numbers, biomass and respiratory metabolism in a beech woodland – Wytham Woods, Oxford. *Oecologia*, **43**: 173–93.
Plesner-Jensen, S. (1993). Temporal changes in the food preferences of wood mice (*Apodemus sylvaticus* L.). *Oecologia*, **94**: 76–82.
Plesner-Jensen, S. (1996). Juvenile dispersal in relation to adult densities in wood mice *Apodemus sylvaticus*. *Acta Theriologica*, **41**: 177–86.
Pontin, A.J. (1961) Population stabilisation and competition between the ants *Lasis flavus* (F.) and *L. niger* (L.). *Journal of Animal Ecology*, **30**: 47–54.
Preston, C.D., Pearman, D.A., and Dines, T.D. (2002a). *New Atlas Of The British And Irish Flora*. Oxford University Press, Oxford.
Preston, C.D., Telfer, M.G., Arnold, H.R. *et al.* (2002b). *The Changing Flora Of The UK*. Defra, London.
Proffitt, F. M. (2002). *Causes of Population Decline of the Bullfinch* Pyrrula pyrrula *in Agricultural Environments*. D. Phil thesis, University of Oxford.
Putman, R.J. (1978). The role of carrion-frequenting arthropods in the decay process. *Ecological Entomology*, **3**: 133–39.
Quinn, J.L., Patrick, S., Wilkin, T.D., and Sheldon, B.C. (2009). Heterogeneous selection on a temperament trait in a variable environment. *Journal of Animal Ecology*, **78**: 1203–1215.
Rackham, O. (1976). *Trees And Woodland In The British Landscape*. J.M. Dent and Sons, London.
Rackham, O. (1980). *Ancient Woodland*. Edward Arnold, London.
Rackham, O. (2003). *Ancient Woodland* (revised edition). Castlepoint Press, Dalbeattie.
Rackham, O. (2006). *Woodlands*. Collins New Naturalist, Harper Collins, London.
Radley, G. (1979). *Changes in herbaceous vegetation during scrub development*. Unpublished D.Phil., Oxford.
Ratcliffe, D.A. (1977). *A Nature Conservation Review*. Volume 1. Cambridge University Press, Cambridge.
Raymond, B. and Hails, R.S. (2007). Variation in plant resource quality and the transmission and fitness of the Winter Moth, *Operophtera brumata* nucleopolyhedrovirus. *Biological Control*, **41**: 237–45.
Read, H. (2000). *The Veteran Tree Management Handbook*. English Nature, Peterborough.
Richards, C.G.J. (1985). The population dynamics of *Microtus agrestis* in Wytham, 1949 to 1978. *Acta Zooligica Fennica*, **173**: 35–38.

Richards, E.G. (2003). *British Forestry in the Twentieth Century: Policy and Achievements*. Brill, Leiden.

Riddington, R. and Gosler, A.G. (1995). Differences in reproductive success and parental qualities between habitats in the Great Tit *Parus major*. *Ibis*: **137**: 371–78.

Roberts, R.R., Hopkins, R., and Morecroft, M. (1999). Towards a predictive description of forest canopies from litter properties. *Functional Ecology*, **13**: 265–72.

Robinson, J.G. (2006). Conservation biology and real-world conservation. *Conservation Biology*, **20**: 658–69.

Robinson, M.A. (1991). The Neolithic and Late Bronze Age insect assemblages. In: S.P. Needham, (ed). *Excavation and Salvage at Runnymede Bridge, 1978: the Late Bronze Age Waterfront Site*, pp. 277–326. British Museum Press: London.

Robinson, M.A. (2001). Insects as palaeoenvironmental indicators. In: D.R. Brothwell, and A.M. Pollard (eds.)., *Handbook Of Archaeological Sciences*, pp. 121–33. Chichester: Wiley and sons.

Rodwell, J. (1991). *British Plant Communities: 1. Woodland And Scrub*. Cambridge University Press, Cambridge.

Rodwell, J.S. (1992). *British Plant Communities. 3. Grassland And Montane Communities*. Cambridge University Press, Cambridge.

Rodwell, J.S. (1995). *British Plant Communities. 4. Aquatic Communities, Swamps And Tall-Herb Fens*. Cambridge University Press, Cambridge.

Root, T.L., Liverman, D., and Newman, C. (2007). Managing biodiversity in the light of climate change: Current biological effects and future impacts. In: D.W. Macdonald and K. Service, (eds). *Key Topics in Conservation*, pp. 85–104. Blackwell Publishing, Oxford.

Rose, F. (1999). Indicators of ancient woodland. *British Wildlife*, **10**: 241–51.

Rosenzweig, C., Casassa, G., Karoly, D.J., Imeson, A., Liu, C., Menzel, A., Rawlins, S., Root, T.L., Seguin, B., Tryjanowski, P., and Hanson, C.E. (2007). Assessment of observed changes and responses in natural and managed systems. In: M.L. Parry, O.F. Canziani, J.P. Palutikof, and P.J. van der Linden, (eds).*Climate Change 2007: Impacts, Adaptation and Vulnerability. Contribution of Working Group II to the Fourth Assessment Report of the Intergovernmental Panel on Climate Change*, pp. 79–131. Cambridge University Press.

Seel, D.C. (1968). *The breeding biology of the House Sparrow* (Passer domesticus). D. Phil thesis, University of Oxford.

Sheail, J. (1995). War and the Development of Nature Conservation in Britain. *Journal of Environmental Management*, **44**: 267–83.

Sheehan, P. (1979). *Management plan for Wytham Woods 1980–1985*. Dissertation, Oxford Forestry Institute.

Skerl, K.L. (1999) Spiders in conservation planning: a survey of US Natural Heritage Programs *Journal of Insect Conservation*, **3**: 341–47.

Smart, S.M., Ashmore, M.R., Hornung, M. *et al.* (2005). Detecting the signal of atmospheric N deposition in recent national-scale vegetation change across Britain. *Water, Air and Soil Pollution: Focus*, **4**: 269–78.

Smeathers R (1939). *Working plan for Wytham Woods*. Unpublished, Oxford Forestry Institute.

Smith, H., Feber, R.E., Johnson, P.J., McCallum, K., Plesner Jensen, S., Younes, M., and Macdonald, D.W. (1993). The conservation management of arable field margins. *English Nature Science* **18**. English Nature, Peterborough.

Smith, H.E. and Macdonald, D.W. (1989). Secondary succession on extended arable field margins: its manipulation for wildlife benefit and weed control. In: *Proceedings of the 1989 Brighton Crop Protection Conference—Weeds*, pp. 1063–68. British Crop Protection Council Publications, Farnham, Brighton.

Smith, H.E., McCallum, K., and Macdonald, D.W. (1997). Experimental comparison of the nature conservation value, productivity and ease of management of a conventional and a more species-rich grass ley. *Journal of Applied Ecology*, **34**: 53–64.

Smith, K. Paviour- and Elbourn, C.A. (1993). A quantitative study of the fauna of small dead and dying wood in living trees in Wytham Woods, near Oxford. pp. 33–57 in: K.J. Kirby, and C.M. Drake, (eds). Dead wood matters. *English Nature Science* 7, English Nature, Peterborough.

SMS (1998). English Nature Site Management Statement. Unpublished.

Snow, D.W. (1958). *A Study of Blackbirds*. George Allen and Unwin, London.

Snow, B. and Snow, D. (1988). *Birds and Berries*. Poyser, Calton.

Solomon, S., Qin, D., Manning, M., Chen, Z., Marquis, M., Avery, K.B., Tignor, M., and Miller H.L. (eds.) (2007). *Climate Change 2007: The Physical Science Basis. Contribution of Working Group I to the Fourth Assessment Report of the Intergovernmental Panel on Climate Change*. Cambridge University Press, Cambridge, United Kingdom and New York.

Sorensen, A.E. (1981). Interactions between birds and fruit in a temperate woodland. *Oecologia*, **50**: 242–49.

Southern, H.N. (1955). A Britain without rabbits. *Discovery*, **16**: 186–89.

Southern, H.N. (1970). The natural control of a population of Tawny owls (*Strix aluco*). *Journal of the Zoological Society of London*, **162**: 197–285.

Southern, H.N. (1979). Population processes in small mammals. In: D.M. Stoddart, (ed). *Ecology Of Small Mammals*, pp. 65–101. Chapman and Hall, London.

Southern, H.N. and Lowe, V.P.W. (1968). The pattern of distribution of prey and predation in tawny owl territories. *Journal of Animal Ecology*, **37**: 75–97.

Southwood, T.R.E. and Clarke, J.R. (1999). Charles Sutherland Elton. *Biographical memoirs of Fellows of the Royal Society*, **45**: 129–146.

Southwood, T.R.E. and Henderson, P.A. (2000). *Ecological Methods* (3rd edn.). Blackwell Science, Oxford.

Southwood, T.R.E., Brown, V.K., and Reader, P.M. (1979). The relationships of plant and insect diversities in succession. *Biological Journal of the Linnaean Society*, **12**: 327–48.

Southwood, T.R.E., Wint, G.R.W., Kennedy, C.E.J., and Greenwood, S.R. (2004). Seasonality, abundance, species richness and specificity of the phytophagous guild of insects on oak (*Quercus*) canopies. *European Journal of Entomology*, **101**: 43–50.

Spears, N. and Clarke, J.R. (1986). Effect of male presence and of photoperiod on the sexual maturation of the field vole (*Microtus agrestis*). *Journal of Reproduction and Fertility*, **78**: 231–38.

Spears, N. and Clarke, J.R. (1987). Effect of nutrition, temperature and photoperiod on the rate of sexual maturation of the field vole (*Microtus agrestis*). *Journal of Reproduction and Fertility*, **80**: 175–81.

Spears, N. and Clarke, J.R. (1988). Selection in field voles (*Microtus agrestis*) for gonadal growth under short photoperiods. *Journal of Animal Ecology*, **57**: 61–70.

Speight, M.C.D. (1989). Saproxylic invertebrates and their conservation. *Nature and Environment Series, No 42*. Council of Europe, Strasbourg.

Speight, M.C.D., Hunter, M.J., and Watt, A.S. (2008). *Ecology of Insects. Concepts and Applications* (2nd Edn.). Wiley-Blackwell, Oxford.

Spencer, J.W. and Kirby, K.J. (1992). An inventory of ancient woodland for England and Wales. *Biological Conservation*, **62**: 77–93.

Staley, J.T., Mortimer, S.R., Masters, G.J., Morecroft, M.D., Brown, V.K., and Taylor, M.E. (2006). Drought stress differentially affects leaf-mining species. *Ecological Entomology*, **31**: 460–69.

Staley, J.T., Mortimer, S.R., Morecroft, M.D., Brown, V.K., and Masters, G.J. (2007a). Summer drought alters plant-mediated competition between foliar- and root-feeding insects. *Global Change Biology*, **13**: 866–77.

Staley, J.T., Hodgson, C.J., Mortimer, S.R., Morecroft, M.D., Masters, G.J., Brown, V.K., and Taylor, M.E. (2007b). Effects of summer rainfall manipulations on the abundance and vertical distribution of herbivorous soil macro-invertebrates. *European Journal of Soil Biology*, **43**: 189–98.

Steel, D. (1984). *Shotover: The History of a Royal Forest*. Pisces Publications, Oxford.

Steele, R.C. and Welch, R.C. (1973). *Monks Wood—a National Nature Reserve Record*. The Nature Conservancy, Huntingdon.

Sterling, P.H. and Hambler, C. (1988). Coppicing for conservation: do hazel communities benefit? In: K.J. Kirby and F.J. Wright, (eds). *Woodland Conservation and Research in the Clay Vale of Oxfordshire and Buckinghamshire*. NCC, Peterborough.

Sterling, P.H., Gibson, C.W.D. and Brown, V.K. (1992). Leaf miner assemblies: effects of plant succession and grazing management. *Ecological Entomology*, **17**: 167–178.

Stern, R. (1993). Obituary of E.W. Jones. *Forestry*, **66**: 221–223.

Sternberg, M., Brown, V.K., Masters, G.J., and Clarke, I.P. (1999). Plant community dynamics in a calcareous grassland under climate change manipulations. *Plant Ecology*, **143**: 29–37.

Stewart, P.D. and Macdonald, D.W. (2003). Badgers and badger fleas: strategies and counter-strategies. *Ethology*, **109**: 751–64.

Stewart, P.D., Bonesi, L. and Macdonald, D.W. (1999). Individual differences in den maintenance effort in a communally dwelling mammal: the Eurasian badger. *Animal Behaviour*, **57**: 153–61.

Stewart, P.D., Ellwood, S.A., and Macdonald, D.W. (1997). Remote video-surveillance of wildlife—an introduction from experience with the European badger *Meles meles*. *Mammal Review*, **27**: 185–204.

Stewart, P.D., Macdonald, D.W., Newman, C., and Tattersall, F.H. (2002). Behavioural mechanisms of information transmission and reception by badgers, *Meles meles*, at latrines. *Animal Behaviour*, **63**: 999–1007.

Stiven, R. (2007). *Managing Sycamore in Semi-Natural Woodland*. Ravine WOODLIFE project, Natural England.

Stockley, P. and Macdonald, D.W. (1998). Why do female common shrews produce so many offspring? *Oikos*, **83**: 560–66.

Stockley, P., Searle, J.B., Macdonald, D.W., and Jones, C.S. (1994). Alternative reproductive tactics in male common shrews: relationships between mate-searching behaviour, sperm production, and reproductive success as revealed by DNA fingerprinting. *Behavioral Ecology and Sociobiology*, **34**: 71–78.

Stockley, P., Searle, J.B., Macdonald, D.W., and Jones, C.S. (1996). Correlates of reproductive success within alternative mating tactics of the common shrew. *Behavioral Ecology*, **7**: 334–40.

Stokes, V.J. (2002). *The impact of microenvironment, leaf development and phenology on annual carbon gain and water loss of two deciduous tree species*. PhD Thesis, University of Essex.

Stokes, V.J., Morecroft, M.D., and Morison, J.I.L. (2006). Boundary layer conductance for contrasting leaf shapes in a deciduous broadleaved forest canopy. *Agricultural and Forest Meteorology*, **159**: 40–54.

Stone, G.N. and Cook, J.M. (1998). The structure of cynipid oak galls: patterns in the evolution of an extended phenotype. *Proceedings of the Royal Society of London Series B*, **265**: 979–88.

Stopka, P. and Macdonald, D.W. (1998). Signal interchange during mating in the wood mouse (*Apodemus sylvaticus*): The concept of active and passive signalling. *Behaviour*, **135**: 231–49.

Stopka, P. and Macdonald, D.W. (1999). The market effect in the wood mouse, *Apodemus sylvaticus*: Selling information on reproductive status. *Ethology*, **105**: 969–82.

Stopka, P. and Macdonald, D.W. (2003). Way-marking behaviour: an aid to spatial navigation in the wood mouse (*Apodemus sylvaticus*). *BMC Ecology*, **3**: doi:10.1186/1472-6785-3-3.

Strong, D.R., Lawton, J.H., and Southwood, T.R.E. (1984). *Insects on Plants: Community Patterns and Mechanisms*. Blackwell Scientific Publications, Oxford.

Swengel, A.B. (2001). A literature review of insect responses to fire, compared to other conservation managements of open habitat, *Biodiversity and Conservation*, **10**: 1141–69.

Sykes, J.M. and Lane, A.M.J. (1996). *The United Kingdom Environmental Change Network: Protocols for Standard Measurements at Terrestrial Sites*. Natural Environment Research Council, H.M.S.O., London.

Szulkin, M., Garant, D., McCleery, R.H., and Sheldon, B.C. (2007) Inbreeding depression along a life-history continuum in the great tit. *Journal of Evolutionary Biology*, **20**: 1531–43.

Tansley, A.G. and Adamson, R.S. (1925). Studies of the vegetation of the English Chalk III. The chalk grasslands of the Hampshire-Sussex border. *Journal of Ecology*, **13**: 177–223.

Tansley, A.G. and Adamson, R.S. (1926). Studies of the vegetation of the English Chalk IV. A preliminary survey of the chalk grasslands of the Sussex Downs. *Journal of Ecology*, **14**: 1–32.

Taylor, G. and Morecroft, M.D. (1997). *Estimating summer 1997 deer populations in Wytham woods, via faecal accumulation method*. Unpublished Internal report to Environmental Change Network, Wytham.

Tew, T.E. and Macdonald, D.W. (1994). Dynamics of space use and male vigour amongst wood mice, *Apodemus sylvaticus*, in the cereal ecosystem. *Behavioral Ecology and Sociobiology*, **34**: 337–45.

The Times (2005). Obituary of Professor Sir Richard Southwood. *The Times, 1st November 2005*.

Thomas, R.C. (1987). *The historical ecology of Bernwood Forest*. Unpublished Ph.D thesis, Oxford Polytechnic (now Oxford Brookes University), Oxford.

Thornton, P.D., Newman, C., Johnson, P.J., Buesching, C.D., Baker, S.E., Slater, D., Johnson, D.D.P., and Macdonald, D.W. (2005). Preliminary comparison of four anaesthetic regimes in badgers (*Meles meles*). *Veterinary Anaesthesia and Analgesia*, **32**: 40–47.

Tsouvalis, J. (2000). *A Critical Geography Of Britain's State Forests*. Oxford University Press, Oxford.

Varley, G.C. (1951). The Winter Moth and other defoliators of oak. *Proceedings of the Royal Entomological Society Series C*, **16**: 36.

Varley, G.C. and Gradwell, G.R. (1968). Population models for the Winter Moth. In: T.R.E. Southwood, (ed). *Insect Abundance*, pp. 377–78. Blackwell Scientific Publications, Oxford.

Varley, G.C., Gradwell, G.R., and Hassell, M.P. (1973). *Insect Population Ecology: An Analytical Approach*. Blackwell Scientific Publications, Oxford.

Vera, F.W.M. (2000). *Grazing Ecology and Forest History*. CABI International, Wallingford.

Verhulst, S., Perrins, C.M., and Riddington, R. (1997). Natal dispersal of Great Tits in a patchy environment. *Ecology*, **78**: 864–72.

Vickery, J.A., Bradbury, R.B., Henderson, I.G., Eaton, M.A., and Grice, P.V. (2004). The role of agri-environment schemes and farm management practices in reversing the decline of farmland birds in England. *Biological Conservation*, **119**: 19–39.

Wade, L.M., Armstrong, R.A., and Woodell, S.R.J. (1981). Experimental studies on the distribution of the sexes of *Mercurialis perennis* L.:1. Field observations and canopy removal experiments. *New Phytologist*, **87**: 431–38.

Walker, J.J. (ed.) (1926). *The Natural History Of The Oxford District*. Oxford University Press, London.

Walther, B. A. and Gosler, A.G. (2001) The effects of food availability and distance to protective cover on the winter foraging behaviour of tits (Aves: *Parus*). *Oecologia*, **129**: 312–20.

Ward, A.I. (2005). Expanding ranges of wild and feral deer in Great Britain. *Mammal Review*, **35**: 165–73.

Ward, J.F., Macdonald, D.W. and Doncaster, C.P. (1997). Responses of foraging hedgehogs to badger odour. *Animal Behaviour*, **53**: 709–20.

Wardell, D.A. (1987). *Management plan for Wytham Woods 1987–1991*. Dissertation, Oxford Forestry Institute.

Waring, P. (1984). Recovery of marked *Catocala nupta* L. (red underwing) 6.5 km from release point. *Entomologists' Record*, **96**: 128–29.

Warren, M.S. and Key, R.S. (1991). Woodlands: past, present and potential for insects. In: N.M. Collins and J.A. Thomas, (eds). *The Conservation of Insects and their Habitats*, pp. 155–211. Academic Press, London.

Warui, C.M., Villet, M.H., Young, T.P., and Jocque, R. (2005). Influence of grazing by large mammals on the spider community of a Kenyan savanna biome. *Journal of Arachnology*, **33**: 269–79.

Waters, E.G.R. (1926). Micro-Lepidoptera. In J.J. Walker, (ed.) *The Natural History of the Oxford District*, pp. 230–47. Oxford University Press.

Waters, T.L. and Savill, P.S. (1991). Ash and sycamore regeneration and the phenomenon of their alternation. *Forestry*, **65**: 417–33.

Watt, T.A. and Gibson, C.W.D. (1988). The effects of sheep grazing on seedling establishment and survival in grassland. *Vegetatio*, **78**: 91–98.

Watt, T.A. and Kirby, K.J. (1983). Wytham Meads SSSI. Unpublished MS, Dept of Agricultural and Forest Sciences, Oxford/Nature Conservancy Council, Huntingdon.

West, C. (1985). Factors underlying the late seasonal appearance of the lepidopterous leaf-mining guild on oak. *Ecological Entomology*, **10**: 111–20.

Wetton, M.N. and Gibson, C.W.D. (1987). Grass flowers in the diet of *Megaloceraea recticornis* (Heteropter: Miridae): plant structural defences and interspecific competition reviewed. *Ecological Entomology*, **12**: 451–57.

Whittaker, J.B. (1969). Quantitative and habitat studies of the frog-hoppers and leaf-hoppers (Homoptera, Auchenorhyncha) of Wytham Woods, Berkshire. *Entomologists Monthly Magazine*, **105**: 27–37.

Whittaker. J.B. (1987). Obituary of H.N.Southern. *Journal of Animal Ecology*, **56**: 715–17.

Whittaker, J.B. (1973). Density regulation in a population of *Philaenus spumarius* (L.) (Homoptera: Cercopidae). *Journal of Animal Ecology*, **42**: 163–172.

Wilkin, T.A., Garant, D., Gosler, A.G., and Sheldon, B.C. (2006). Density effects on life-history traits in a wild population of the great tit *Parus major*: analyses of long-term data with GIS techniques. *Journal of Animal Ecology*, **75**: 604–15.

Williamson, M.H. (1959). The separation of molluscs from woodland leaf-litter. *Journal of Animal Ecology*, **28**: 153–55.

Wilson, C. and Reid, C. (1995). *Oxfordshire inventory of ancient woodland*. Unpublished report. Nature Conservancy Council, Peterborough.

Wint, G.R.W. (1983). The effect of foliar nutrients upon the growth and feeding of a Lepidopteran larva. In: J.A. Lee, S. McNeill, and I.H. Rorison, (eds). *Nitrogen as an Ecological Factor. 22nd Symposium of the British Ecological Society, Oxford, 1981*, pp. 301–20. Blackwell Scientific, Oxford.

Wong, J., Stewart, P.D. and Macdonald, D.W. (1999). Vocal repertoire in the European badger (*Meles meles*): structure, context and function. *Journal of Mammalogy*, **80**: 570–88.

Wood, M.J., Cosgrove, C.L., Wilkin, T.A., Knowles, S.C.L., Day, K.P., and Sheldon, B.C. (2007). Within-population variation in prevalence and lineage distribution of avian malaria in blue tits *Cyanistes caeruleus*. *Molecular Ecology*, **16**: 3263–73.

Wood, T.G. and Lawton, J.H. (1973). Experimental studies on the respiratory rates of mites (Acari) from beech-woodland leaf litter. *Oecologia*, **12**: 169–91.

Woodcock, B.A., Potts, S.G., Pilgrim, E., Ramsay, A.J., Tscheulin, T., Parkinson, A., Smith, R.E.N., Gundrey, A.L., Brown, V.K., and Tallowin, J.R. (2007). The potential of grass field margin management for enhancing beetle diversity in intensive livestock farms. *Journal of Applied Ecology*, **44**, 60–69.

Woodell, S.R.J. and Steel, J. (1990). *Changes in the Seed Bank and Vegetation of an Abandoned Arable Field at Wytham, Oxfordshire, between 1982 and 1989*. Unpublished report to the Nature Conservancy Council, Haslemere, BIOSCAN (UK) Ltd.

Woodland Trust (2007). *Back on the Map: an Inventory of Ancient and Long-Established Woodland for Northern Ireland*. Woodland Trust, Grantham.

Woodroffe, R. and Macdonald, D.W. (2000). Helpers provide no detectable benefits in the European badger (*Meles meles*). *Journal of Zoology*, **250**: 113–19.

Wu, J., Yu, X.-D. and Zhou, H.-Z. (2008). The saproxylic beetle assemblage associated with different host trees in Southwest China. *Insect Science*, **15**: 251–61.

Yalden, D. (1999). *The History of British Mammals*. Poyser, London.

Young, R.P., Davison, I., Trewby, I.D., Wilson, G.J., Delahey, R.J., and Doncaster, C.P. (2006). Abundance of hedgehogs (*Erinaceus europaeus*) in the relation of the density and distribution of badgers (*Meles meles*). *Journal of Zoology*, **269**: 349–56.

Zulka, K.P., Milasowszky, N., and Lethmayer, C. (1997). Spider biodiversity potential of an ungrazed and a grazed inland salt meadow in the National Park 'Neusiedler See-Seewinkel' (Austria): implications for management (Arachnida: Araneae). *Biodiversity and Conservation*, **6**: 75–88.

# Index

Abingdon, Earls of 8, 34–5, 197
   fifth *see* Bertie, Montagu
Abingdon Abbey 8, 29, 32, 33, 70, 197
*Accipter nisus* – *see* sparrowhawk
*Acer campestre* – *see* field maple
*Acer pseudoplatanus* – *see* sycamore
acid rain 28, 217
acidification 27–8
*Adarrus multinotatus* 104
*Adarrus ocellaris* 104
*Adoxa moschatellina* – *see* moschatel
Adult Aggression Hypothesis 177
Aegithalidae 169
*Aegithalos caudatus* – *see* long-tailed tit
*Aesculus hippocastanum* – *see* horse chestnut
agro-chemicals 217
*Agrostis stolonifera* – *see* creeping bent
*Agrotis cenerea* – *see* light-feathered rustic moth
*Aix galericulata* – *see* mandarin duck
*Alca torda* – *see* razorbill
*Alcedo atthis* – *see* kingfisher
allo-marking 194
alloparental behaviour 188
ammonia 28, 217
*Anacamptis pyramidalis* – *see* pyramidal orchid
ancient woodland
   indicators 77, 79
   inventories 44, 46
*Anas crecca* – *see* teal
*Anas platyrhynchos* – *see* mallard
animal diversity, vegetation and 48
Animal Kingdom course 126
*Anisantha sterilis* – *see* barren brome
Ant Reserve 55
*Anthus trivialis* – *see* tree pipit
*Apodemus sylvaticus* – *see* wood mouse
Apparency Theory 118, 144
arable land 34, 52
archaeology 5
*Archanara geminipuncta* – *see* twin-spotted wainscot
*Ardea cinerea* – *see* grey heron
*Argynnis aglaia* – *see* dark green fritillary
*Arrhenatherum elatius* – *see* false oat-grass
ash (*Fraxinus excelsior*) 5, 54, 204, 205
   climate change sensitivity 228
   ground flora under 80
   growth 224
   numbers trends 62, 64–5, 67–8
   planting 58
   seeds 167

   squirrel damage immunity 210
   sycamore and 62, 67, 205
   timber sale 35–6
aspen 67
Australia 126
avian malaria 164
aviaries, studies in 162–3

badgers (*Meles meles*) 131, 185, 186, 187–95
   census 195–6
   competition with other species 194
   numbers 191–3
   road traffic accidents 193
   sociality 187–9, 190–1, 193–4
   territoriality 189–91
Bagley Wood 70, 120, 175, 176
banded snails (*Cepaea* spp.) 112, 117
bank vole (*Clethrionomys glareolus*) 165, 174, 179–80, 184
barn owl (*Tyto alba*) 149, 179
barren brome (*Anisantha sterilis*) 97, 98
barren strawberry (*Potentilla sterilis*) 79
bats 214
Beacon Hill 41
Bean Wood 20, 68, 147, 171, 203
beech (*Fagus sylvatica*) 5, 47, 54, 204
   climate change sensitivity 228
   die-offs 65
   ground flora under 80
   numbers trends 65
   planting 37–8, 58
   seeds 153
   squirrel damage 195, 209
   veteran trees 73, 205
beech bark disease 65
beetles (Coleoptera) 104, 221
   *Agriotes* sp. 223
   *Cis bilamellatus* 113
   in crop pest reduction 122
   *Dromius quadrimaculatus* 140
   *Feronia madida* 141–2
   *Philonthus decorus* 141–2
behavioural ecology, origins 121–2
behavioural response 118
*Bembicia scopigera* – *see* six-belted clearwing moth
Bernwood 45
Bertie, Montagu, fifth Earl of Abingdon 8, 36–8, 197
*Betula* spp. – *see* birch
Bialowieza Forest 4, 33
biodiversity monitors 195–6

## Index

Biological Advisory Committee 201
biological reserves 201, 213
birch (*Betula* spp.) 65–6, 228
bird predation 112
birds 145–71
 foraging 121–2
 species in Wytham 145–6
  changes in 146–9
 studies in Wytham 165–71 *see also* Wytham Tit Study
birdsfoot trefoil (*Lotus corniculatus*) 96
*Biston betularia* – *see* peppered moth
black ant (*Lasius niger*) 117
black bryony (*Tamus communis*) 79
black hairstreak butterfly (*Satyrium pruni*/*Strymonidia pruni*) 106, 133, 208
black medick (*Medicago lupulina*) 97
blackbird (*Turdus merula*) 170
blackcap (*Sylvia atricapilla*) 168
blackthorn (*Prunus spinosa*) 4, 67, 70, 106, 133, 208
blue tit (*Cyanistes caeruleus*) 148–50, 153–4, 156, 163–4, 166, 184 *see also* Wytham Tit Study
*Carex flacca* 96
bluebell (*Hyacinthoides non-scripta*) 54, 75, 219
 frequency 81–2
 in old-field succession 98
bone-skipper fly 133
botanical records 75–6
boundary layer conductance 225
Bowling Alley 95, 102–4, 109, 114
*Brachypodium pinnatum* – *see* tor grass
*Brachypodium sylvaticum* – *see* wood false-brome
bracken (*Pteridium aquilinum*) 75
 frequency 81–2
 spread 92
Bradfield Woods 88
bramble (*Rubus fruticosus*) 54, 55, 75, 211
 frequency 81–2, 84
 as key species 87
Brasenose Woods 125
breeding seasons 178–9
British Deer Society (BDS) 183
British Ecological Society (BES) 174, 198
British woodland
 distribution 43
 ground flora changes 83–8
 history 3–7
 types 54
Broad Oak 146
Brogdens Belt 37, 69
brome grasses (*Bromus* spp.) 98
*Brometalia* grasslands 100
*Bromus erectus* – *see* erect brome; upright brome
*Bromus hordeaceus hordeaceus* – *see* soft brome
*Bromus* spp – *see* brome grasses
Bronze Age 6
Brown soils 25–6
bullfinch (*Pyrrula pyrrula*) 167

Bureau of Animal Population 110, 112, 114, 174, 176, 199, 201
*Buteo buteo* – *see* buzzard
butterflies 48, 49, 106, 214
 climate change and 126, 220, 221
 in field margins 122, 123
Buxton 222
buzzard (*Buteo buteo*) 146, 148–9, 179

cabbage white butterfly 144
Calcareous grit 24
calcium 165
*Callophrys rubi* – *see* green hairstreak
Cammoor Copse 20
*Campanula trachelium* – *see* nettle-leaved bell-flower
Canada 143, 175
canopy cover 62–8
 and ground flora richness 79–81
canopy height 68
canopy walkway 21, 224
*Capreolus capreolus* – *see* roe deer
*Caprimulgus europaeus* – *see* nightjar
Carabidae *see* ground beetles
carbon cycle 223–4, 227
carbon fluxes 227
*Carex* spp. – *see* sedges
carrion, mechanical breakdown 113
carrion beetle (*Nicrophorus germanicus*) 133
caterpillars
 and food plants 144
 on oak 131
 as prey 156–7, 167
*Catocala nupta* – *see* red underwing moth
cattle-grazing 49, 212
*Centaurea nigra* – *see* common knapweed
centipedes 113
Central Mining 10
Centre for Ecology and Hydrology (CEH) 219, 222, 226, 227
*Cepaea* spp. – *see* banded snails
*Cerastium pumilum* – *see* small mouse-ear
chalcid wasps 120
*Cervus elaphus* – *see* red deer
chalet 10–11
*Chamerion angustifolium* – *see* rosebay willowherb
change, nature of 91–2
cherry (*Prunus avium*) 210
cherry pickers 227
chiffchaff (*Phylloscopus collybita*) 168
Chilterns 45, 46, 47, 48, 149
Chitty, Dennis 174, 175, 178
*Circaea lutetiana* – *see* enchanter's nightshade
*Circus aeruginosus* – *see* marsh harrier
*Cirsium eriophorum* – *see* woolly thistle
citizen science 195
Clarke, John 177, 178–9, 181

*Clethrionomys glareolus* – *see* bank vole
climate 21–4
climate change 157–8, 202, 215, 217–28
   and badgers 192–3
   and biodiversity 126
   and birds 157
   and flora 88 – 9
   and trees 74, 223–7, 228
   Upper Seeds experiment 221–3 *see also* Environmental Change Network
*Clinopodium vulgare* – *see* wild basil
*Clubiona caerulescens* – *see* woodland spider
Clubionids 105
clutch size 151, 156–7, 170
coal 7, 32
coal tit (*Periparus ater*) 162, 163
coccidiosis 192
Coccoidea 223
*Coccothraustes coccothraustes* – *see* hawfinch
co-existence 117–20
cohort analysis 183
*Colchicum autumnale* – *see* meadow saffron
Coleoptera *see* beetles
colonization 95–7
*Columba palumbus* – *see* wood pigeon
*Centaurium erythraea* 96
common knapweed (*Centaurea nigra*) 55
Common Piece 147
common reed (*Phragmites australis*) 50–1, 56
common shrew (*Sorex araneus*) 182
common vetch (*Vicia sativa*) 98
competition 117–18
conifers 7, 14, 41
   planting 58
   removal 66, 204
conservation biology 124, 202
conservation management 197–215
   before 1900 197–8
   by Oxford University 200–1
   climate change and 215
   deer 206–9
   effectiveness monitoring 213–14
   Elton, Charles 123 – 4
   fen 212
   future challenges 214–15
   grasslands 211–12
   grey squirrels 209–11
   invertebrates and 123–6
   last 60 years 202–3
   ponds 212
   public access promotion 213
   research and 213
   rides 212
   twentieth century 198–200
   woodland management 203–5
   Wytham's role in national policy 201–2
coppice, sale 35–6

coppicing 32, 33, 72, 203, 211
Coral Rag limestone 24, 25
corn bunting (*Emberiza calandra*) 148
*Cornus sanguinea* – *see* dogwood
*Corvus corax* – *see* raven
*Corvus frugilegus* – *see* rook
*Corvus monedula* – *see* jackdaw
*Corylus avellana* – *see* hazel
Cothill Fen 50
Cotswolds 44, 47, 48
Countrywide Stewardship 53, 215
Countrywide Surveys 84–5
cover
   importance 114
   studies 114, 116
cowslip (*Primula veris*) 96, 98
*Crataegus monogyna* – *see* hawthorn
*Cratichneumon culex* – *see* ichneumon wasp
creeping bent (*Agrostis stolonifera*) 97, 98, 105
*Crepis capillaris* – *see* smooth goat's beard
crested dog's tail (*Cynosurus cristatus*) 55
*Cucullia verbasci* – *see* mullein moths
Cumnor parish, tithings 33
Cumnor Woods 8, 32, 70
curlew (*Numenius arquata*) 146, 147
cut-leaved cranesbill (*Geranium dissectum*) 97
*Cyanistes caeruleus* – *see* blue tit
*Cynosurus cristatus* – *see* crested dog's tail
*Cyzenis albicans* – *see* flies

*Dama dama* – *see* fallow deer
Dark Ages 6
dark green fritillary (*Argynnis aglaia*) 106
Dartington Hall 149, 150
Dawkins, Colyear 62, 63
Dawkins Plots 62, 64, 77–9, 204, 205, 207
deadwood 73, 123–4, 205, 214
dead-wood beetles 132
deer 15–16, 70–2, 182–3, 206–9
   confinement to Wytham Woods 15, 72, 87, 180, 183, 206
   conservation management 206–9
   exclosure plots 87
   historically 36, 70
   impact 70–2, 146, 164, 168, 183, 206, 212
   numbers control 228
   population targets 208–9
   species 15, 70, 182–3
Dell, The 112
demesne 34
denatonium benzoate 187
density-dependent processes 116–17, 137–8
density-independent processes 137–8, 143
*Deschampsia cespitosa* – *see* tufted hair-grass
detritivores 121
Dietric Vacuum sampler (D-vac) 129
*Digitalis purpurea* – *see* foxglove

Diptera (true flies) 50, 51
diversity, and stability 120–1
dog's mercury (*Mercurialis perennis*) 54, 75, 76–7, 81, 221
　frequency 81–2, 85
dogwood (*Cornus sanguinea*) 70
Domesday Book 6, 32
Down Field 34
drought 219, 220, 221–3, 225–6
Druce, G.C. 40
ducks 146
Duffey, Eric 114–16, 124–5, 128–9
Duke of Burgundy fritillary (*Hamearis lucina*) 106, 133
Dutch elm disease 5, 16, 65

Eadwig, King of Wessex 29
Earthwatch Institute (Europe) 53, 195–6, 211, 227
earthworms 188, 189, 194
East African savannah 126
ecological energetics 121
ecological genetics, pioneers 111–13
ecology 198–9
　foundation 2
　pioneers 110–13
ectoparasites 190
edge effects 228
Edward Grey Institute of Field Ornithology (EGI) 145, 149–51, 153, 202, 207
eggshell 164–5
elaiosomes 95
elder 47, 55, 70, 221
Ellenberg scores 76
elm
　decline 5, 16, 31
　numbers trends 62, 65
　timber sale 35–6
Elton, Charles 41, 48, 92
　biographical details 176–7
　in conservation science 123–4
　on diversity and stability 120
　ecological research 110, 112
　and mammals 173, 174, 175
　and woodland structure 72, 73
　and Wytham Ecological Survey 113, 202–3
*Emberiza calandra* – see corn bunting
emergence traps 129, 140
*Ena montana* – see mountain bulin snail
enchanter's nightshade (*Circaea lutetiana*), frequency 81–2
Enclosure Acts 7, 8, 36, 197
English Channel 3
English Nature see Natural England
Environmental Change Network (ECN) 23, 25, 67, 205, 215, 217–21
　measurements 218

meteorological observation 21, 218, 219
　sites 23, 218–19
　species monitoring 126, 214
　tree growth monitoring 224
Environmental Impact Assessment 128, 129
Environmental Stewardship schemes 53
Environmentally Sensitive Areas 53
*Equisetum sylvaticum* – see wood horsetail
*Equisetum telmateia* – see great horsetail
erect brome (*Bromus erectus*) 48, 49, 55
*Erithacus rubecula* – see robin
*Erinaeus europaeus* – see hedgehog
Essex University 225
Eulipotyphla 181
*Eupithecia puchellata* – see foxglove pug
evapotranspiration 69
evidence-based management 125
exines 3

*Fagus* – see beech
*Falco peregrinus* – see peregrine
fallow deer (*Dama dama*) 15, 70–2, 182–3, 208
false oat-grass (*Arrhenatherum elatius*) 55, 98, 99–102, 105, 211
farm wildlife 122–3
farmed landscape 5–6
Farmoor Reservoir 27
*Feltria romijni* – see water mites
fens 50–1, 212
　types 56
ffennell, Hazel 12–13
ffennell, Raymond 9–11, 13, 38, 40, 200
*Ficedula hypoleuca* – see pied flycatcher
field madder (*Sherardia arvensis*) 97
field maple (*Acer campestre*) 49, 54, 70
field margins 122–3
field sports 198, 212
field vole (*Microtus agrestis*) 175, 177, 178–80, 184
*Filipendula ulmaria* – see meadow-sweet
finches 146, 147, 166, 167–8
fire, impacts 126
firecrest (*Regulus ignicapillus*) 148
Fisher, R.A. 111
flies (*Cyzenis albicans*) 138–9, 142, 143
flight ponds 38
flocking 162
flooding 27, 219
floodplain 34, 219
flora
　changes 1974–99 81–3
　　compared to whole country 83–8
　climate change and 221, 222
　species occurrence 76–9
　and stand dynamics 79–81
flux tower 226–7
foraging, optimal 121
Ford, Edmund B. (Henry) 111, 112

forest clearance 5–6
Forest of Dean 61
Forest Stewardship Council certification 211
Forestry Commission 7, 14, 41, 43, 204, 212
　establishment 57
　Woodland Dedication Scheme 60
　Woodland Grant Scheme (WGS) 60, 214
forestry management 57–60
fox *see* red fox
foxglove (*Digitalis purpurea*) 81
foxglove pug (*Eupithecia puchellata*) 81
*Fraxinus excelsior* – *see* ash
frog hoppers 223
Further Clay Hill 44, 147

*Galium aparine* – *see* goosegrass
*Galium pumilum* – *see* slender bedstraw
*Galium verum* – *see* lady's bedstraw
*Gallinago gallinago* – *see* snipe
Game Act 198
*Gammarus* – *see* water shrimp
gannet (*Morus bassanus*) 145, 146
genetic polymorphism 111
Geographic Information System (GIS) 213
geology 24
*Geranium dissectum* – *see* cut-leaved cranesbill
Gibb, John 151
Gibson, C.W.D. 79, 116, 203, 211, 213
*Glechoma hederacea* – *see* ground ivy
*Glomeris marginata* – *see* pill millipede
Godfray, Charles 131
Godstow Nunnery 32
goosander (*Mergus merganser*) 145
goosegrass (*Galium aparine*), frequency 82, 84, 87
Gosler, Andy 152
Gradwell, George 136, 137, 138–40, 142–3, 144
graphical key factor analysis 142
grasses, frequency 81, 84 – 7
grasshoppers 113
grassland 48–50, 211–12, 221
　types 55
grazing
　decline 49
　plant susceptibility to 87
　treatments studies 93–5
great horsetail (*Equisetum telmateia*) 50
great tit (*Parus major*) 131, 149–51, 153–7, 159–65, 166, 184
　climate change and 221
　personalities 155–6
　recent studies 163–5
　song 161–2 *see also* Wytham Tit Study
Great Wood
　birds 171
　flora 77, 84 – 7
　flux tower 226–7
　mammals 179

　replanting 58
　trees 227
green hairstreak (*Callophrys rubi*) 106
green tortrix (*Tortrix viridana*) 135–6, 156
grey heron (*Ardea cinerea*) 145
grey squirrel (*Sciurus carolinensis*) 15, 65, 195, 209–11
Grimsbury Wood 226
ground beetles (Carabidae) 141, 220, 221
　*Notiophilus biguttatus* 117
　*Notiophilus rufipes* 117
ground ivy (*Glechoma hederacea*) 54
　frequency 82
groundsel (*Senecio vulgaris*) 97
groundwater gley soils 26
guelder rose (*Viburnum opulus*) 70

Habitat Saturation Hypothesis 177
Hagley pool 51
hairy violet (*Viola hirta*) 96, 98, 101, 106
haloing 205
*Hamearis lucina* – *see* Duke of Burgundy fritillary
Harvard Forest 92
hawfinch (*Coccothraustes coccothraustes*) 148
hawthorn (*Crataegus monogyna*) 47, 54–5, 70, 98
Hayley Wood 95
hazel (*Corylus avellana*) 3, 67, 70
*Hedera helix* – *see* ivy
hedgehog (*Erinaeus europaeus*) 194–5
hedges, removal 52
*Helianthemum nummularium* – *see* rockrose
Herb Paris (*Paris quadrifolia*) 76–9
Higgins Copse 20, 33, 151
Hill End 40
Hill End Camp 10, 13, 171
*Hippocrepis comosa* – *see* horseshoe vetch
*Hirundo rustica* – *see* swallow
Hither Clay Hill 146
*Holcus lanatus* – *see* Yorkshire fog
Holies, the 95, 102–3
Holly Hill 31, 38
Home Farm 23, 219
horse chestnut (*Aesculus hippocastanum*) 197
horseshoe vetch (*Hippocrepis comosa*) 49
house sparrow (*Passer domesticus*) 171, 174
hoverflies 73
HSBC Bank 227
Hudson Bay Company 173
*Hyacinthoides non-scripta* – *see* bluebell
hydrology 26–7, 225
*Hypericum perforatum* – *see* St John's Wort

ice caps 3
ichneumon wasp (*Cratichneumon culex*) 142, 143
Ideal Flea Distribution Hypothesis 190
inbreeding depression 164
Indian Forest Service 59
indicator groups 129

insect–plant interactions 144
introgression 112
invertebrates 109–44
   climate change and 221–2
   and conservation management 123–6
   ecology in teaching 126–8
   habitat specialisms 114
   of limestone grassland 107
   notable species in Wytham 132–4
   present work 129–31
   succession studies 114, 116
   in Upper Seeds 104–8 *see also* winter moth
Iron Age 6
Island Biogeography theory 51
ivy (*Hedera helix*) 54–5

jackdaw (*Corvus monedula*) 166–7
Jews Harp 68
Jones, Eustace 4, 60–1

Kettlewell, Bernard 112
key factor analysis 116, 142
King, Carolyn (Kim) 184–5
kingfisher (*Alcedo atthis*) 147
kittiwake (*Rissa tridactyla*) 145–6
Kluijver (Kluyver), H.N. 149, 150
knopper gall 16
Krebs, John 121, 151, 161
Kruuk, Hans 187–90

Lack, David 149–51
Lady Park Wood 61, 65
lady's bedstraw (*Galium verum*) 96
lapwing (*Vanellus vanellus*) 146
larch, planting 37
Lardon Chase 95
*Lasius flavus* – *see* yellow meadow ant
*Lasius niger* – *see* black ant
lay-date 156, 157
leaf-miners 105–6, 118, 223
   *Phytomyza nigra* 105
   *Phytomyza ranunculi* 105
leafhoppers (Auchenorryncha) 104–5, 106, 223
   *Ledra aurita* 132
leatherjackets 121, 167
*Lecanopsis formicarum* – *see* scale insects
lemming 173
Lepidoptera 51, 111, 131
*Lepthyphantes tenuis* – *see* money spider
life tables 137–43
light-feathered rustic moth (*Agrotis cenerea*) 133
lime (*Tilia* spp.) 3, 5, 8, 31
   past dominance 74
*Linum catharticum* 96
Little Ash Hill 205
little grebe (*Tachybaptus ruficollis*) 146, 147
Little Wittenham 125, 130, 131

Lloyd, Philip 62, 92
lobster-pot traps 138–9
*Lolium perenne* – *see* ryegrass
long-tailed tit (*Aegithalos caudatus*) 162, 169–70
*Lotus corniculatus* – *see* birdsfoot trefoil
Lower Seeds 34, 101
Lower Seeds Field 146
*Lullula arborea* – *see* woodlark
*Luscinia megarhynchos* – *see* nightingale

Macdonald, David 173, 185–7, 188
*Macrosteles laevis* 105
Magdalene Hill Down 95, 103–4
mallard (*Anas platyrhynchos*) 146
Mammal Monitoring Project 195
Mammal Society 173
mammals 173–96 *see also* small mammals
mandarin duck (*Aix galericulata*) 146
maple 70
Marlborough Downs 117
Marley Fen 50–2, 56, 91, 212
Marley Plantation 171
Marley Wood 26, 31, 197, 207
   birds 146, 147, 151, 153
   flora 77
   invertebrates 129
   mammals 184
   shrub layer 70
   tree heights 68
marsh harrier (*Circus aeruginosus*) 145
marsh tit (*Poecile palustris*) 163
masting 153
meadow saffron (*Colchicum autumnale*) 76
meadow-sweet (*Filipendula ulmaria*) 56, 77
*Medicago lupulina* – *see* black medick
medieval period 7, 32
*Megaloceraea recticornis* 102
*Meles meles* – *see* badgers
*Mercurialis perennis* – *see* dog's mercury
*Mergus merganser* – *see* goosander
meta-population dynamics 117
mice 174, 176
microbe stripping 112
*Microsphaera alphitoides* 16
*Microtus agrestis* – *see* field vole
migrants, arrival dates 174
*Milvus milvus* – *see* red kite
minimum intervention 60
molluscs 128
money spider (*Lepthyphantes tenuis*) 122
Monks Wood 66, 68, 88, 114, 131
*Morus bassanus* – *see* gannet
moschatel (*Adoxa moschatellina*) 79
moths
   climate change and 221
   *Cochylis flaviciliana* 133
   *Coleophora silenella* 133

*Cydia pallifrontana* 133
*Stephensia brunnichella* 223
Mount Wood 38, 197
mountain bulin snail (*Ena montana*) 132
movement, of species 51–2
mullein (*Verbascum* spp.) 81
mullein moths (*Cucullia verbasci*) 81
*Muntiacus reevesi* – *see* muntjac
muntjac (*Muntiacus reevesi*) 15, 71–2, 183, 208
*Muscicapa striata* – *see* spotted flycatcher
*Mustela erminea* – *see* stoat
*Mustela nivalis* – *see* weasel
My Ladies Common Piece 44, 68
mycorrhizae 101
*Mocydia crocea* 104
*Mythimna straminea* – *see* southern wainscot
myxomatosis 16, 41, 54, 92, 133, 203, 211

narrow-leaved meadow-grass (*Poa angustifolia*) 96
National Nature Reserves 125, 199
national parks 199–200
National Soil Resources Institute 24
National Vegetation Classification 53–6
Natural England (English Nature) 203, 204, 205, 214
Natural Environment Research Council 221
Nature Conservancy 111, 177, 202
Nature Conservancy Council 44, 125, 203
nature conservation *see* conservation management
Nealings Copse 33, 35, 38, 147
near-natural stand development 60
nematodes 177, 182
  *Heligmosomoides polygyrus* 177
  *Porrocaecum* sp. 182
Neolithic period 5
nestboxes 149–52
net radiation 226–7
nettle (*Urtica dioica*) 47, 54, 55, 56, 123
  frequency 81–2, 85
nettle-leaved bell-flower (*Campanula trachelium*) 76
New Forest 65
Newton, Ian 167
*Nicrophorus germanicus* – *see* carrion beetle
nightingale (*Luscinia megarhynchos*) 12–13, 145, 146–7, 148
nightjar (*Caprimulgus europaeus*) 146
nitrate concentrations 27, 220
nitrogen deposition 27–8, 217
nitrogen levels 87, 101
nitrogen mineralization 222
Norreys, Lord 8
North Field 34
Northfield Farm 23, 40–1, 219
*Numenius arquata* – *see* curlew
numerical response 118

oak (*Quercus*) 16–17, 204, 205
  and bird breeding 163–4
  climate change and 221, 228
  ground flora under 80
  numbers trends 62, 64–5, 67
  photosynthesis 224–5
  planting 58
  squirrel damage 210
  timber sale 35–6
  veteran trees 73
oak galls 118, 120
Odum, Eugene 112
old-field succession *see* Upper Seeds case study
old growth 72–4
open habitats 91 *see also* Upper Seeds case study
open space 72–4
*Operophtera brumata* – *see* winter moth
*Oryctolagus cuniculus* – *see* rabbit
Osmaston, F.C. 59
Otmoor 48
owls 146, 165–6
Oxford Botanic Garden 170
Oxford Clay 24, 26
Oxford Ornithological Society 13, 146
Oxford Rare Plants Group 212
Oxford Scientific Films 126
Oxford University
  agricultural school 40
  climate change project collaborations 227
  Department of Zoological Field Studies 174
  Forestry Department 14, 58, 60, 200
  University Farm 45–6, 56, 170–1
  Zoology Department 58
Oxfordshire woodland 43

parapox virus 195
*Pararge aegeria* – *see* speckled wood butterfly
parasites
  *Babesia missiroli* 192
  *Eimeria melis* 192
  *Isospora melis* 192
  *Trypanosoma pestanai* 192
parasitic wasp (*Phaenoserphus vexator*) 117
parasitism 117–20
*Paris quadrifolia* – *see* Herb Paris
partridge 199
*Parus major* – *see* great tit
*Passer domesticus* – *see* house sparrow
*Passer montanus* – *see* tree sparrow
Passive Range Exclusion hypothesis 191
Pasticks 37, 41, 58, 184, 211
*Pastinaca sativa* – *see* white parsnip
peppered moth (*Biston betularia*) 112
peregrine (*Falco peregrinus*) 158
*Periparus ater* – *see* coal tit
Perrins, Chris 151
Perturbation Hypothesis 187
phenology 219, 222, 223
pheromones 179

## Index

*Philaenus spumarius* – *see* spittlebug
phosphorus levels 101
photosynthesis 69, 224–5
*Phaenoserphus vexator* – *see* parasitic wasp
*Phragmites australis* – *see* common reed
pied flycatcher (*Ficedula hypoleuca*) 149
pill millipede (*Glomeris marginata*) 112
pitfall traps 128, 129
Pixey Meads 48
*Phoenicurus phoenicurus* – *see* redstart
*Phylloscopus collybita* – *see* chiffchaff
plant hoppers 223
planting, nineteenth century 37
*Platycerus caraboides* – *see* stag beetle
Plumers Coppice (Copse) 29, 33
*Poa angustifolia* – *see* narrow-leaved meadow-grass
*Poa trivialis* – *see* rough-stalked meadow-grass
*Poecile montanus* – *see* willow tit
*Poecile palustris* – *see* marsh tit
pollarding 32, 33
pollards, sale 36
pollen record 3, 4, 5, 7–8, 29
pollution 27–8
polymorphism 174
polyphenols 144
ponds, management 212
population cycles 175–9
Port Meadow 48
potassium levels 101
*Potentilla sterilis* – *see* barren strawberry
precipitation 21, 219, 220, 222–3
primrose (*Primula vulgaris*) 54, 75, 219
*Primula veris* – *see* cowslip
*Primula vulgaris* – *see* primrose
privet 70
*Prunus avium* – *see* cherry
*Prunus spinosa* – *see* blackthorn
*Pteridium aquilinum* – *see* bracken
public access 213
purple emperor 133
pygmy shrew (*Sorex minutus*) 182
pyramidal orchid (*Anacamptis pyramidalis*) 96
*Pyrrula pyrrula* – *see* bullfinch

Quarry, the (Wytham) 102
quarrying 34–5

rabbit (*Oryctolagus cuniculus*) 7, 174
　grazing 49, 55, 109
　impact 7, 70, 72, 200
　numbers decline 16, 41, 133, 211, 212
　as prey 187
rabies control 185, 187
Radbrook Common 27, 32, 37, 44
　planting 41, 58, 68, 146
　warfarin use 211
Radcliffe Meteorological Observatory 21

ragwort (*Senecio jacobaea*) 98
rainfall 21, 219, 220, 222–3
*Ranunculus parviflorus* – *see* small-flowered buttercup
rats 174
raven (*Corvus corax*) 149
razorbill (*Alca torda*) 145, 146
red deer (*Cervus elaphus*) 70
red fox (*Vulpes vulpes*) 16, 179, 185–7
red kite (*Milvus milvus*) 149
red squirrel (*Sciurus vulgaris*) 195
red underwing moth (*Catocala nupta*) 51
redshank (*Tringa totanus*) 146
redstart (*Phoenicurus phoenicurus*) 145, 147
reed *see* common reed
reforestation 3
*Regulus ignicapillus* – *see* firecrest
relaxation 132
remote sensing techniques 68
Rendzina soil 25
research, conservation management and 213
research estate development 38–41
resilience 228, 229
Resource Dispersion Hypothesis (RDH) 187, 189
restoration ecology 123
ride system 73
rides 212
*Rissa tridactyla* – *see* kittiwake
robin (*Erithacus rubecula*) 149, 150
rockrose (*Helianthemum nummularium*) 106
roe deer (*Capreolus capreolus*) 15, 70, 72, 182–3, 208
Roman times 31
Romano-British artefacts 31–2
rook (*Corvus frugilegus*) 166–7, 174
*Rosa canina* agg. – *see* wild rose
rosebay willowherb (*Chamerion angustifolium*), frequency 85
Rough Common 34–5, 146, 149
rough-stalked meadow-grass (*Poa trivialis*) 54, 56
　frequency 81–2, 85
　in old-field succession 97, 98
royal forests 44, 45
RSPB 148, 167, 171
*Rubus fruticosus* – *see* bramble
ruderals 221, 222
ryegrass (*Lolium perenne*) 98

St John's wort (*Hypericum perforatum*) 223
salad burnet (*Sanguisorba minor*) 96
sampling methods 127, 128–9
*Sanguisorba minor* – *see* salad burnet
sanicle (*Sanicula europaea*) 77
*Sanicula europaea* – *see* sanicle
*Satyrium pruni*/*Strymonidia pruni* – *see* black hairstreak butterfly
*Saxicola torquata* – *see* stonechat
scale insects, *Lecanopsis formicarum* 223
scarlet tiger moths 112

*bimiculata* morph 112
Schumacher, Raymond *see* ffennell, Raymond
*Sciurus carolinensis – see* grey squirrel
*Sciurus vulgaris – see* red squirrel
*Scolopax rusticola – see* woodcock
scrub 47–8, 211
    types 54–5
Seacourt 8, 33
Seacourt Hill 147
Seacourt Stream 27
sedges (*Carex* spp.) 50, 77
    frequency 81–2, 84
seed-dressings 158
seepage 24, 27, 31
*Senecio jacobaea – see* ragwort
*Senecio vulgaris – see* groundsel
sheep-grazing 49, 93, 101, 133, 181, 211–12
Sheldon, Ben 153
*Sherardia arvensis – see* field madder
shooting 198, 212
Shotover 45
shredding 32
shrews 181–2
shrub layer, composition and structure 70
Silwood Park 115
Singing Way 1, 34–5
    as private drive 37
Sites of Special Scientific Interest (SSSIs) 14, 125, 201, 202–3
six-belted clearwing moth (*Bembicia scopigera*) 133
slender bedstraw (*Galium pumilum*) 91
small-flowered buttercup (*Ranunculus parviflorus*) 97
small mammals 173–82
    habitat management effects 179–82
    interactions with other species 182
small mouse-ear (*Cerastium pumilum*) 91
small tortoiseshell butterfly 123
Smithsonian Institution 227
smooth goat's beard (*Crepis capillaris*) 221
smooth tare (*Vicia tetrasperma*) 98
snipe (*Gallinago gallinago*) 146
soft brome (*Bromus hordeaceus hordeaceus*) 98
Soil Association 205
soil seed bank 81, 96
soil temperature 22–3
soil types 24–6, 76–8, 85
solar radiation 24, 227
Somerford Mead 49
*Sorex araneus – see* common shrew
*Sorex minutus – see* pygmy shrew
Southern, H.N. (Mick) 173, 174, 179
southern wainscot (*Mythimna straminea*) 51
Southwood, Sir Richard (Dick) 115, 120, 127–8
sparrowhawk (*Accipiter nisus*) 145, 158–60, 166
species pools 96
Specific Leaf Area 69

speckled wood butterfly (*Pararge aegeria*) 121, 221
spiders 104–5, 106, 114, 116, 122
    climate change and 126
    *Erigone* spp. 122
    as indicator group 129
    *Oedothorax* spp. 122
    species representation 113–14, 130–1
    *Syedra gracilis* 122
    *Tuberta maerens* 133–4 *see also* money spider
spindle 70
Spitsbergen 176
spittlebug (*Philaenus spumarius*) 105
spotted flycatcher (*Muscicapa striata*) 147
stability, diversity and 120–1
stag beetle (*Platycerus caraboides*) 133
stand composition studies 61–9
stand maps 67
Staphilinidae 141
starling (*Sturnus vulgaris*) 166, 171
Stimpson's Copse 20
stoat (*Mustela erminea*) 16, 179
stone 34–5
stonechat (*Saxicola torquata*) 146
*Streptopelia turtur – see* turtle dove
*Strix aluco – see* tawny owl
Stroud Copse 20, 33
Stroud tithing 33
*Sturnus vulgaris – see* starling
sub-fossil remains 133
succession
    studies 114, 116 *see also* Upper Seeds case study
Summerford Meadow 34
Sundays Hill 38, 95, 102–4, 114
surface water gley soils 26
swallow (*Hirundo rustica*) 171
Swinford Toll Bridge 1
sycamore 3–4
    ash and 62, 67, 205
    climate change sensitivity 228
    ground flora under 80
    growth 224
    leaf loss 221
    as native vs non-native 66
    numbers trends 62, 64–8
    photosynthesis 224–5
    planting 197
    regeneration 200
    removal 204
    squirrel damage 195, 210
*Sylvia atricapilla – see* blackcap
Sydlings Copse 50

*Tachybaptus ruficollis – see* little grebe
*Tamus communis – see* black bryony
tannins 144
tawny owl (*Strix aluco*) 165, 174, 179, 184

teal (*Anas crecca*) 146
temperature 21–3, 219
Ten Acre Copse 21–3, 33, 68, 147
tenant farms 40, 41
Terrestrial Initiative in Global Environmental Research (TIGER) 126, 221, 223
territoriality 189–91
Thames, River 1, 51
Thomas, Rachel 35
Thorneycroft 38
Three Pines 171
thrushes 146, 147, 166
Tickner, George 13
TIGER 126, 221, 223
*Tilia* spp. – *see* lime
timber 35–6, 200–1
   old 123–4
tits (Paridae) 147, 166 *see also* Wytham Tit Study
tor grass (*Brachypodium pinnatum*) 48, 49, 55, 92, 107, 211–12
*Tortrix viridana* – *see* green tortrix
toxins 144
transpiration 69, 224, 225
tree growth 223–6
tree-hole fauna 113
tree physiology 69
tree pipit (*Anthus trivialis*) 145, 146, 147, 148
tree sparrow (*Passer montanus*) 148, 171
trenches 38
*Trifolium repens* – *see* white clover
*Tringa totanus* – *see* redshank
*Trisetum flavescens* – *see* yellow oat-grass
*Trochosa spinipalpis* – *see* wetland spider
*Troglodytes troglodytes* – *see* wren
true bugs (Heteroptera) 104
true flies (Diptera) 50, 51
tufted hair-grass (*Deschampsia cespitosa*) 54, 56
   frequency 81–2
*Turdus merula* – *see* blackbird
Turkey oak (*Quercus cerrisi*) 197
turtle dove (*Streptopelia turtur*) 147
twin-spotted wainscot (*Archanara geminipuncta*) 51
*Tyto alba* – *see* barn owl

Uganda 63
UK Biodiversity Action Plan 49, 92–3
UK Woodland Assurance Scheme 203
Upper Field (Upfield) 34, 37
Upper Radbrook pond 212
Upper Seeds
   archaeology 32
   birds 146
   flora 77, 221
   meteorological recording 21–3
   restoration to grassland 49–50, 212

Upper Seeds case study 92–108
   colonization 95–7
   experimental system 92–5
   false oat-grass invasion and decline 99–102
   invertebrates 104–8
   successional communities
      early and mid 97–8
      late 102–4
Upper Seeds climate change experiment 221–3
upright brome (*Bromus erectus*) 101
*Urtica dioica* – *see* nettle

*Vanellus vanellus* – *see* lapwing
Varley, George 116, 127, 134–6, 137–40, 142–4
vegetation patterns 76–9
Vera hypothesis 4, 124
*Verbascum* spp. – *see* mullein
veteran trees 73, 205, 214
*Viburnum lantana* – *see* wayfaring tree
*Viburnum opulus* – *see* guelder rose
*Vicia sativa* – *see* common vetch
*Vicia tetrasperma* – *see* smooth tare
*Viola hirta* – *see* hairy violet
virgin forest 60
voles 173, 176 *see also* bank vole; field vole
*Vulpes vulpes* – *see* red fox
Vyrnwy, Lake 175, 178

warblers 146, 147, 168
warfarin 195, 210–11
wasps
   larvae 223
   *Syntretus lyctaea* 114
water fluxes 225–7
water mites (*Feltria romijni*) 118
water shrimp (*Gammarus*) 116
water traps 140
wayfaring tree (*Viburnum lantana*) 55, 70
weasel (*Mustela nivalis*) 16, 151, 179, 184–5
Weigall, Hope 10
weight cycle 160
wetland spider (*Trochosa spinipalpis*) 133
wheat 97
Wheatley stone 34
white clover (*Trifolium repens*) 98
wild basil (*Clinopodium vulgare*) 98, 223
wild parsnip (*Pastinaca sativa*) 50, 55, 98, 105, 222
wild rose (*Rosa canina* agg.) 98
Wildlife Conservation Research Unit (WildCRU) 122, 173, 193, 195–6, 202, 214
Wildlife and Countryside Act 203
Williams, Lord, of Thame 8
willow herbs, frequency 81–2
willow tit (*Poecile montanus*) 147–8, 163
wind speeds 24
winter disappearance 142–3

winter moth (*Operophtera brumata*) 116–17, 118, 134–43, 156
  climate change and 221
  life tables 142–3
wireworms 223
Woburn Park 183
wood, as commodity 32
wood false-brome (*Brachypodium sylvaticum*) 54, 75
  frequency 81–2, 84
wood horsetail (*Equisetum sylvaticum*) 75
wood melick 54
wood mouse (*Apodemus sylvaticus*) 165, 173–5, 177, 179–81, 184
wood pigeon (*Columba palumbus*) 167, 174
wood white 133
woodcock (*Scolopax rusticola*) 146
woodland *see* ancient woodland; British woodland; Oxfordshire woodland
woodland census 57
woodland flora *see* flora
Woodland Grant Scheme (WGS) 60, 214
woodland management 203–5
woodland spider (*Clubiona caerulescens*) 133
woodland structure
  changes 70–2 *see also* stand composition
woodlark (*Lullula arborea*) 146
woodlice 112
woodpeckers 146, 148, 151
woolly thistle (*Cirsium eriophorum*) 98
Wormstall 38, 212
wren (*Troglodytes troglodytes*) 168–9
Wychwood Forest 131, 132

Wye valley 61
Wytham Atlas 44
Wytham Ecological Survey 113–14, 124, 131, 176, 201–2
Wytham Park 43, 205
  creation 33–4, 38, 197
Wytham Tit Study 149–63
  breeding season 156–8, 217
  flocking 162
  great tit song 161–2
  populations 153–6
  sparrowhawk 158–60
  storing seeds 163
  survival 153
  weights 160–1
Wytham Village 171
Wytham Woods
  bequest to Oxford University 13, 200
  history 7–9
  location 1–2, 19–20
  management 13–15
  role in national conservation policy 201–2
  and surroundings 52–3
  time line 30–1
  topography 21 *see also* conservation management

Yarnton Meads 48
yellow meadow ant (*Lasius flavus*) 117
yellow oat-grass (*Trisetum flavescens*) 96
Yorkshire fog (*Holcus lanatus*) 55, 56, 98, 105

zoology 199